Ordinary Differential Equations

Many important real-life problems in mathematics, physics, chemistry, biology, engineering, economics, sociology and psychology are modelled using the tools and techniques of ordinary differential equations (ODEs). This book on ODE discusses the relevant topics including first and second order linear equations, initial value problems and qualitative theory. The text covers two-point boundary value problems for second order linear and nonlinear equations. Using two linearly independent solutions, a Green's function is also constructed for given boundary conditions.

The authors emphasize the use of calculus concepts in justification and analysis of equations to get solutions in explicit form. While discussing first order linear systems, tools from linear algebra are used and the importance of these tools is clearly explained. Real-life applications are interspersed throughout. Additiotnally readers can find the methods and tricks to solve numerous mathematical problems with sufficient derivations and explanations. The first few chapters can be used for an undergraduate course on ODE, and later chapters can be used at graduate level.

A. K. Nandakumaran is a Professor at the Department of Mathematics, Indian Institute of Science, Bangalore. His areas of interest are partial differential equations, homogenization, control and controllability problems, inverse problems and computations.

P. S. Datti retired from the Centre for Applicable Mathematics, Tata Institute of Fundamental Research, Bangalore after serving for over 35 years. His research interests include nonlinear hyperbolic equations, hyperbolic conservation laws, ordinary differential equations, evolution equations and boundary layer phenomenon.

Raju K. George is Senior Professor and Dean (R&D) at the Indian Institute of Space Science and Technology (IIST), Thiruvananthapuram. His research areas include functional analysis, mathematical control theory, soft computing, orbital mechanics and industrial mathematics.

CAMBRIDGE–IISc SERIES

Cambridge–IISc Series aims to publish the best research and scholarly work on different areas of science and technology with emphasis on cutting-edge research.

The books will be aimed at a wide audience including students, researchers, academicians and professionals and will be published under three categories: research monographs, centenary lectures and lecture notes.

The editorial board has been constituted with experts from a range of disciplines in diverse fields of engineering, science and technology from the Indian Institute of Science, Bangalore.

IISc Press Editorial Board:

G. K. Ananthasuresh, *Professor, Department of Mechanical Engineering*
K. Kesava Rao, *Professor, Department of Chemical Engineering*
Gadadhar Misra, *Professor, Department of Mathematics*
T. A. Abinandanan, *Professor, Department of Materials Engineering*
Diptiman Sen, *Professor, Centre for High Energy Physics*

Titles in print in this series:

- *Continuum Mechanics: Foundations and Applications of Mechanics* by C. S. Jog
- *Fluid Mechanics: Foundations and Applications of Mechanics* by C. S. Jog
- *Noncommutative Mathematics for Quantum Systems* by Uwe Franz and Adam Skalski
- *Mechanics, Waves and Thermodynamics* by Sudhir Ranjan Jain

Cambridge-IISc Series

Ordinary Differential Equations
Principles and Applications

A. K. Nandakumaran
P. S. Datti
Raju K. George

CAMBRIDGE
UNIVERSITY PRESS

Shaftesbury Road, Cambridge CB2 8EA, United Kingdom

One Liberty Plaza, 20th Floor, New York, NY 10006, USA

477 Williamstown Road, Port Melbourne, VIC 3207, Australia

314 to 321, 3rd Floor, Plot No.3, Splendor Forum, Jasola District Centre, New Delhi 110025, India

103 Penang Road, #05–06/07, Visioncrest Commercial, Singapore 238467

Cambridge University Press is part of the University of Cambridge.

It furthers the University's mission by disseminating knowledge in the pursuit of education, learning and research at the highest international levels of excellence.

www.cambridge.org
Information on this title: www.cambridge.org/9781108416412

First published 2017
Reprint 2019, 2024

Printed in India by Avantika Printers Pvt. Ltd.

A catalogue record for this publication is available from the British Library

Library of Congress Cataloging-in-Publication Data
NAMES: Nandakumaran, A. K., author. | Datti, P. S., author. | George, Raju K., author.
TITLE: Ordinary differential equations : principles and applications / A.K. Nandakumaran, P.S. Datti, Raju K. George.
DESCRIPTION: Delhi, India : Cambridge University Press, 2017. | Includes bibliographical references and index.
INTENTIFIERS: LCCN 2017022841 | ISBN 9781108416412 (hardback : alk. paper)
SUBJECTS: LCSH: Differential equations.
CLASSIFICATION: LCC QA372 .N36 2017 | DDC 515/.352–DC23 LC record available at https://lccn.loc.gov/2017022841

ISBN 978-1-108-41641-2 Hardback

Additional resources for this publication at www.cambridge.org/9781108416412

We would like to dedicate the book to our parents
who brought us to this wonderful world.

Contents

Figures

Preface

Many interesting and important real life problems are modeled using ordinary differential equations (ODE). These include, but are not limited to, physics, chemistry, biology, engineering, economics, sociology, psychology etc. In mathematics, ODE have a deep connection with geometry, among other branches. In many of these situations, we are interested in understanding the future, given the present phenomenon. In other words, we wish to understand the time evolution or the dynamics of a given phenomenon. The subject field of ODE has developed, over the years, to answer adequately such questions. Yet, there are many important intriguing situations, where complete answers are still awaited. The present book aims at giving a good foundation for a beginner, starting at an undergraduate level, without compromising on the rigour.

We have had several occasions to teach the students at the undergraduate and graduate level in various universities and institutions across the country, including our own institutions, on many topics covered in the book. In our experience and the interactions we have had with the students, we felt that many students lack a clear notion of ODE including the simplest integral calculus problem. For other students, a course on ODE meant learning a few tricks to solve equations. In India, in particular, the books which are generally prescribed, consist of a few tricks to solve problems, making ODE one of the most uninteresting subject in the mathematical curriculum. We are of the opinion that many students at the beginning level do not have clarity about the essence of ODE, compared to other subjects in mathematics.

While we were still contemplating to write a book on ODE, to address some of the issues discussed earlier, we got an opportunity to present a video course on ODE, under the auspices of the National Programme for Technology Enhanced Learning (NPTEL), Department of Science and Technology (DST), Government of India, and our course is freely available on the NPTEL website (see www.nptel.ac.in/courses/ 111108081). In this video course, we have presented several topics. We have also tried to address many of the doubts that students may have at the beginning level and the misconceptions some other students may possess.

Many in the academic fraternity, who watched our video course, suggested that we write a book. Of course, writing a text book, that too about a classical subject at a beginning level, meant a much bigger task than a video course, involving choosing and presenting the material in a very systematic way. In a way, the video course may supplement the book as it gives a flavour of a classroom lecture. We hope that in this way, students in remote areas and/or places where there is lack of qualified teachers, benefit from the book and the video course, making good use of the modern technology available through the Internet. The teachers of undergraduate courses can also benefit, we hope, from this book in fine tuning their skills in ODE.

We have written the present book with the hope that it can also be used at the undergraduate level in universities everywhere, especially in the context of Indian universities, with appropriately chosen topics in Chapters 1, 2 and 3. As the students get more acquainted with basic analysis and linear algebra, the book can be introduced at the graduate level as well and even at the beginning level of a research programme.

We now briefly describe the contents of the book. The book has a total of ten chapters and one appendix.

Chapter 1 describes some important examples from real life situations in the field of physics to biology to engineering. We thought this as a very good motivation for a beginner to undertake the study of ODE; in a rigorous course on ODE, often a student does not see a good reason to study the subject. We have observed that this has been one of the major concerns faced by students at a beginning level.

As far as possible, we have kept the prerequisite to a minimum: a good course on calculus. With this in mind, we have collected, in Chapter 2, a number of important results from analysis and linear algebra that are used in the main text. Wherever possible, we have provided proofs and simple presentations. This makes the book more or less self contained, though a deeper knowledge in analysis and linear algebra will enhance the understanding of the subject.

First and second order equations are dealt with in Chapter 3. This chapter also contains the usual methods of solutions, but with sufficient mathematical explanation, so that students feel that there is indeed rigorous mathematics behind these methods. The concept behind the exact differential equation is also explained. Second order linear equations, with or without constant coefficients, are given a detailed treatment. This will make a student better equipped to study linear systems, which are treated in Chapter 5.

Chapter 4 deals with the hard theme of existence, non-existence, uniqueness etc., for a single equation and also a system of first order equations. We have tried to motivate the reader to wonder why these questions are important and how to deal with them. We have also discussed other topics such as continuous dependence on initial data, continuation of solutions and the maximal interval of existence of a solution.

Linear systems are studied in great detail in Chapter 5. We have tried to show the power of linear algebra in obtaining the phase portrait of 2×2 and general systems. We have also included a brief discussion on Floquet theory, which deals with linear systems with periodic coefficients.

In the case of a second order linear equation with variable coefficients, it is not possible in general, to obtain a solution in explicit form. This has been discussed at length in Chapter 3. Chapter 6 deals with a class of second order linear equations, whose solutions may be written explicitly, although in the form of an infinite series. This method is attributed to Frobenius.

Chapter 7 deals with the regular Sturm–Lioville theory. This theory is concerned with boundary value problems associated with linear second order equations with smooth coefficients, in a compact interval on the real, involving a parameter. We, then, show the existence of a countable number of values of the parameter and associated non-trivial solutions of the differential equation satisfying the boundary conditions. There are many similarities with the existence of eigenvalues and eigenvectors of a matrix, though we are now in an infinite dimensional situation.

The qualitative theory of nonlinear systems is the subject of Chapter 8. The contents may be suitable for a senior undergraduate course or a beginning graduate course. This chapter does demand for more prerequisites and these are described in Chapter 2. The main topics of the chapter are equilibrium points or solutions of autonomous systems and their stability analysis; existence of periodic orbits in a two-dimensional system. We have tried to make a presentation of these important notions so that it can be easily understood by any student at a senior undergraduate level. The proofs of two important theorems on the existence of periodic orbits are given in the Appendix.

Chapter 9 considers the study of two point boundary value problems for second order linear and nonlinear equations. The first dealing with linear equations fully utilises the theory developed in Chapter 3. Using two linearly independent solutions, a Green's function is constructed for given boundary conditions. This is similar to an integral calculus problem. For nonlinear equations, we no longer have the luxury of two linearly independent solutions. A result which gives a taste of delicate analysis is proved. It is also seen through some examples how phase plane analysis can help in deciding whether a given boundary value problem has a solution or not.

In Chapter 10, we have attempted to show how the methods of ODE are used to find solutions of first order partial differential equations (PDE). We essentially describe the method of characteristics for solving general first order PDE. As very few books on ODE deal with this topic, we felt like including this, as a student gets some benefit of studying PDE and (s)he can later pursue a course on PDE.

We have followed the standard notations. Vectors in Euclidean faces and matrices are in boldface.

Acknowledgement

We wish to express our sincere appreciation to Gadadhar Misra and others at the IISc Press for suggesting to publish our book through the joint venture of IISc Press and Cambridge University Press. We also would like to thank Gadadhar Misra for all the help in this regard. We wish to acknowledge the support we received from our respective institutions and the moral support from our colleagues, during the preparation of the manuscript. We thank our academic fraternity, who have made valuable suggestions after reading through the various parts of the book. We would like to thank the students who attended our lectures at various places and contributed in a positive way. Over the years, we have had the opportunity to deliver talks in various lecture programs conducted by National Programme in Differential Equations (NPDE), India and the Indian Science Acadamies; our sincere thanks to them. We also wish to thank the anonymous referees for their constructive criticism and suggestions, which have helped us in improving the presentation. The illustrations have been drawn using the freely available software packages tikz and circuitikz. We are also thankful to the CUP team for their coordination from the beginning and their excellent production. Last but not the least, we wish to thank our family members for their patience and support during the preparation of this book.

1

Introduction and Examples: Physical Models

1.1   A Brief General Introduction

The beginning of the study of ordinary differential equations (ODE) could perhaps be attributed to Newton and Leibnitz, the inventors of differential and integral calculus. The theory began in the late 17th century with the early works of Newton, Leibnitz and Bernoulli. As was customary then, they were looking at the fundamental problems in geometry and celestial mechanics. There were also important contributions to the development of ODE, in the initial stages, by great mathematicians – Euler, Lagrange, Laplace, Fourier, Gauss, Abel, Hamilton and others. As the modern concept of function and analysis were not developed at that time, the aim was to obtain solutions of differential equations (and in turn, solutions to physical problems) in terms of elementary functions. The earlier methods in this direction are the concepts of *integrating factors* and method of separation of variables.

In the process of developing more systematic procedures, Euler, Lagrange, Laplace and others soon realized that it is hopeless to discover methods to solve differential equations. Even now, there are only a handful of sets of differential equations, that too in a simpler form, whose solutions may be written down in explicit form. It is in this scenario that the *qualitative analysis – existence, uniqueness, stability properties, asymptotic behaviour* and so on – of differential equations became very important. This qualitative analysis depends on the development of other branches of mathematics, especially analysis. Thus, a second phase in the study of differential equations started from the beginning of the 19th century based on a more rigorous approach to calculus via the mathematical analysis. We remark that the first existence theorem for first order differential equations is due to Cauchy in 1820. A class of differential equations known as *linear differential equations*, is much easier to handle. We will analyse linear equations and linear systems in more detail and see the extensive use of linear algebra; in particular, we will see how the nature of eigenvalues of a given matrix influences the stability of solutions.

After the invention of differential calculus, the question of the existence of antiderivative led to the following question regarding differential equation: Given a function f, does there exist a function g such that $\dot{g}(t) = f(t)$? Here, $\dot{g}(t)$ is the derivative of g with respect to t. This was the beginning of integral calculus and we refer to this problem as an *integral calculus problem*. In fact, Newton's second law of motion describing the motion of a particle having mass m states that the rate change of momentum equals the applied force. Mathematically, this is written as $\frac{d}{dt}(mv) = -F$, where v is the velocity of the particle. If $x = x(t)$ is the position of the particle at time t, then $v(t) = \dot{x}(t)$. In general, the applied force F is a function of t, x and v. If we assume F is a function of t, x, we have a second order equation for x given by $m\ddot{x} = -F(t,x)$. If F is a function of x alone, we obtain a conservative equation which we study in Chapter 8. If on the other hand, F is a function of t alone, then the second law leads to two integral calculus problems: namely, first solve for the momentum $p = mv$ by $\dot{p} = -F(t)$ and then solve for the position using $m\dot{x} = p$. This also suggests that one of the best ways to look at a differential equation is to view it as a *dynamical system*; namely, the motion of some physical object. Here t, the *independent variable* is viewed as time and x is the unknown variable which depends on the independent variable t, and is known as the *dependent variable*.

A large number of physical and biological phenomena can be modelled via differential equations. Applications arise in almost all branches of science and engineering–radiation decay, aging, tumor growth, population growth, electrical circuits, mechanical vibrations, simple pendulum, motion of artificial satellites, to mention a few.

In summary, real life phenomena together with physical and other relevant laws, observations and experiments lead to mathematical models (which could be ODE). One would like to do mathematical analysis and computations of solutions of these models to simulate the behaviour of these physical phenomena for better understanding.

Definition 1.1.1

An ODE is an equation consisting of an independent variable t, an unknown function (dependent variable) $y = y(t)$ and its derivatives up to a certain order. Such a relation can be written as

$$f\left(t, y, \frac{dy}{dt}, \cdots, \frac{d^n y}{dt^n}\right) = 0. \tag{1.1.1}$$

Here, n is a positive integer, known as the order of the differential equation.

For example, first and second order equations, respectively, can be written as

$$f\left(t, y, \frac{dy}{dt}\right) = 0 \text{ and } f\left(t, y, \frac{dy}{dt}, \frac{d^2 y}{dt^2}\right) = 0. \tag{1.1.2}$$

We will be discussing some special cases of these two classes of equations. It is possible that there will be more than one unknown function and in that case, we will have a system of differential equations. A higher order differential equation in one unknown function may be reduced into a system of first order differential equations. On the other hand, if there are more than one independent variable, we end up with partial differential equations (PDEs).

1.2   Physical and Other Models

We begin with a few mathematical models of some real life problems and present solutions to some of these problems. However, methods of obtaining such solutions will be introduced in Chapter 3, and so are the terminologies like linear and nonlinear equations.

1.2.1   Population growth model

We begin with a linear model. If $y = y(t)$, represents the population size of a given species at time t, then the rate of change of population $\frac{dy}{dt}$ is proportional to $y(t)$ if there is no other species to influence it and there is no net migration. Thus, we have a simple linear model [Bra78]

$$\frac{dy}{dt} = ry(t), \tag{1.2.1}$$

where r denotes the difference between birth rate and death rate. If $y(t_0) = y_0$ is the population at time t_0, our problem is to find the population for all $t > t_0$. This leads to the so-called *initial value problem (IVP)* which will be discussed in Chapter 3. Assuming that r is a constant, the solution is given by

$$y(t) = y_0 e^{r(t-t_0)} \tag{1.2.2}$$

Note that, if $r > 0$, then as $t \to \infty$, the population $y(t) \to \infty$. Indeed, this linear model is found to be accurate when the population is small and for small time. But it cannot be a good model as no population, in reality, can grow indefinitely. As and when the population becomes large, there will be competition among the population entities for the limited resources like food, space etc.

This suggests that we look for a more realistic model which is given by the following logistic nonlinear model. The statistical average of the number of encounters of two members per unit time is proportional to y^2. Thus, a better model would be

$$\frac{dy}{dt} = ay - by^2, \ y(t_0) = y_0. \tag{1.2.3}$$

Here a, b are positive constants. The negative sign in the quadratic term represents the competition and reduces the growth rate. This is known as the *logistic law of population growth*. It was introduced by the Dutch mathematical biologist Verhulst in 1837. It is also known as the Malthus law.

Practically, b is small compared to a. Thus, if y is not too large, then by^2 will be negligible compared to ay and the model behaves similar to the linear model. However, when y becomes large, the term by^2 will have a considerable influence on the growth of y, as can be seen from the following discussion.

The solution of (1.2.3) is given by[1]

$$\frac{1}{a} \log \frac{|y|}{|y_0|} \left| \frac{a - by_0}{a - by} \right| = t - t_0, \ t > t_0. \tag{1.2.4}$$

Note that $y \equiv 0$ and $y \equiv \frac{a}{b}$ are solutions to the nonlinear differential equation in (1.2.3) with the initial condition $y(t_0) = 0$ and $y(t_0) = \frac{a}{b}$,

---

[1]The reader, after getting familiarised with the methods of solutions in Chapter 3, should work out the details for this and the other examples in this chapter.

respectively. Hence, if the initial population y_0 satisfies $0 < y_0 < \frac{a}{b}$, then the solution will remain in the same interval for all time. This follows from the existence and uniqueness theory, which will be developed in Chapter 4. A simplification of (1.2.4) gives

$$y(t) = \frac{ay_0}{by_0 + (a - by_0)e^{-a(t-t_0)}}. \tag{1.2.5}$$

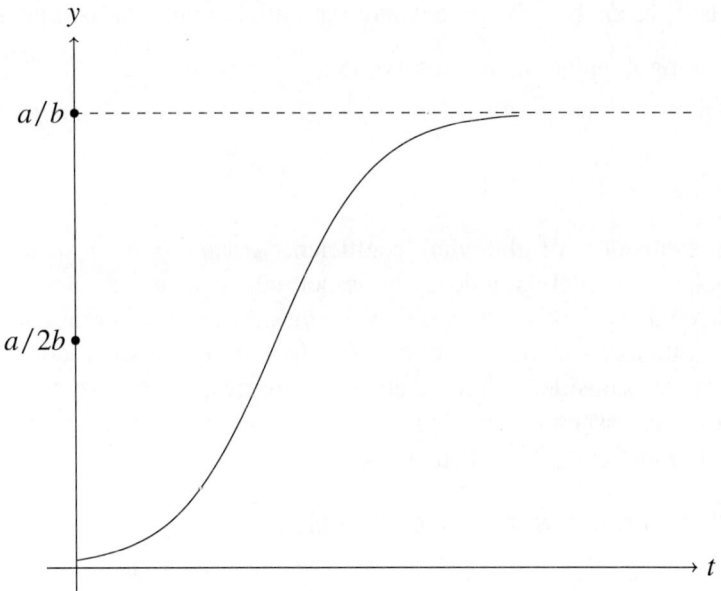

Fig. 1.1   Logistic map

In case $0 < y_0 < \dfrac{a}{b}$, the curve $y(t)$ is depicted as in Fig. 1.1. This curve is called the *logistic* curve; it is also called an *S-shaped curve*, because of its shape. Note that $\dfrac{a}{b}$ is the limiting population, also known as capacity of the ecological environment. In this case, the rate of population $\dfrac{dy}{dt}$ is positive and hence, y is an increasing function. Since $\dfrac{d^2y}{dt^2} = (a - 2by)\dfrac{dy}{dt}$, we immediately see that it is positive if the population is between 0 and half the limiting population, namely, $\dfrac{a}{2b}$, whereas, it is negative when the population crosses the half way mark $\dfrac{a}{2b}$. This indicates that if the initial population is less than half the limiting population, then there is an accelerated growth $\left(\dfrac{dy}{dt} > 0, \dfrac{d^2y}{dt^2} > 0\right)$, but after reaching half the population, the population still grows $\left(\dfrac{dy}{dt} > 0\right)$, but it has now a decelerated growth $\left(\dfrac{d^2y}{dt^2} < 0\right)$.

When we analyse the case where the initial population is bigger than the limiting population, we observe that $\dfrac{dy}{dt} < 0$ and $\dfrac{d^2y}{dt^2} < 0$. Thus, the population *decreases with a decelerated growth* to the limiting population.

Remark 1.2.1

The estimation of the vital coefficients a and b in a particular population model is indeed an important issue which has to be updated in a period of time as they are influenced by other parameters like pollution, sociological trends, etc. In a more realistic model, one needs to consider more than one species, their interactions, unforeseen issues like epidemics, natural disasters, etc., which may lead to more complicated equations.

1.2.2   An atomic waste disposal problem

The dumping of tightly sealed drums containing highly concentrated radioactive waste in the sea below a certain depth (say 300 feet) from the surface is a very sensitive issue as it could be environmentally hazardous. The drums could break due to the impact of their velocity exceeding a certain limit, say 40 ft/sec. Our problem is to compute the velocity by using Newton's second law of motion and assess the level of safety involved in the process. Let $y(t)$ denote the position, at time t, of the object, the drum, (considered as a particle) measured from the sea surface (indicating $y = 0$) as a positive quantity. The total force acting on the object is given by

$$F = W - B - D,$$

where the weight $W = mg$ is the force due to gravity, B is the buoyancy force of water acting against the forward movement and $D = cV$ is the drag exerted by water (it is a kind of resistance), where $V = \dfrac{dy}{dt}$, the velocity of the object and $c > 0$ is a constant of proportionality. Thus, we have the differential equation

$$\frac{d^2y}{dt^2} = \frac{1}{m}F = \frac{1}{m}(W - B - cV) = \frac{g}{W}(W - B - cV), \; y(0) = 0. \quad (1.2.6)$$

Equivalently,

$$\frac{dV}{dt} + \frac{cg}{W}V = \frac{g}{W}(W - B), \; V(0) = 0. \quad (1.2.7)$$

Equation (1.2.7) can be solved to get

$$V(t) = \frac{W - B}{c}\left(1 - e^{-\frac{cg}{W}t}\right). \quad (1.2.8)$$

Thus, $V(t)$ is increasing and tends to $\frac{W-B}{c}$ as $t \to \infty$ and the value (practically) of $\dfrac{W - B}{c} \approx 700$.

The limiting value 700 ft/sec of velocity is far above the permitted critical value. Thus, it remains to ensure that $V(t)$ does not reach 40 ft/sec by the time it reaches the sea bed. But it is not possible to compute t at which time the drum hits the sea bed and one needs to do further analysis.

Analysis: The idea is to view the velocity $V(t)$ not as a function of time, but as a function of position y. Let $v(y)$ be the velocity at height y measured from the surface of the sea downwards. Then, clearly, $V(t) = v(y(t))$ so that $\dfrac{dV}{dt} = \dfrac{dv}{dy}\dfrac{dy}{dt} = v\dfrac{dv}{dy}$. Hence, (1.2.7) becomes

$$\begin{cases} \dfrac{v}{W - B - cv}\dfrac{dv}{dy} = \dfrac{g}{W}, \\[2mm] v(0) = 0. \end{cases} \quad (1.2.9)$$

This is a first order non-homogeneous nonlinear equation for the velocity v. Indeed, the equation is more difficult, but it is in a variable separable form and can be integrated easily. We can solve this equation to obtain the solution in the form

$$\frac{gy}{W} = -\frac{v}{c} - \frac{W - B}{c^2}\log\frac{W - B - cv}{W - B}. \quad (1.2.10)$$

Of course, v cannot be explicitly expressed in terms of y as it is a nonlinear equation. However, it is possible to obtain accurate estimates for the velocity $v(y)$ at height y and it is estimated that $v(300) \approx 45$ ft/sec and hence, the drum could break at a depth of 300 feet.

Tail to the Tale: This problem was initiated when environmentalists and scientists questioned the practice of dumping waste materials by the Atomic Energy Commission of USA. After the study, the dumping of atomic waste was forbidden, in regions of sea not having sufficient depths.

1.2.3   Mechanical vibration model

The fundamental mechanical model, namely *spring-mass-dashpot system (SMD)* has applications in shock absorbers in automobiles, heavy guns, etc. An object of mass m is attached to an elastic spring of length l which is suspended from a rigid horizontal body. This is a spring–mass system. Elastic spring has the property that when it is stretched or compressed by a small length Δl, it will exert a force of magnitude proportional to Δl, say $k\Delta l$ in the opposite direction of stretching or compressing. The positive constant k is called *spring constant* which is a measure of stiffness of the spring. We then obtain an SMD system when this spring–mass is immersed in a medium like oil which will also resist the motion of the spring–mass. In a simple situation, we may assume that the force exerted by the medium on the spring–mass is proportional to the velocity of the mass and in the opposite direction of the movement of mass. It is also similar to a seismic instrument used to obtain a seismograph to detect the motion of the earth's surface.

Let $y(t)$ denote the position of mass at time t, $y = 0$ being the position of the mass at equilibrium and let us take the downward direction as positive. There are four forces acting on the system, that is, $F = W + R + D + F_0$, where $W = mg$, the force due to gravity; $R = -k(\Delta l + y)$, the restoring force; D, the damping or drag force and F_0, the external applied force, if any. Drag force is the kind of resistance force which the medium exerts on the mass and hence, it will be negative. It is usually proportional to the velocity, that is, $D = -c\dfrac{dy}{dt}$. At equilibrium, the spring has been stretched a length Δl and so $k\Delta l = mg$. Applying Newton's second law, we get

$$m\frac{d^2y}{dt^2} = -ky - c\frac{dy}{dt} + F_0(t). \tag{1.2.11}$$

That is,

$$m\frac{d^2y}{dt^2} + c\frac{dy}{dt} + ky = F_0(t), \ m,c,k \geq 0. \tag{1.2.12}$$

This is a second order non-homogeneous linear equation with constant coefficients and we study such equations in detail in Chapter 3. Such a system also arises in electrical circuits, which we discuss next.

1.2.4   Electrical circuit

A basic LCR electrical circuit is shown in Fig. 1.2, and is described as follows:

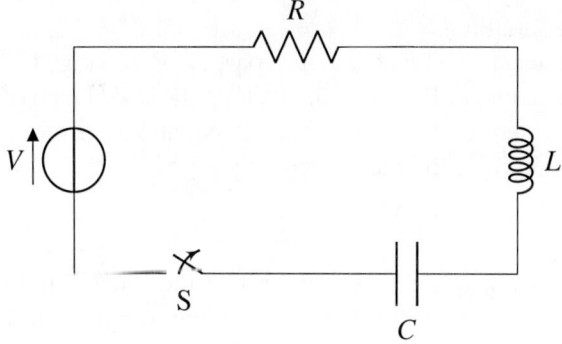

Fig. 1.2   A basic LCR circuit

By Kirchoff's second law, the impressed voltage in a closed circuit equals the sum of the voltage drops in the rest of the circuit. Let $E(t)$ be the source of electro motive force (emf), say a battery, $I = \dfrac{dQ}{dt}$ be the current flow, $Q(t)$ the charge on the capacitor at time t. Then, the voltage drops across inductance (L), resistance (R) and capacitance (C), respectively, are given by $L\dfrac{dI}{dt} = L\dfrac{d^2Q}{dt^2}$, $RI = R\dfrac{dQ}{dt} + \dfrac{Q}{c}$. Thus, we obtain a similar equation for Q as in (1.2.12):

$$L\frac{d^2Q}{dt^2} + R\frac{dQ}{dt} + \frac{Q}{c} = E(t). \tag{1.2.13}$$

More often, the current $I(t)$ is the physical quantity of interest; by differentiating (1.2.13) with respect to t, the equation satisfied by I is

$$L\frac{d^2I}{dt^2} + R\frac{dI}{dt} + \frac{1}{c}I = \frac{dE}{dt}(t). \qquad (1.2.14)$$

Mathematically, the equation is exactly same as the equation obtained in the spring–mass–dashpot system. We can also see the similarity between various quantities: inductance corresponding to mass, resistance corresponding to damping constant and so on.

1.2.5 Satellite problem

Consider an artificial satellite of mass m orbiting the earth. We assume that the satellite has thrusting capacity with radial thrust u_1 and a thrust u_2 which is applied in a direction perpendicular to the radial direction. The thrusters u_1 and u_2 are considered as the external force F or control inputs applied to the satellite.

The satellite can be considered as a particle P moving around the earth in the equatorial plane. If (x, y) is the rectangular coordinate of the particle P of mass m, then by Newton's law, the equations of motion along the rectangular coordinate axes are given by

$$m\ddot{x} = F_x, \quad m\ddot{y} = F_y \qquad (1.2.15)$$

where, F_x and F_y denote the components of the force F in the directions of the axes (see Fig. 1.3). It will be convenient to represent the motion in polar coordinates (r, θ), where,

$$x = r\cos\theta, \ y = r\sin\theta$$

We will resolve the velocity, acceleration and force of the particle into components along the radial direction and the direction perpendicular to it. Denote by u, v; a_1, a_2 and F_r, F_θ the components of velocity, acceleration and force, respectively in the new coordinate system. The resultant of u and v is also equal to the resultant of the components of $\dot{x}$ and $\dot{y}$. Therefore, by resolving parallel to the x-axis, we get

$$\dot{x} = u\cos\theta - v\sin\theta \qquad (1.2.16)$$

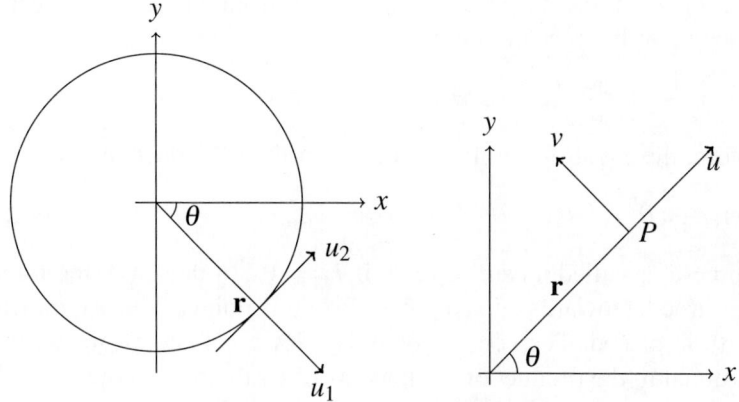

Fig. 1.3   Satellite problem

Since $x = r\cos\theta$, differentiating with respect to time t,

$$\dot{x} = \dot{r}\cos\theta - r(\sin\theta)\dot{\theta} \qquad (1.2.17)$$

From (1.2.16) and (1.2.17), we have

$$u\cos\theta - v\sin\theta = \dot{r}\cos\theta - r(\sin\theta)\dot{\theta} \qquad (1.2.18)$$

Comparing coefficients of $\cos\theta$ and $\sin\theta$ from (1.2.18), we have

$$u = \dot{r}, \quad v = r\dot{\theta}. \qquad (1.2.19)$$

Then, by resolving the acceleration parallel to the x- and y-axes, we get

$$\ddot{x} = a_1\cos\theta - a_2\sin\theta.$$

By differentiating (1.2.17), we obtain

$$\ddot{x} = \frac{d\dot{x}}{dt} = \frac{d}{dt}(\dot{r}\cos\theta - r(\sin\theta)\dot{\theta}) \qquad (1.2.20)$$

$$= \ddot{r}\cos\theta - \dot{r}(\sin\theta)\dot{\theta} - \dot{r}(\sin\theta)\dot{\theta} - r(\cos\theta)\dot{\theta}^2 - r(\sin\theta)\ddot{\theta} \qquad (1.2.21)$$

$$= (\ddot{r} - r\dot{\theta}^2)\cos\theta - (2\dot{r}\dot{\theta} + r\ddot{\theta})\sin\theta \qquad (1.2.22)$$

Equating the expressions of $\ddot{x}$ obtained here, we get

$$a_1\cos\theta - a_2\sin\theta = (\ddot{r} - r\dot{\theta}^2)\cos\theta - (2\dot{r}\dot{\theta} + r\ddot{\theta})\sin\theta \quad (1.2.23)$$

Comparing coefficients of $\cos\theta$ and $\sin\theta$ from (1.2.23), we get the components of the acceleration as

$$a_1 = \ddot{r} - r\dot{\theta}^2, \quad a_2 = 2\dot{r}\dot{\theta} + r\ddot{\theta} \tag{1.2.24}$$

Therefore, the equations of motion of the particle P reduce to

$$m(\ddot{r} - r\dot{\theta}^2) = F_r, \quad 2m\dot{r}\dot{\theta} + mr\ddot{\theta} = F_\theta, \tag{1.2.25}$$

The force F is called a *central force* if $F_\theta = 0$. In this case, the force is always directed towards a fixed point. Take this point as the origin, where the earth is placed. The central force, by Newton's law of gravitation is proportional to the product of the mass M of earth, mass m of the satellite and inversely proportional to the square of the distance between them. Thus, $F_r = -\dfrac{GMm}{r^2}$, where G is the gravitational constant. Let $k = GMm$. Now, the equations of motion are given by

$$m(\ddot{r} - r\dot{\theta}^2) = -\frac{k}{r^2} \tag{1.2.26}$$

$$mr\ddot{\theta} + 2\dot{r}\dot{\theta}m = 0. \tag{1.2.27}$$

Newton derived Kepler's laws of planetary motion using these equations. The interested reader can refer to [Sim91]. Note that $r(t) = \sigma$, $\theta(t) = \omega t$, where σ, ω are appropriate constants, is a special solution to the aforementioned equations which corresponds to a circular orbit.

Assume that the mass is equipped with the ability to exert a thrust u_1 in the radial direction and u_2 in the direction perpendicular to the radial direction. Then, under the presence of these external forces (known as controls), the equations of motion become

$$m\ddot{r} - mr\dot{\theta}^2 + \frac{k}{r^2} = u_1$$

$$2m\dot{r}\dot{\theta} + r\ddot{\theta}m = u_2. \tag{1.2.28}$$

By scaling the time variable, we may assume that the mass m of the satellite is 1. Then, the motion of the satellite is described by a pair of second order nonlinear differential equations:

$$\frac{d^2r}{dt^2} = r(t)\left(\frac{d\theta}{dt}\right)^2 - \frac{k}{r^2(t)} + u_1(t). \tag{1.2.29}$$

$$\frac{d^2\theta}{dt^2} = -\frac{2}{r(t)} \frac{d\theta}{dt} \frac{dr}{dt} + \frac{u_2(t)}{r(t)}. \tag{1.2.30}$$

In applications, when a satellite is injected into an orbit, it usually drifts from its prescribed orbit due to the influence of other cosmic forces. The thrusters (controls) are activated to maintain the desired orbit of the satellite.

1.2.6   Flight trajectory problem

We consider an aeroplane which departs from an airport located at point $(a,0)$ and intends to reach an airport located at $(0,0)$ in the western direction from the departure airport. Assume that the constant wind velocity in the northern direction is w and the plane travels with constant speed v_0 relative to the wind. Assume that the plane's pilot maintains its heading directly towards the origin $(0,0)$.

The ground velocities of the plane in the direction of the x-axis and the y-axis are given by

$$\frac{dx}{dt} = -v_0 \cos\theta = -\frac{v_0 x}{\sqrt{x^2 + y^2}}$$

$$\frac{dy}{dt} = -v_0 \sin\theta + w = -\frac{v_0 y}{\sqrt{x^2 + y^2}} + w,$$

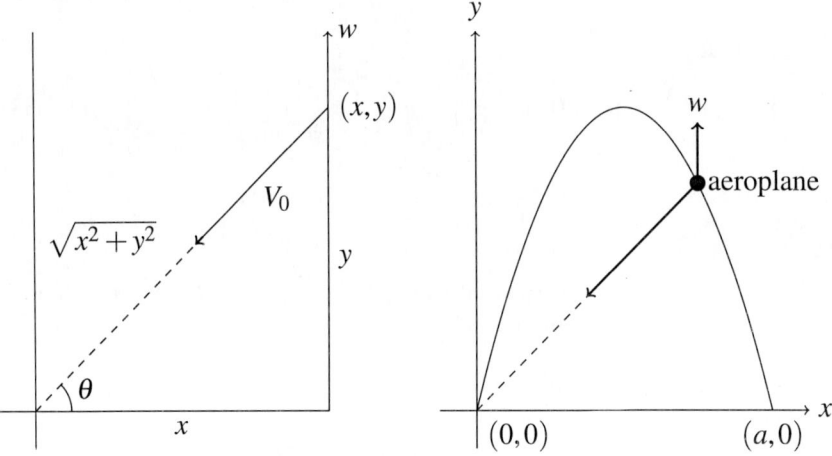

Fig. 1.4   Flight trajectory

The path $\{(x(t), y(t)), t \geq 0\}$ is called the orbit or trajectory of the aircraft in the xy plane (Figure 1.4). These equations can be implicitly written as

$$\frac{dy}{dx} = \frac{1}{v_0 x}(v_0 y - w \sqrt{x^2 + y^2}).$$

1.2.7 Other examples

Example 1.2.2

[Unforced Duffing equation or oscillator] This is a second order equation, named after Georg Duffing, and is given by

$$\ddot{x} - \alpha x + \beta x^3 + \delta \dot{x} = 0. \tag{1.2.31}$$

Here α, β are nonzero real numbers and $\delta \geq 0$. This equation, referred to as a nonlinear oscillator, is a perturbation of the usual linear oscillator, namely (1.2.31) with $\beta = \delta = 0$ and $\alpha < 0$. The nonlinear equation (1.2.31) has a cubic nonlinearity and a linear damping term. It models more complicated dynamics of a spring pendulum whose spring stiffness does not exactly obey Hooke's law. The case of $\delta = 0$ and a periodic forcing term was extensively studied by Duffing.

By dilating the variables x to ax and t to bt, for suitable constants a, b, we can write (1.2.31) in the following standard form

$$\ddot{x} - x + x^3 + \delta \dot{x} = 0, \tag{1.2.32}$$

if $\alpha \beta > 0$ and

$$\ddot{x} + x + x^3 + \delta \dot{x} = 0, \tag{1.2.33}$$

if $\alpha \beta < 0$.

Example 1.2.3

[Unforced van der Pol equation or oscillator] This is also a second order nonlinear equation given by

$$\ddot{x} - \mu(x^2 - 1)\dot{x} + x = 0, \mu \in \mathbb{R}. \tag{1.2.34}$$

This equation, apparently first introduced in 1896 by Lord Rayleigh, was extensively studied both theoretically and experimentally using electrical circuits by the Dutch engineer van der Pol when he was working for the Philips company (in the Netherlands) around 1920. He also studied this equation with forced periodic term $A \sin \omega t$ and observed the phenomenon, which in the current literature is termed as *chaos*. A detailed mathematical analysis of this equation was done by Cartwright and Littlewood [CL45] and by Levinson [Lev49]; their study revealed the existence of the paradoxical combination of randomness and structure, which is also called *deterministic chaos* in the current literature; see Example 1.2.5, Lorenz equations.

The van der Pol equation is also used to model certain situations in physical and biological sciences. For example, in seismology, it is used to model the motion of two plates in a geological fault; in biology, it is used to model the action potential of neurons.

Example 1.2.4

[Pendulum equation or Nonlinear oscillator] Again, this is a second order equation given by

$$\ddot{x} + k\sin x = 0, k > 0. \tag{1.2.35}$$

When x is small, we have $\sin x \approx x$ and one obtains the linear pendulum equation.

Example 1.2.5

[Lorenz equations] The Lorenz system is given by

$$
\begin{aligned}
\dot{x} &= -\sigma x + \sigma y \\
\dot{y} &= Rx - y - xz \\
\dot{z} &= -bz + xy,
\end{aligned}
\tag{1.2.36}
$$

where R, σ, b are fixed parameters.

Motivated by the meteorological problem of weather prediction, Lorenz derived these equations as a much simplified model of Rayleigh–Bernard convection in fluids, which provided a first specific example of chaotic dynamics persisting for all time. Later, in Japan, Ueda studied the steady state chaotic behaviour in the context of the periodically forced Duffing oscillator.

Lorenz's work was not noticed by the mathematical community when it was published, perhaps due to its appearance in a non-mathematical journal. Once it was noticed, a host of papers appeared announcing similar phenomenon in the context of other equations. All these works have been put together in book form and the interested reader may consult [Hao84].

Example 1.2.6

[Lotka–Volterra Prey–Predator Model]

The dynamical behaviour governing the growth, decay and general evolution of two interacting biological species (the case of a single species was discussed earlier using the logistic model) is modelled by the Lotka–Volterra prey–predator equations, which are given by

$$\dot{x} = ax - bxy$$
$$\dot{y} = -cy + dxy. \tag{1.2.37}$$

Here x denotes the population of the prey, say rabbits at a given time and y denotes the population of the predator, say foxes. The constants a, b, c, d are all positive and represent the growth and decay rates of the prey and the predator. In (1.2.37), there is no *competition* within the same species. If we incorporate this also in the model, the dynamics changes; see, the exercises in Chapter 8.

Example 1.2.7

[Mathematical Epidemiology] We now describe the basic SIR model of mathematical epidemiology and an extension of it. The host population (humans for example) is divided into a small number of compartments, each containing individuals that are identical in terms of their status with respect to the disease (tuberculosis, for example) in question. In the SIR model, we have the following three compartments:

1. *Susceptible (S)*: individuals who have no immunity to the infectious agent, so might become infected if exposed.

2. *Infectious (I)*: individuals who are currently infected and can transmit the infection to susceptible individuals who they contact.

3. *Removed (R)*:   individuals who are immune to the infection, and consequently do not affect the transmission dynamics in any way when they contact other individuals.

The total host population is $N = S + I + R$. When N is very large, we may treat the variables in question as continuous variables and the dynamics of S, I variables is described by the following differential equations:

$$\dot{S} = -\beta SI$$
$$\dot{I} = \beta SI - \gamma I. \qquad (1.2.38)$$

Here, the **transmission rate** (per capita) is β and the **recovery rate** is γ (so the **mean infectious period** is $\frac{1}{\gamma}$). Since (1.2.38) do not contain R, no equation for R is written; the appropriate equation for R is $\dot{R} = \gamma I$ (outflow from I compartment goes into the R compartment). But, once the I is determined from these equations, R can easily be determined.

If we now expand the *SIR* model to include B, the births per unit time and a natural mortality rate μ (per capita), then these equations become

$$\dot{S} = B - \beta SI - \mu S$$
$$\dot{I} = \beta SI - \gamma I - \mu I. \qquad (1.2.39)$$

The timescale for substantial changes in birth (decades) is much larger than, say, a measles epidemic (a few months), so one may assume that the total population is constant and that $B = \mu N$. Therefore, there is essentially only one new parameter here.

Example 1.2.8

[Hamiltonian system] This is an even dimensional system described in terms of a smooth function (a **Hamiltonian**) $H(x, y)$ of $2n$ variables, where $H : \mathbb{R}^n \times \mathbb{R}^n \to \mathbb{R}$. The system is described by the following set of equations

$$\dot{x}_j = \frac{\partial H}{\partial y_j}, \quad \dot{y}_j = -\frac{\partial H}{\partial x_j} \qquad (1.2.40)$$

for $j = 1, 2, \cdots, n$.

If $(x(t), y(t))$ is a solution of (1.2.40), it is easy to see that $H(x(t), y(t)) \equiv$ constant; in particular, if $H(x(t_0), y(t_0)) = 0$ for some t_0, then $H(x(t), y(t)) = 0$ for all t.

A conservative system of order n is a system of second order equations

$$\ddot{x}_j + \frac{\partial V}{\partial x_j}(x_1, \cdots, x_n) = 0, \tag{1.2.41}$$

for $j = 1, \cdots, n$. Here $V : \mathbb{R}^n \to \mathbb{R}$ is a smooth function called a **potential function**. By introducing new variables $y_j = \dot{x}_j$, we immediately see that a conservative system becomes a Hamiltonian system with the corresponding Hamiltonian given by the total energy

$$H(x_1, \cdots, x_n, y_1, \cdots, y_n) = \frac{1}{2}(y_1^2 + \cdots + y_n^2) + V(x_1, \cdots, x_n).$$

We will recall these examples in Chapter 8 where a detailed analysis of them will be done.

Example 1.2.9

[Discrete Dynamical System]

When the time (independent) variable t varies discretely instead of varying continuously, we have a *discrete dynamical system*. In this scenario, we deal with a *difference equation* instead of a differential equation. A difference equation of nth order is given by

$$x_{n+1+k} = f(x_{1+k}, \cdots, x_{n+k}), \tag{1.2.42}$$

for $k = 0, 1, \cdots$ and $n = 1, 2, \cdots$, for an appropriate function f defined on a subset of $\mathbb{R}^n$. For a given set of arbitrary $x_1, \cdots, x_n$, the given equation (1.2.42) will generate $x_{n+k}, k = 1, 2, \cdots$ and we need to study the sequence $\{x_m\}$, for its boundedness, convergence, etc.

A very familiar example of a difference equation is the *Fibonacci sequence* generated by any given arbitrary real numbers x_1, x_2 and satisfying the difference equation

$$x_{n+2} = x_n + x_{n+1},$$

for $n = 1, 2, \cdots$. This is a second order equation. Another example is given by the logistic map:

$$x_{n+1} = ax_n(1 - x_n),$$

for $n = 1, 2, \cdots$. This is a first order equation. The constant $a \in [0,4]$. Thus, if $x_1 \in [0,1]$, then $x_n \in [0,1]$ for all $n > 1$. The logistic map has been studied extensively and it reveals many surprising properties of the sequence $\{x_n\}$ for a certain range of values of a.

We will not pursue this subject in this book, but the interested reader may look into, for example in [Hao84, Wig90].

1.3 Exercises

1. Consider the initial value problem[2]

$$\frac{dy}{dt} = ay(t) - by^2(t), \ y(t_0) = y_0$$

where $a, b > 0, t_0, y_0 \in \mathbb{R}$. Assume the unique existence of the (local) solution $y = y(t)$ in the interval (t_1, t_2) with $t_0 \in (t_1, t_2)$.

   (a) Without attempting an explicit representation of the solution, show that y satisfies sign $(y(t)(y(t) - a/b)) = \text{sign}(y_0(y_0 - a/b)))$.

   (b) Now solve the IVP to get the implicit form

$$\log \frac{|y|}{|y_0|} \frac{|a - by_0|}{|a - by|} = t - t_0.$$

   (c) Use the first part to obtain the solution y in the explicit form

$$y(t) = \frac{ay_0}{by_0 + (a - by_0)e^{-a(t-t_0)}}$$

   (d) In each of the cases of the first part, describe the maximal interval (t_*, t^*), where the solution y is defined. This is referred to as the maximal interval of existence, which will be discussed in detail in Chapter 4. (Note that t^* can be $+\infty$ or t_* can be $-\infty$). Further, compute the limits

$$\lim_{t \downarrow t_*} y(t) \quad \text{and} \quad \lim_{t \uparrow t^*} y(t).$$

---

[2]The methods of solutions are described in Chapter 3

(e) In each of the cases, find $\frac{dy}{dt}$, $\frac{d^2y}{dt^2}$ and analyse the shape of the curve.

(f) Find the conditions on y_0 so that $t_* = -\infty$ and/or $t^* = +\infty$.

(g) Plot the graphs of the solutions y in the ty plane for different values of y_0.

(h) Let $y = y(t)$ be the solution as earlier and $z = z(t)$ be the solution to the initial value problem:

$$\frac{dz}{dt} = az(t) - bz^2(t), \; z(t_1) = y_0.$$

Represent z in terms of y. Sketch with different initial times. Do you observe any property? Describe the observed properties for the general problem

$$\frac{dy}{dt} = f(y(t)), \; y(t_0) = y_0.$$

2. Consider the modified population model with a real parameter λ, namely

$$\frac{dy}{dt} = ay(t) - by^2(t) - \lambda, \; y(t_0) = y_0.$$

Do a similar analysis for various values of the parameter. More precisely, show that there is a critical value λ_{cr} such that for $\lambda > \lambda_{cr}$, the behaviour is exactly similar, but for $\lambda < \lambda_{cr}$, the behaviour of the solution is completely different.

3. Consider the linear model of the atomic waste disposal problem:

$$\frac{dV}{dt} + \frac{cg}{W}V(t) = \frac{g}{W}(W - B), \; V(0) = 0,$$

where $V = V(t)$ is the velocity at time t.

(a) Find the solution V and find the limit $\lim_{t \to \infty} V(t)$.

(b) Now derive the non linear model:

$$\frac{v}{W - B - cv}\frac{dv(y)}{dy} = \frac{g}{W}, \; v(0) = 0$$

where $v = v(y)$ is the velocity at the distance y, and solve the same to obtain the solution in the implicit form:

$$\frac{gy}{W} = -\frac{v}{c} - \frac{W-B}{C^2}\log\frac{W-B-cv}{W-B}.$$

4. Obtain (1.2.32) and (1.2.33) from (1.2.31) by suitable dilations.

1.4 Notes

We have presented a few real world problems to highlight the importance of modelling using ODEs and their analysis. Of course, the examples are not exhaustive; in fact, one can find several text books devoted to a particular topic, for example, mathematical biology, mechanical systems, etc. We have seen through the atomic waste disposal problem (Section 1.2.3) that through the simple linear model, we can solve the problem explicitly, but incomplete answer to the question set out therein. However, a little reformulation gives us a non-linear equation, which in general is hard to solve, yet gives us a complete answer to the question. This exhibits the importance of correct modelling and its analysis even if the solution is not available in explicit form. Such phenomena can be observed in other models like population growth (Sections 1.2.1 and 1.2.2). One should bear in mind such peculiarities arising in the analysis of ODEs. In general, it is hard to obtain explicit or implicit or even a representation of a solution leading to the necessity of analysing the solution in the absence of such forms.

A large number of real life examples are available in Martin Brown [Bra78, Bra75]. See also [AMR95, TS86]

2

Preliminaries

2.1 Introduction

In this chapter, we present some topics from linear algebra and analysis which are extensively used in the subsequent chapters of the book. Our discussion will only be brief and more details and longer proofs may be found in the references cited at the end of this chapter. Quite often, the explicit solution may not be available and we may appeal to the analysis to derive the qualitative nature of the solution, which in turn may help us to arrive at conclusions about the behaviour of the physical or biological problems modelled through ODE. Even when the explicit solution is known, it may be hard to draw significant conclusions regarding the global behaviour of the system. We have therefore emphasized the importance of analysis and linear algebra throughout this book, with the hope that the beginner starts appreciating the essential role of these subjects in the study of ODE.

2.2 Preliminaries from Real Analysis

First, we recall the concepts of pointwise convergence, uniform convergence, etc. of a sequence of functions of one variable and Lipschitz continuity of functions.

2.2.1 Convergence and uniform convergence

Let I be any interval in $\mathbb{R}$. Let $f_n : I \to \mathbb{R}$, $n = 1, 2, \ldots$, be a sequence of functions. We say that the sequence $\{f_n\}$ converges pointwise to a function $f : I \subset \mathbb{R} \to \mathbb{R}$, if the numerical sequence $\{f_n(x)\}$ converges to $f(x)$ for every $x \in I$.

Consider the sequence

$$f_k(x) \;=\; \frac{kx}{kx+1}, \; x \in [0,1], \; k = 1,2,\cdots \qquad\qquad (2.2.1)$$

Clearly $f_k(0) = 0$ for all k and hence $f_k(0) \to 0$, whereas it is easy to see that for $x \in (0,1]$, the sequence $f_k(x) \to 1$. Thus, the limit function is

$$f(x) = \begin{cases} 0 & \text{if} \;\; x = 0 \\ 1 & \text{if} \;\; x \in (0,1]. \end{cases}$$

Notice that each function f_k defined here is continuous on $I = [0,1]$, but the limit function f is discontinuous at $x = 0$. We thus lost the important property of continuity under the pointwise convergence. Therefore, we now discuss a stronger convergence under which the continuity property is preserved. This is the notion of *uniform convergence*.

Definition 2.2.1

[Uniform Convergence] Let $f_k, f : I \to \mathbb{R}$, $k = 1,2,\ldots$ be functions defined on I. Then, the sequence f_k is said to converge to f uniformly in I if for any given $\varepsilon > 0$, there exists $N = N(\varepsilon) \in \mathbb{N}$ such that $|f_k(x) - f(x)| < \varepsilon$ for all $k \geq N$ and for all $x \in I$.

We remark that in the pointwise convergence, N depends both on ε and x, whereas in uniform convergence N depends only on ε. This uniformity in x has far reaching consequences including preservation of continuity, interchange of the limit and the integral.

Theorem 2.2.2

Let $f_k : I \to \mathbb{R}$, $k = 1,2,\ldots$ be a sequence of continuous functions that converges uniformly to a function $f : I \to \mathbb{R}$. Then f is continuous.

This theorem immediately shows that the sequence of functions given by (2.2.1) does not converge uniformly as f is not continuous on $[0,1]$.

Consider the example $f_k(x) = x^k, x \in [0,1]$. It is easy to see that $f_k(x) \to f(x)$ point wise, where

$$f(x) = \begin{cases} 0, & 0 \le x < 1 \\ 1, & x = 1, \end{cases}$$

which is discontinuous at $x = 1$. The reader can directly verify that the aforementioned convergence is not uniform without appealing to the theorem. However, it is not hard to see that the convergence is uniform in $[0, 1 - \eta]$ for any $0 < \eta < 1$.

On the other hand, the sequence f_k defined by $f_k(x) = \dfrac{kx^2}{k|x| + 1}$ converges to $f(x) = |x|$, pointwise in $\mathbb{R}$; however, the convergence is uniform only in bounded intervals. Note that the limit function is also continuous in $\mathbb{R}$.

We remind the reader that the converse of Theorem 2.2.2 may not be true. That is, $f_k \to f$ pointwise, f_k, f are continuous may not imply that the convergence is uniform.

Example 2.2.3

Consider $g_k : [0, 1] \to \mathbb{R}$ defined by

$$g_k(x) = \begin{cases} kx, & 0 \le x \le \dfrac{1}{k} \\ k\left(\dfrac{2}{k} - x\right), & \dfrac{1}{k} \le x \le \dfrac{2}{k} \\ 0, & \dfrac{2}{k} \le x \le 1. \end{cases}$$

The reader can easily verify that $g_k \to g \equiv 0$ pointwise and g_k, g are continuous, but the convergence is not uniform.

Theorem 2.2.4

Let $f_k : [a, b] \to \mathbb{R}$ be a sequence of functions converging uniformly to a function $f : [a, b] \to \mathbb{R}$. Assume f_n, f are Riemann integrable on $[a, b]$, then

$$\lim_{k\to\infty}\int_a^b f_k(t)\,dt = \int_a^b f(t)\,dt.$$

In particular, if each f_k is continuous, the integrability condition is automatically satisfied. Again, we remark that uniform convergence may not be a necessary condition for the validity of Theorem 2.2.4. This can be seen from Example 2.2.3 because $\int_0^1 g_k \to 0 = \int_0^1 0$. At the same time, in general, we may not be able to interchange the limit and integral signs if the convergence is not uniform.

In view of Theorem 2.2.2 and Theorem 2.2.4, for a given sequence of functions, extracting a uniformly convergent subsequence is very important in analysis. In this direction, we need to have conditions under which one can derive uniformly convergent subsequences. A well-known theorem is the *Arzela–Ascoli theorem*. Before stating this result, we introduce some more concepts.

We discuss the convergence and uniform convergence of series of functions. Let $\{u_k\}$ be a sequence of functions defined on I. Consider the sequence of partial sums $f_k = \sum_{i=1}^k u_i$. If the sequence f_k converges pointwise (respectively, uniformly) to a function u on I, then we say that the infinite series, denoted by $\sum_{k=1}^\infty u_k$ converges pointwise (respectively, uniformly) to u on I.

Theorem 2.2.2 and Theorem 2.2.4 are valid for series under appropriate hypotheses.

Theorem 2.2.5

[Weierstrass M-Test] Let M_k be a sequence of positive constants such that the series $\sum_{k=1}^\infty M_k$ converges. Suppose $\{u_k\}$ is a sequence of functions defined on I such that $|u_k(x)| \le M_k$, for all $x \in I$. Then, the series $\sum_{k=1}^\infty u_k$ converges uniformly to a function u on I.

For example, the series $\sum_{k=1}^\infty \dfrac{\sin kx}{k^2}$, $x \in \mathbb{R}$ converges uniformly since the series $\sum_{k=1}^\infty \dfrac{1}{k^2}$ is convergent.

Definition 2.2.6

[Uniform Boundedness] A sequence of functions $\{f_k\}$ defined on I is said to be uniformly bounded if there exists a constant $M > 0$ such that $|f_k(x)| \leq M$, for all $x \in I$, for all $k \in \mathbb{N}$.

Definition 2.2.7

[Equicontinuity] A sequence of functions $\{f_k\}$ defined on I is said to be equicontinuous on I, if for every $\varepsilon > 0$, there exists $\delta > 0$ such that $|f_k(x) - f_k(y)| < \varepsilon$ whenever $x, y \in I$ and $|x - y| < \delta$, and for all k.

Clearly, if the family of functions is equicontinuous, then each member in the family is uniformly continuous. So, in Definition 2.2.7, we require the existence of uniform δ which works for all the members of the family.

Choose $f \in C(\mathbb{R})$ with compact support, that is, f is identically zero outside a compact set. Then, the sequence $f_k(x) = \dfrac{1}{k} f(x)$ is a family of equicontinuous functions. On the other hand, it is not difficult to verify that the family of functions $f_k(x) = x^k, 0 \leq x \leq 1, k = 1, 2, \ldots$ is not an equicontinuous family; however, each member of the family is a uniformly continuous function, as it is a continuous function on a closed and bounded interval.

Denote by $C[a, b]$, the space of all continuous functions $f : [a, b] \to \mathbb{R}$.

Theorem 2.2.8

[Arzela–Ascoli] Let $\{f_k\}$ be a sequence of functions in $C[a, b]$ which is uniformly bounded and equicontinuous. Then, there exists a subsequence $\{f_{k_n}\}$ of $\{f_k\}$ such that $\{f_{k_n}\}$ converges uniformly to a function $f \in C[a, b]$.

A proof can be found in several books, see, for instance, [Rud76, CL72].

Definition 2.2.9

[Lipschitz Continuity] A function $f : D \subseteq \mathbb{R} \to \mathbb{R}$, is said to be locally Lipschitz in D if for any $x_0 \in D$, there exists a neighbourhood N_{x_0} of x_0 and an $\alpha = \alpha(x_0) > 0$ such that

$$|f(x) - f(y)| \leq \alpha |x - y|, \text{ for all } x, y \in N_{x_0}.$$

The function $f : D \subset \mathbb{R} \to \mathbb{R}$ is said to be Lipschitz (or globally Lipschitz) in D if there exists $\alpha > 0$ such that

$$|f(x) - f(y)| \leq \alpha |x - y|, \text{ for all } x, y \in D. \tag{2.2.2}$$

The smallest α satisfying (2.2.2) is called the *Lipschitz constant* of f. If f is Lipschitz (globally), then it is uniformly continuous. We also have the following result giving a sufficient condition for Lipschitz continuity; the proof trivially follows from the mean value theorem.

Theorem 2.2.10

Suppose D is an open interval in $\mathbb{R}$ and $f : D \to \mathbb{R}$ is differentiable on D and $\alpha = \sup_{x \in D} |f'(x)| < \infty$. Then, f is Lipschitz with a Lipschitz constant less than or equal to α.

Thus, the class of Lipschitz continuous functions is quite large and includes all our familiar functions: polynomial functions, polynomials of sine and cosine functions, exponential functions, etc. The definition of Lipschitz continuity of a function of one variable given in Definition 2.2.9 can be analogously extended to any vector valued map $\mathbf{f} : (a,b) \times D \to \mathbb{R}^n$, where (a,b) is an interval in $\mathbb{R}$ and D is a domain in $\mathbb{R}^n$. In Chapter 4, which includes a discussion on the existence of solutions, we need to introduce the Lipschitz continuity of $\mathbf{f}(t, \mathbf{y})$ with respect to the second argument $\mathbf{y}$ keeping t as a parameter. While defining Lipschitz continuity in the single variable case, we used the usual absolute value to denote the distance between any two real numbers. We need a similar concept of distance in the several variables case, which we introduce now.

   In $\mathbb{R}^n$, we define a notion of distance of a vector from the origin, in terms of a function called *norm*.

Definition 2.2.11

[Norm] A norm, denoted by $\|.\|$ on $\mathbb{R}^n$ is a mapping from $\mathbb{R}^n \to \mathbb{R}$ that satisfies:

1. $\|\mathbf{x}\| \geq 0$; $\|\mathbf{x}\| = 0$ if and only if $\mathbf{x} = \mathbf{0}$,
2. $\|a\mathbf{x}\| = |a| \|\mathbf{x}\|$,

3. (Triangle inequality) $\|\mathbf{x}+\mathbf{y}\| \le \|\mathbf{x}\| + \|\mathbf{y}\|$,

for all $\mathbf{x},\mathbf{y} \in \mathbb{R}^n$ and scalars a.

See also Definition 2.4.2 for a general definition of a normed linear space. The following are some examples of norms in $\mathbb{R}^n$. For $\mathbf{x} = (x_1, x_2, \cdots, x_n) \in \mathbb{R}^n$, define, for $\mathbf{x} \in \mathbb{R}^n$

$$\|\mathbf{x}\|_p = \left(\sum_{i=1}^{n} |x_i|^p \right)^{1/p}, 1 \le p < \infty \quad \text{and} \quad \|\mathbf{x}\|_\infty = \max_{1 \le i \le n} |x_i|.$$

Usually, it is the third property of the norm that does not follow in an obvious way and needs proof. In the context of $\mathbb{R}^n$, it is called *Minkowski's inequality*. When $p = 2$, it is the usual Euclidean norm (or distance).

It is convenient to take $p = 1$ for the discussion in Chapter 4 and we write $\|\cdot\|_1 = |\cdot|$ and state the definition of Lipschitz continuity now in terms of this $1-$norm.

Definition 2.2.12

A function $\mathbf{f}(t,\mathbf{y}) : (a,b) \times D \to \mathbb{R}^n$ is said to be Lipschitz continuous (globally) with respect to $\mathbf{y}$ if there exists $\alpha > 0$ such that

$$|\mathbf{f}(t,\mathbf{y}_1) - \mathbf{f}(t,\mathbf{y}_2)| \le \alpha |\mathbf{y}_1 - \mathbf{y}_2|$$

for all $(t,\mathbf{y}_1)$ and $(t,\mathbf{y}_2)$ in $(a,b) \times D$.

The smallest such constant α is known as the Lipschitz constant of $\mathbf{f}$. We can also define local Lipschitzness analogously.

We will now see some examples of Lipschitz continuous functions.

Example 2.2.13

(i) $f(t,y) = t + 3y$ is globally Lipschitz continuous in $\mathbb{R}$ with respect to y.

(ii) $f(t,y) = t^2 + y^2$ is Lipschitz continuous on any bounded domain D in $\mathbb{R}^2$.

(iii) $f(t,y) = t^2 y^3$ is Lipschitz continuous on any bounded domain in $\mathbb{R}^2$.

(iv) For the function $f(t,y) = e^{-t^2}y^2 \sin t$ on $D = \{(t,y) : 0 \le y \le 2,$ $t \in \mathbb{R}\}$, we have

$$|f(t,y_1) - f(t,y_2)| = |e^{-t^2}\sin t||y_1 + y_2||y_1 - y_2| \le 4|y_1 - y_2|$$

for any $(t,y_1), (t,y_2)$ in D. Thus, $f(t,y)$ is Lipschitz continuous on the strip D.

(v) $f(t,y) = t\sqrt{y}$ on the rectangle $D = \{(t,y) : 0 \le t \le 1, 0 \le y \le 1\}$

Note that

$$|f(1,y) - f(1,0)| = \sqrt{y} = \frac{1}{\sqrt{y}}|y - 0|$$

and $\dfrac{1}{\sqrt{y}} \to +\infty$ as $y \to 0^+$. Hence, the function f is not Lipschitz continuous on the rectangle D, but is continuous on D.

Here we state a sufficient condition for Lipschitz continuity of $\mathbf{f}(t,\mathbf{y})$ with respect to $\mathbf{y}$.

Theorem 2.2.14

Let $\mathbf{f} : (a,b) \times D \to \mathbb{R}^n$ be a C^1 vector valued function, where D is a convex domain in $\mathbb{R}^n$ such that

$$\sup_{(t,\mathbf{y}) \in (a,b) \times D} \left| \frac{\partial f_i}{\partial y_j}(t,\mathbf{y}) \right| = \alpha < \infty,$$

for $i, j = 1, 2, \cdots, n$. Then, $\mathbf{f}(t,\mathbf{y})$ is Lipschitz continuous on $(a,b) \times D$ with respect to $\mathbf{y}$ having a Lipschitz constant less than or equal to a multiple of α.

The convexity assumption means, by definition, that the line segment joining any two points of D lies inside D.

Example 2.2.15

Let $f(t,y) = t + y^2$ on $D : |t| \le a, |y| \le b$. Then, $\frac{\partial f}{\partial y} = 2y$ which implies $\left| \frac{\partial f}{\partial y}(t,y) \right| = |2y| \le 2b$. Thus, $f(t,y)$ is Lipschitz continuous on D with Lipschitz constant $\alpha = 2b$.

Example 2.2.16

Let $f(t,y) = t|y|$ on $D : |t| \leq a, |y| \leq b$. Then,

$$|f(t,y_1) - f(t,y_2)| = |t|y_1| - t|y_2|| \leq |t||y_1 - y_2| \leq a|y_1 - y_2|.$$

Note that $f(t,y)$ is Lipschitz continuous, but $\dfrac{\partial f}{\partial y}$ does not exist at any point $(t,0) \in D$ for which $t \neq 0$. Thus, Lipschitz continuity is a smoothness property stronger than continuity, but weaker than differentiability, locally.

A few more results from analysis, namely calculus lemma, differentiation under integral sign and Taylor's formula are presented here.

Lemma 2.2.17

[Calculus Lemma] Let (a,b) be a finite or infinite interval and $h : (a,b) \to \mathbb{R}$ satisfy either

(i) h is bounded above and non-decreasing or

(ii) h is bounded below and non-increasing,

then, $\lim\limits_{t \to b} h(t)$ exists.

The lemma is important in the sense that the boundedness alone will not give the existence of limit. For example, $f(x) = \sin\dfrac{1}{x}$ is bounded above and below for $0 < x < 1$, but $\lim\limits_{x \to 0} \sin\dfrac{1}{x}$ does not exist.

Proof: If h satisfies (i), then $-h$ satisfies (ii) and vice versa. Hence, it is enough to prove with one of the assumptions. So assume (i) holds true. Put $\alpha = \sup\limits_{t \in (a,b)} h(t) < \infty$ as h is bounded above. For any $\varepsilon > 0$, there exists $t_0 \in (a,b)$ such that $\alpha - \varepsilon < h(t_0) \leq \alpha$. If $t \in (a,b), t > t_0$, then $\alpha - \varepsilon \leq h(t_0) \leq h(t) \leq \alpha$ which implies $0 \leq \alpha - h(t) \leq \varepsilon$ for all $t \geq t_0$, that is, $\lim\limits_{t \to b} h(t) = \alpha$. $\qquad\square$

We know that if $f : [a,b] \to \mathbb{R}$ is continuous, then $F(t) = \displaystyle\int_a^t f(s)\, ds$ is differentiable, $F'(t) = f(t)$ and F is referred to as an *antiderivative* of f.

Often in integral calculus, we employ the method of substitution, which is also called the *change of variable formula*, to evaluate integrals. This is also the basis for the method of separable variables to solve certain first order ODE. The formula is an easy consequence of the chain rule used in differentiation.

Let f be a (Riemann) integrable function over an interval $[a,b]$ in $\mathbb{R}$. If F is an anti derivative of f, then F is differentiable and

$$\int_a^b f(t)\,dt = F(b) - F(a),$$

provided f is continuous. If u is a differentiable function defined on some interval J in $\mathbb{R}$, then by chain rule, we have

$$(F \circ u)'(t) = F'(u(t))u'(t) = f(u(t))u'(t),$$

for $t \in [a,b]$, provided that the compositions of the functions are well-defined; here $'$ denotes the derivative of the function in question. Therefore, we obtain the formula

$$\int_a^b f(u(t))u'(t)\,dt = F(u(b)) - F(u(a)) = \int_{u(a)}^{u(b)} f(s)\,ds.$$

The foregoing discussion on change of variable is stated in the following theorem.

Theorem 2.2.18

[Change of Variable Formula]

Let $g : [c,d] \to \mathbb{R}$ be a C^1 function, that is, g is continuously differentiable and let $[a,b]$ be any interval containing the image of g, that $g[c,d] \subset [a,b]$. If $f : [a,b] \to \mathbb{R}$ is a continuous function, then

$$\int_c^d f(g(t))g'(t)\,dt = \int_{g(c)}^{g(d)} f(x)\,dx.$$

The change of variable formula is very important in many occasions and it is equally important to understand certain symbolic notations we use frequently. For example, to compute $\int y^3\,dy$ by substitution, say $y^2 = t$, we quite often write $2y\,dy = dt$ which in turn gives $\int y^3\,dy = \dfrac{1}{2}\int t\,dt$.

How do we interpret the symbolic notation $2y\,dy = dt$? This can be done via the change of variable formula; take $f(y) = y^3$ and $g(t) = \sqrt{t}$, then, by Theorem 2.2.18, $\displaystyle\int y^3\,dy = \int (\sqrt{t})^3 \frac{1}{2\sqrt{t}}\,dt = \frac{1}{2}\int t\,dt.$

We now discuss an important result known as *differentiation under the integral sign*.

Theorem 2.2.19

[Generalized Leibnitz Formula]

Let $\alpha, \beta : [a,b] \to \mathbb{R}$ be differentiable functions and c, d be real numbers satisfying

$$c \leq \alpha(t), \beta(t) \leq d, \quad \text{for all } t \in [a,b].$$

Let $f : [a,b] \times [c,d] \to \mathbb{R}$ be a continuous function such that $\frac{\partial f}{\partial t}(t,s)$ is also continuous. Define

$$F(t) = \int_{\alpha(t)}^{\beta(t)} f(t,s)\,ds.$$

Then, F is differentiable and

$$\frac{dF}{dt} = \int_{\alpha(t)}^{\beta(t)} \frac{\partial f}{\partial t}(t,s)\,ds + f(t,\beta(t))\frac{d\beta}{dt} - f(t,\alpha(t))\frac{d\alpha}{dt}. \qquad (2.2.3)$$

We mention two particular cases of (2.2.3) which are often used. The first one is obtained by taking $\alpha(t) = a$, a constant and $\beta(t) = t$. Then, we obtain

$$\frac{d}{dt}\left(\int_a^t f(t,s)\,ds\right) = \int_a^t \frac{\partial f}{\partial t}(t,s)\,ds + f(t,t).$$

For the other, we take $f = f(t)$, a one variable function, $\alpha(t) = a$, a constant and $\beta(t) = t$ and obtain the fundamental theorem of calculus:

$$\frac{d}{dt}\left(\int_a^t f(s)\,ds\right) = f(t).$$

Theorem 2.2.20

[Taylor's Formula]

Let $f : (a,b) \to \mathbb{R}$ be a C^2 function, that is twice continuously differentiable function and $x_0 \in (a,b)$. Then

$$f(x_0 + y) = f(x_0) + f'(x_0)y + \frac{1}{2}f''(\zeta)y^2$$

for some point ζ between x_0 and $x_0 + y$. Here $'$ denotes differentiation with respect to x.

More generally, in the several variables case, let $f : D \to \mathbb{R}$ be a C^2 function, where D is an open set in $\mathbb{R}^n$. Let $\mathbf{x}_0 \in D$ and $r > 0$ be such that $B(\mathbf{x}_0, r) \subset D$. For $\mathbf{y} \in B(\mathbf{x}_0, r)$, define

$$F(t) = f(\mathbf{x}_0 + t\mathbf{y}),$$

for $0 \leq t \leq 1$ and $\mathbf{x}_0 + t\mathbf{y} \in B(\mathbf{x}_0, r)$. Applying the one variable Taylor formula, we get (in the following $'$ denotes differentiation with respect to t)

$$F(1) - F(0) = F'(0) + \frac{1}{2}F''(\zeta)$$

for some $\zeta \in [0,1]$, that is,

$$f(\mathbf{x}_0 + \mathbf{y}) - f(\mathbf{x}_0) = F'(0) + \frac{1}{2}F''(\zeta).$$

By chain rule, $F'(t) = \sum_{j=1}^{n} \frac{\partial f}{\partial x_j}(\mathbf{x}_0 + t\mathbf{y})y_j$. Hence,

$$F'(0) = \nabla f(\mathbf{x}_0) \cdot \mathbf{y},$$

where

$$\nabla f(\mathbf{x}_0) = \left(\frac{\partial f}{\partial x_1}(\mathbf{x}_0), \cdots, \frac{\partial f}{\partial x_n}(\mathbf{x}_0) \right)$$

is the gradient of f at $\mathbf{x}_0$. Doing a further differentiation of $F'(t)$, we get

$$F''(t) = \sum_{i,j=1}^{n} \frac{\partial^2 f}{\partial x_i \partial x_j}(\mathbf{x}_0 + t\mathbf{y})y_i y_j.$$

Thus, $F''(\zeta) = O(|\mathbf{y}|^2)$ contains terms of quadratic and higher orders in $\mathbf{y}$. Hence, it follows that

$$f(\mathbf{x}_0 + \mathbf{y}) = f(\mathbf{x}_0) + \nabla f(\mathbf{x}_0) \cdot \mathbf{y} + O(|\mathbf{y}|^2).$$

We can extend this result further to multi-valued cases. Let $\mathbf{f} : D \subset \mathbb{R}^n \to \mathbb{R}^n$. Hence, $\mathbf{f} = (f_1, \cdots, f_n)$ is considered as a column vector and each $f_i : D \to \mathbb{R}$. Thus, the formula holds for each f_i. Using the matrix notation,

$$D\mathbf{f}(\mathbf{x}_0) = \begin{bmatrix} \nabla f_1(\mathbf{x}_0) \\ \cdot \\ \cdot \\ \cdot \\ \nabla f_n(\mathbf{x}_0) \end{bmatrix}$$

which is an $n \times n$ matrix and we may write

$$\mathbf{f}(\mathbf{x}_0 + \mathbf{y}) = \mathbf{f}(\mathbf{x}_0) + D\mathbf{f}(\mathbf{x}_0)\mathbf{y} + O(|\mathbf{y}|^2).$$

Note that $\nabla f(\mathbf{x}_0) \cdot \mathbf{y}$ is the dot product, whereas $D\mathbf{f}(\mathbf{x}_0)\mathbf{y}$ is the action of the matrix $D\mathbf{f}(\mathbf{x}_0)$ on the vector $\mathbf{y}$.

2.3   Fixed Point Theorem

We now introduce the notion of a metric space.

Definition 2.3.1

[Metric Space] Let X be a non-empty set. A metric d is a mapping $d : X \times X \to \mathbb{R}$ which satisfies the following properties:

1. $d(x,y) \geq 0$; $d(x,y) = 0$ if and only if $x = y$,
2. (Symmetry) $d(x,y) = d(y,x)$,
3. (Triangle inequality) $d(x,y) \leq d(x,z) + d(z,y)$,

for all $x, y, z \in X$. The set X together with the metric d is called a *metric space*. We refer to (X, d) as a metric space; when the metric in the context is clear, we say X is a metric space.

The n dimensional real and complex Euclidean spaces, $\mathbb{R}^n$ and $\mathbb{C}^n$ are examples of metric spaces, where the metric d is defined by $d(\mathbf{x}, \mathbf{y}) =$

$$\left(\sum_{i=1}^{n} |x_i - y_i|^2 \right)^{1/2} \quad \text{for all } \mathbf{x} = (x_1, \cdots, \mathbf{x}_n), \mathbf{y} = (y_1, \cdots, y_n) \in \mathbb{R}^n \text{ or } \mathbb{C}^n.$$

This is the standard Euclidean metric or distance. There are many other metrics we can introduce on $\mathbb{R}^n$ and $\mathbb{C}^n$. We will see more examples later.

Let (X, d) be a metric space. A sequence $\{x_k\}$ in X is said to converge to a point $x \in X$ if for given $\varepsilon > 0$, there exists $N \in \mathbb{N}$ such that $d(x_k, x) < \varepsilon$ for all $k \geq N$. This statement may also be written as $d(x_k, x) \to 0$ or $x_k \to x$ as $k \to \infty$. It is easy to see, from the triangle inequality, that if $x_k \to x$ and $x_k \to y$, then $x = y$.

A sequence $\{x_n\} \subset X$ is said to be a *Cauchy sequence*, if $d(x_n, x_m) \to 0$ as $n, m \to \infty$. A metric space (X, d) is said to be a *complete metric space* if every Cauchy sequence in X converges in X. A normed linear space which is a complete metric space (metric induced by the norm) is called a *Banach space* (see, Definition 2.4.2).

In particular, if $X = \mathbb{R}^n, \mathbf{u}_k \in \mathbb{R}^n$ converges to $\mathbf{u} \in \mathbb{R}^n$ if $|\mathbf{u}_k - \mathbf{u}| \to 0$ as $k \to \infty$ and $\mathbb{R}^n$ is a Banach space. It is also a Banach space under the norms

$$\|\mathbf{x}\|_p = \left(\sum_{i=1}^{n} |x_i|^p \right)^{1/p}, 1 \leq p < \infty,$$

and

$$\|\mathbf{x}\|_\infty = \max_{1 \leq i \leq n} |x_i|,$$

for $\mathbf{x} \in \mathbb{R}^n$. The function space $C[0, 1]$ or, more generally, $C[a, b]$ with sup norm is a Banach space. However, it is *not* a complete space with respect to $\| \cdot \|_1$ introduced earlier. The completeness plays a crucial role in the fixed point theorem to be studied later.

It is also easy to check that for a sequence $\{f_n\} \subset C[a, b]$, the statement $f_n \to f$ in sup norm is equivalent to saying that f_n converges uniformly to f.

Suppose (X, d) is a metric space, $x \in X$, $r > 0$. The set $B_r(x) \equiv \{y \in X : d(x, y) < r\}$ is called an *open ball* of radius r centred at x. The collection $\{B_r(x) : x \in X, r > 0\}$ forms a *basis* for a *topology* in X. This is referred to as the *topology induced by the metric d in X*. An *open set* in X is, by definition, an arbitrary union of open balls. A subset of X is *closed* if its complement is open in X.

If $Y \subset X$, then (Y,d) is also a metric space. If (X,d) is a complete metric space and Y is a *closed* (referring to the metric topology) subset of X, then (Y,d) is also a complete metric space. In particular, the *closed* balls $\bar{B}_r(x) \equiv \{y \in X : d(x,y) \le r\}$ are complete metric spaces in a complete metric space (X,d).

A subset A of X is said to be *bounded*, if there exists an $M > 0$ such that $d(x,y) \le M$ for all $x,y \in A$. The *smallest* such M is called the *diameter* of A. A subset A of X is said to be *compact* [1] if given any sequence $\{x_n\} \subset A$, there is a subsequence $\{x_{n_k}\}$ which converges to some element *in* A. If $A \subset X$ is compact, it is easy to see that A is closed and bounded. However, the converse may not be true. In $\mathbb{R}^n$, A is compact if and only if A is closed and bounded. This is the *Heine–Borel theorem*.

In this section, we present the Banach fixed point theorem which is used in the existence result. Suppose $T : X \to X$ is a mapping. A point $x^* \in X$ is said be a *fixed point* of T if $Tx^* = x^*$.

Theorem 2.3.2

[Banach Fixed Point Theorem] Suppose (X,d) is a complete metric space and $T : X \to X$ is a contraction, that is, there exists an $\alpha \in (0,1)$ such that

$$d(Tx,Ty) \le \alpha \, d(x,y) \tag{2.3.1}$$

for all $x,y \in X$. Then, T has a unique fixed point $x^* \in X$. Further, the sequence $\{x_k\}$ defined by $x_k = Tx_{k-1}$, $x_0 \in X$ is arbitrary and $k = 1,2,\ldots$, converges to x^*.

We remark that many interesting problems in mathematics can be formulated in terms of finding fixed points of appropriate maps. There are different types of fixed point theorems in different contexts, but probably Theorem 2.3.2 is the easiest one to apply and prove as well. Indeed, the condition (2.3.1) is stringent and may not be found easily.

In general, if we omit either the completion or contraction condition, we may not get a fixed point. For example, consider $f : \mathbb{R} \to \mathbb{R}$ defined by $f(x) = x + 1$. Then $d(f(x),f(y)) = |f(x) - f(y)| = |x - y|$. Hence, $\alpha = 1$ and f does not have any fixed point. On the other hand, the function $f : (0,1) \to (0,1)$ defined by $f(x) = mx$, for some m, $0 < m < 1$, is a

---

[1] There are other notions of compactness in general topological spaces. It turns out that all these are equivalent to the one given here for a metric space.

contraction with $\alpha = m$, but has no fixed point. In the second example, $(0,1)$ is not complete.

Proof of Theorem 2.3.2:   We will sketch the proof. Choose any $x_0 \in X$ and define the sequence $x_1 = Tx_0$, $x_2 = T^2x_0$, $\cdots$, $x_k = T^kx_0$, $\cdots$. The sequence $\{x_k\}$ is a Cauchy sequence in X. To see this, observe that

$$d(x_{n+1},x_n) = d(Tx_n, Tx_{n-1}) \leq \alpha d(x_n, x_{n-1})$$

and by induction, we get

$$d(x_{n+1},x_n) \leq \alpha^n d(x_1,x_0).$$

Next, for any $m < n$, using triangle inequality, we have

$$d(x_n,x_m) \leq d(x_n,x_{n-1}) + \cdots + d(x_{m+1},x_m) \leq \alpha^m \frac{1 - \alpha^{n-m-1}}{1 - \alpha} d(x_1,x_0)$$

and the right-most term tends to 0 as $n,m \to \infty$. This proves x_k is a Cauchy sequence and by completeness, there exists an $x^* \in X$ such that $x_k \to x^* \in X$. By continuity of T, we get $Tx_k \to Tx^*$. But $Tx_k = x_{k+1} \to x^*$. Thus, $Tx^* = x^*$. The uniqueness can also be proved easily.   □

The interesting advantage in this proof is that it is constructive. It gives us a method to get the fixed point and to obtain an approximate one to any desired accuracy. The second and more useful fact is that we can start at any point in X as the initial guess. Quite often, in numerics, finding a suitable initial point itself is a big challenge.

Corollary 2.3.3

Let $T : X \to X$ be such that T^k is a contraction for some $k \geq 1$. Then, T has a unique fixed point.

The corollary follows from the theorem. Let x^* be the unique fixed point of T^k, that is, $T^kx^* = x^*$. Applying T, we get $T^k(Tx^*) = Tx^*$ and hence, Tx^* is also a fixed point of T^k. By uniqueness, $Tx^* = x^*$ and thus, the unique fixed point of T^k is also a fixed point of T. If x_1 is another fixed point of T, that is, $Tx_1 = x_1$, then by repeated application of T, we see that $T^kx_1 = x_1$. By uniqueness of the fixed point of T^k, we have $x_1 = x^*$ as required.

2.4   Some Topics in Linear Algebra

To motivate the use of tools of linear algebra in the study of differential equations, we consider a first order system of ODE with constant coefficients written in the form

$$\dot{\mathbf{x}} = \mathbf{A}\mathbf{x}, \tag{2.4.1}$$

where $\mathbf{A}$ is a given real $n \times n$ constant matrix. It is not easy, in general, to describe any qualitative property of a solution to (2.4.1) just by looking at $\mathbf{A}$. Hence, we try to reduce (2.4.1) to an *equivalent system*

$$\dot{\mathbf{y}} = \mathbf{B}\mathbf{y} \tag{2.4.2}$$

where, $\mathbf{B}$ is a matrix *similar* to $\mathbf{A}$. If we can obtain $\mathbf{B}$ in a very simple form so that (2.4.2) is completely or partially decoupled, then it may be possible to describe the qualitative behaviour of a solution $\mathbf{y}$ of (2.4.2); this in turn will describe the nature of a solution $\mathbf{x}$ of (2.4.1).

Our main aim is to describe a procedure to reduce (2.4.1) to (2.4.2). This is done using the *Jordan canonical form*, which requires many important tools developed in a systematic course on linear algebra! We content ourselves with the description of several steps of this reduction. We now introduce the concept of a vector space. Though a vector space may be defined over any field, we only consider the field of real numbers $\mathbb{R}$.

Definition 2.4.1

A real vector space V is a non-empty set possessing a binary operation, called *addition*: $u + v \in V$ whenever $u, v \in V$ and a *scalar multiplication*: $au \in V$, whenever $u \in V$ and $a \in \mathbb{R}$, satisfying the following axioms:

1. $<V, +>$ is an abelian group:

    (i) $u + v = v + u$ for all $u, v \in V$. (commutativity)

    (ii) $(u + v) + w = u + (v + w)$ for all $u, v, w \in V$. (associativity)

    (iii) there exists an additive identity, called *zero vector* and is denoted by $0 \in V$ such that $u + 0 = 0 + u = u$ for all $u \in V$.

    (iv) to each $u \in V$, there exists its *additive inverse*, denoted by $-u \in V$ such that $u + (-u) = (-u) + u = 0$.

2. (Associativity of the scalar multiplication) $a(bu) = (ab)u$ for all $u \in V$ and $a, b \in \mathbb{R}$.

3. (Distributive property) $(a+b)u = au + bu$ and $a(u+v) = au + av$ for all $u, v \in V$ and $a, b \in \mathbb{R}$.

4. $1u = u$ for all $u \in V$; 1 is the multiplicative identity in $\mathbb{R}$.

The elements in a vector V are referred to as *vectors*. It is easy to see that 0 is unique; so is $-u$ for each $u \in V$. The associative property enables us to define the sum $u_1 + \cdots + u_k$ unambiguously for any vectors $u_1, \cdots, u_k \in V$. A vector space is also called a *linear space*.

Examples of vector spaces are given here.

1. Let X be any non-empty set and let $V = \{f : X \to \mathbb{R}\}$ be the set of all real valued functions defined on X. Define, for $f, g \in V$ and $a \in \mathbb{R}$, the following

$$(f+g)(t) = f(t) + g(t), \, t \in X,$$

$$(af)(t) = af(t), \, t \in X,$$

where on the right are the usual addition and multiplication of real numbers. Thus, $f + g, af \in V$ whenever $f, g \in V$ and $a \in \mathbb{R}$. It is easy to check that V is a vector space with these operations. The additive identity in V is the zero function: $0(t) = 0$ for all $t \in X$ and the additive inverse of $f \in V$ is the function $-f$ defined by $(-f)(t) = -f(t), \, t \in X$.

2. If we take $X = \{1, 2, \cdots, n\}$ (n, a given positive integer) in Example 1, then we identify the vector space V with $\mathbb{R}^n$.

3. If instead we take X as an interval in $\mathbb{R}$, then we may consider the subsets of V consisting of polynomial functions, continuous functions, continuously differentiable functions, etc. It is easy to verify that all these are examples of real vector spaces. A continuously differentiable function is one which is differentiable and its derivative is also continuous. Higher order continuously differentiable functions are defined in a similar way.

We now define some important concepts such as linear dependence and independence of vectors, linear span, basis and dimension.

A finite collection of vectors $u_1, \cdots, u_k$ in a vector space V, $k \geq 1$, are said to be *linearly independent* if whenever

$$a_1 u_1 + \cdots + a_k u_k = 0, \tag{2.4.3}$$

for $a_1, \cdots, a_k \in \mathbb{R}$, we have $a_1 = \cdots = a_k = 0$. Otherwise, $u_1, \cdots, u_k$ are said to be *linearly dependent*. Thus, $u_1, \cdots, u_k$ are linearly dependent if there exists $a_1, \cdots, a_k \in \mathbb{R}$, not all zero, such that (2.4.3) holds. The left side of (2.4.3) is referred to as a *linear combination* of $u_1, \cdots, u_k$.

Suppose S is subset of V. The *span of S*, denoted by span(S), is the set of all finite linear combinations of vectors in S:

$$\text{span}(S) = \{a_1 u_1 + \cdots + a_k u_k : k \geq 1, u_1, \cdots, u_k \in S, a_1, \cdots, a_k \in \mathbb{R}\}.$$

A subset S of V is, by definition, said to be *linearly independent* if any finite number of vectors in S are linearly independent. Otherwise, S is *linearly dependent*. The following are immediate from the definition.

- If S is linearly independent, then any non-empty subset of S is also linearly independent.

- If S is linearly dependent, then any superset of S is also linearly dependent.

A non-empty subset M of V is a *subspace* of V if M itself is a vector space with the same addition and scalar multiplication as in V. If M is a non-empty subset of V, it is easy to see that M is a subspace of V if and only if the conditions $u + v \in M$ and $au \in M$ whenever $u, v \in M$ and $a \in \mathbb{R}$ are satisfied.

Note that $\{0\}$ and V are always subspaces of V; these are the *trivial* subspaces of V. For any subset S of V, $span(S)$ is a subspace of V; it is the *smallest* subspace of V containing S, that is, if M is any subspace of V containing S, then M also contains span(S).

The vector space V is said to be infinite dimensional if for each integer $k \geq 1$, there is a linearly independent set $\{u_1, \cdots, u_k\}$ consisting of k vectors in V. If the space V is not infinite dimensional, then we say V is finite dimensional. If V is finite dimensional, then there is an integer $n \geq 1$ such that *every* subset of V containing more than n vectors is linearly dependent. Then, there is a smallest such n. This n is called the *dimension* of V and V is called an n dimensional vector space. It is easy to see that if V is n dimensional, then there is a subset $S = \{u_1, \cdots, u_n\}$ such that S is linearly independent and span$(S) = V$. Such a set S is called a *basis* of V. If $S = \{u_1, \cdots, u_n\}$ is basis, so is $\tilde{S} = \{u_1 + u_2, u_2 + u_3, \cdots, u_{n-1} + u_n\}$. Thus, a basis is not unique.

Definition 2.4.2

[Normed Linear Space] A norm, denoted by $\|.\|$ on a vector space or a linear space X is a mapping from $X \to \mathbb{R}$ that satisfies:

1. $\|x\| \geq 0$; $\|x\| = 0$ if and only if $x = 0$,
2. $\|ax\| = |a| \|x\|$,
3. (Triangle inequality) $\|x + y\| \leq \|x\| + \|y\|$,

for all $x, y \in X$ and scalar a.

Every normed linear space X is a metric space, where the metric is given by $d(x, y) = \|x - y\|$, for all $x, y \in X$. The Euclidean space $\mathbb{R}^n$ and $\mathbb{C}^n$ are normed linear spaces. The space of continuous functions $C[0, 1]$ is a normed space with the norm (known as *sup norm*) given by $\|f\| = \sup\limits_{x \in [0,1]} |f(x)|$, $f \in C[0, 1]$. We can also give different norms in the same space and it is important to give appropriate norms as needed in the applications. The space $C[0, 1]$ can also be equipped with the integral norm $\|f\|_1 = \int_0^1 |f(x)| dx$. These norms are fundamentally different in the sense that $C[0, 1]$ equipped with the sup norm is complete, whereas with the integral norm, it is not complete.

2.4.1   Euclidean space $\mathbb{R}^n$

Any point or vector $\mathbf{x} \in \mathbb{R}^n$, is denoted as a row vector: $\mathbf{x} = (x_1, \cdots, x_n)$, $x_i \in \mathbb{R}$. However, for convenience, we also treat it as a column vector. The set $\mathbb{R}^n$ is a vector space (linear space) over the field $\mathbb{R}$ with the addition and scalar multiplication defined, respectively, by $\mathbf{x} + \mathbf{y} = (x_1 + y_1, \cdots, x_n + y_n)$ and $\alpha \mathbf{x} = (\alpha x_1, \cdots, \alpha x_n)$, for all $\mathbf{x} = (x_1, \cdots, x_n), \mathbf{y} = (y_1, \cdots, y_n) \in \mathbb{R}^n$ and $\alpha \in \mathbb{R}$. The Euclidean norm, metric and inner product (usual dot or scalar product) are, respectively, defined by

$$|\mathbf{x}|^2 = \sum_{i=1}^n x_i^2, \ d(\mathbf{x}, \mathbf{y}) = |\mathbf{x} - \mathbf{y}|, \ \text{ and } \ (\mathbf{x}, \mathbf{y}) \equiv \mathbf{x} \cdot \mathbf{y} = \sum_{i=1}^n x_i \, y_i.$$

for all $\mathbf{x}, \mathbf{y} \in \mathbb{R}^n$. With this inner product, it can be shown that $\mathbb{R}^n$ is a complete innerproduct space, that is, $\mathbb{R}^n$ is a Hilbert space.

2.4.2   Points versus vectors

A point $\mathbf{x} \in \mathbb{R}^n$ can also be viewed as a vector given by the position vector. The vector gives direction and magnitude. Where the initial position (now it is the origin) of the vector is, is immaterial. This has a great advantage in visualization. For example, consider a particle moving along a curve which is at $\mathbf{x}(t) \in \mathbb{R}^n$ at time t. So for a fixed t, we view the position $\mathbf{x}(t)$ as a point, whereas the velocity $\dot{\mathbf{x}}(t)$ is also a point in $\mathbb{R}^n$, but we view it as a vector which is positioned at $\mathbf{x}(t)$.

Similarly, if $\mathbf{A}$ is an $n \times n$ matrix, we can see it as a linear mapping $\mathbf{A} : \mathbb{R}^n \to \mathbb{R}^n$ by the correspondence $\mathbf{x} \mapsto \mathbf{A}\mathbf{x}$. Now, for every $\mathbf{x} \in \mathbb{R}^n$, the point $\mathbf{A}\mathbf{x} \in \mathbb{R}^n$ can be viewed as a vector at $\mathbf{x}$. Thus, $\mathbf{A} \in M_n(\mathbb{R})$, generates a vector field in $\mathbb{R}^n$, where $M_n(\mathbb{R})$ denotes the set of all real $n \times n$ matrices. We will see this in the study of linear systems in Chapter 5.

With these notations, $\mathbb{R}^n$ is n dimensional and a standard basis can be chosen as $\{\mathbf{e_1}, \cdots, \mathbf{e_n}\}$, where $\mathbf{e_1} = (1, 0, \cdots, 0), \mathbf{e_2} = (0, 1, 0, \cdots, 0), \cdots \mathbf{e_n}$ $= (0, \cdots 0, 1)$. Further, any $\mathbf{x} \in \mathbb{R}^n$ can be written as $\mathbf{x} = \sum_{i=1}^{n} x_i \mathbf{e_i} = \sum_{i=1}^{n}$ $(\mathbf{x} \cdot \mathbf{e_i}) \mathbf{e_i}$, where x_i is the coordinate of $\mathbf{x}$ in the direction of $\mathbf{e_i}$ and is given by the dot product of $\mathbf{x}$ and $\mathbf{e_i}$.

2.4.3   Linear operators

Let X and Y be finite dimensional vector spaces. A mapping $T : X \to Y$ is said to be a *linear operator* or *linear transformation* if $T(\alpha x + \beta y) = \alpha\, Tx + \beta\, Ty$ for all $x, y \in X$ and for all scalars α, β. The set of all linear operators is denoted by $\mathscr{L}(X, Y)$ and if $Y = X$, we write $\mathscr{L}(X, X) = \mathscr{L}(X)$. If $Y = \mathbb{R}$, then T is called a *linear functional* and the set $X^* = \mathscr{L}(X, \mathbb{R})$, is called the dual space of X.

Suppose $\dim(X) = k$ and $\dim(Y) = m$. Given bases $\{u_1, \cdots, u_k\}$ for X and $\{v_1, \cdots, v_m\}$ of Y, we can represent $T \in \mathscr{L}(X, Y)$ by an $m \times k$ matrix $\mathbf{A}$ whose columns are given by the coefficients of $T(u_i) = \sum_{j=1}^{m} a_{ij}\, v_j, i = 1, 2, \cdots, k$. This essentially means that X, Y, respectively can be viewed as $\mathbb{R}^k, \mathbb{R}^m$ as sets and complete identification is done if X is equipped with an inner product. In particular, if T is a linear operator and $\mathbf{A}$ is the $m \times k$ matrix associated with the standard bases, then $T\mathbf{x} = \mathbf{A}\mathbf{x}$ for all $\mathbf{x} \in \mathbb{R}^k$. The spectral analysis of T is essentially the same as that of $\mathbf{A}$.

We notice that $M_n(\mathbb{R})$ corresponds to the set of all linear operators from $\mathbb{R}^n$ to $\mathbb{R}^n$. Given a norm $|\cdot|$ in $\mathbb{R}^n$, we can introduce the following *induced norm* in $M_n(\mathbb{R})$ as follows: For $\mathbf{A} \in M_n(\mathbb{R})$, define

$$|\mathbf{A}| = \sup_{\mathbf{x} \in \mathbb{R}^n, \, |\mathbf{x}|=1} |\mathbf{Ax}| \tag{2.4.4}$$

which is same as

$$|\mathbf{A}| = \sup_{\mathbf{x} \in \mathbb{R}^n, \, \mathbf{x} \neq \mathbf{0}} \frac{|\mathbf{Ax}|}{|\mathbf{x}|} = \sup_{\mathbf{x} \in \mathbb{R}^n, \, |\mathbf{x}| \leq 1} |\mathbf{Ax}|.$$

Note that for identity matrix $\mathbf{I}$, we have $|\mathbf{I}| = 1$. Using the properties of $|\cdot|$ in $\mathbb{R}^n$, it is not hard to verify the following:

1. $|\mathbf{A}| \geq 0$; $|\mathbf{A}| = 0$ if and only if $\mathbf{A} = 0$,

2. $|a\mathbf{A}| = |a||\mathbf{A}|$,

3. (Triangle inequality) $|\mathbf{A} + \mathbf{B}| \leq |\mathbf{A}| + |\mathbf{B}|$,

4. $|\mathbf{AB}| \leq |\mathbf{A}||\mathbf{B}|$

for all $\mathbf{A}, \mathbf{B} \in M_n(\mathbb{R})$ and scalars a.

   Any mapping from $M_n(\mathbb{R}) \to \mathbb{R}$ satisfying the aforementioned four properties is termed as a *matrix norm*. Note that a matrix norm satisfies Property 4 in addition to the usual first three properties satisfied by any norm. It should be noted that matrix norm need not be an induced norm. For example, if $\mathbf{A} = [a_{ij}] \in M_n(\mathbb{R})$, define $|\mathbf{A}|_F^2 = \sum_{i,j=1}^{n} |a_{ij}|^2$. Then, $|\mathbf{A}|_F$ is a matrix norm (but not an induced norm, why?) and is known as the *Frobenius norm*.

   The norm induces a metric: $d(\mathbf{A}, \mathbf{B}) = |\mathbf{A} - \mathbf{B}|$ in $M_n(\mathbb{R})$. This makes it a complete metric space: every Cauchy sequence $\{\mathbf{A}_k\} \subset M_n(\mathbb{R})$ converges to some $\mathbf{A} \in M_n(\mathbb{R})$, that is, $|\mathbf{A}_k - \mathbf{A}| \to 0$ as $k \to \infty$.

2.5   Matrix Exponential $e^{\mathbf{A}}$ and its Properties

We next proceed to define the exponential of a matrix. Let $\mathbf{A} \in M_n(\mathbb{R})$, define the sequence of matrices

$$\mathbf{S}_k = \mathbf{I} + \mathbf{A} + \frac{\mathbf{A}^2}{2!} + \cdots + \frac{\mathbf{A}^k}{k!}.$$

Here, $\mathbf{I}$ is the identity matrix and $\mathbf{A}^2 = \mathbf{AA}, \mathbf{A}^3 = \mathbf{A}^2\,\mathbf{A}, \cdots$. Using the properties of a matrix norm, it is easy to see that, for $k > l$,

$$|\mathbf{S}_k - \mathbf{S}_l| \leq \sum_{j=l+1}^{k} \frac{|\mathbf{A}|^j}{j!} \to 0 \text{ as } l, k \to \infty.$$

Note that the term on the right side is a partial sum of the tail of the (scalar) exponential $e^{|\mathbf{A}|}$. Thus, $\{\mathbf{S}_k\}$ is a Cauchy sequence and consequently converges to some $\mathbf{S} \in M_n(\mathbb{R})$.

Definition 2.5.1

Given $\mathbf{A} \in M(\mathbb{R}^n)$, the *exponential* of $\mathbf{A}$, denoted by $e^{\mathbf{A}}$ or $\exp(\mathbf{A})$, is defined by

$$e^{\mathbf{A}} = \mathbf{S},$$

where $\mathbf{S} = \lim\limits_{k \to \infty} \sum\limits_{j=0}^{k} \dfrac{\mathbf{A}^j}{j!}$.

We also write $e^{\mathbf{A}} = \sum\limits_{j=0}^{\infty} \dfrac{\mathbf{A}^j}{j!}$. Note that $e^{\mathbf{A}} \in M_n(\mathbb{R})$. Clearly $|e^{\mathbf{A}}| \leq e^{|\mathbf{A}|}$, which is an interesting inequality. The computation of $e^{\mathbf{A}}$ is not easy. However, if $\mathbf{A} = \text{diag}\,(\lambda_1, \cdots, \lambda_n)$ is a diagonal matrix, that is, the main diagonal entries are $\lambda_1, \cdots, \lambda_n$ and all other elements are zero, then $\mathbf{A}^k$ is also a diagonal matrix with diagonal entries $\lambda_1^k, \cdots, \lambda_n^k$ (show this by induction) and hence, $e^{\mathbf{A}} = \text{diag}\,(e^{\lambda_1}, \cdots, e^{\lambda_n})$.

Here are a couple of important observations:

1. Suppose that the matrix $\mathbf{A}$ is similar to a matrix $\mathbf{B}$, that is, there exists a non-singular matrix $\mathbf{P}$ such that $\mathbf{B} = \mathbf{PAP}^{-1}$. Then,

$$\mathbf{B}^2 = (\mathbf{PAP}^{-1})(\mathbf{PAP}^{-1}) = \mathbf{PA}(\mathbf{P}^{-1}\mathbf{P})\mathbf{AP}^{-1} = \mathbf{PA}^2\mathbf{P}^{-1},$$

and, by induction, we get $\mathbf{B}^k = \mathbf{PA}^k\mathbf{P}^{-1}$ for any $k = 1, 2, \cdots$. This implies that

$$e^{\mathbf{B}} = \mathbf{P}e^{\mathbf{A}}\mathbf{P}^{-1} \text{ and } e^{\mathbf{A}} = \mathbf{P}^{-1}e^{\mathbf{B}}\mathbf{P} \qquad (2.5.1)$$

Thus, $e^{\mathbf{A}}$ and $e^{\mathbf{B}}$ are also similar.

2. Suppose **A** is represented as a block diagonal matrix:

$$\mathbf{A} = \mathrm{diag}\,(\mathbf{A}_1, \cdots, \mathbf{A}_k) = \begin{bmatrix} \mathbf{A}_1 & \mathbf{O} & \cdots & \mathbf{O} \\ \mathbf{O} & \mathbf{A}_2 & \cdots & \mathbf{O} \\ \cdot & \cdot & \cdots & \cdot \\ \mathbf{O} & \mathbf{O} & \cdots & \mathbf{A}_k \end{bmatrix}$$

with square matrices on the diagonal (may be of different orders) and the rest are **O**, the zero matrix. Then, it can be easily seen that $e^{\mathbf{A}}$ is also a block diagonal matrix and is given by

$$e^{\mathbf{A}} = \mathrm{diag}\,\left(e^{\mathbf{A}_1}, \cdots, e^{\mathbf{A}_k}\right). \tag{2.5.2}$$

Further, it is not hard to show that $|\mathbf{A}| \leq \max\{|\mathbf{A}_1|, \cdots, |\mathbf{A}_k|\}$ and hence, $|e^{\mathbf{A}}| \leq \max\{e^{|\mathbf{A}_1|}, \cdots, e^{|\mathbf{A}_k|}\}$. (Equality holds for the Euclidean norm)

These observations motivate us to look for an invertible matrix **P** so that **B** is diagonal and hence, $e^{\mathbf{A}}$ can be computed easily. If such a matrix **P** exists, then we say the matrix **A** is diagonalizable. If **A** is not diagonalizable, we next look for a **P** so that $\mathbf{PAP}^{-1}$ is a block diagonal matrix, with easily computable $e^{\mathbf{A}_i}$.

Diagonalizability is related to eigenvalues and eigenvectors of the given matrix. To get an idea, suppose $\mathbf{\Lambda} = \mathrm{diag}(\lambda_1, \cdots, \lambda_n)$ and $\mathbf{x} = (0, \cdots, 0, x_i, 0, \cdots, 0)$ be a vector in the ith coordinate direction. Then, the action of **A** on **x** is $\mathbf{Ax} = (0, \cdots, 0, \lambda_i x_i, 0, \cdots, 0)$ and therefore, **Ax** also lies on the same coordinate direction. Further, **Ax** is a scaled version of **x** by the factor λ_i. The coodinate axes are invariant under **A**. Thus, for a general matrix **A**, the idea is to look for n directions, if possible and **A** acts invariantly along each of these directions and each vector in any of these directions is a multiple of itself. Essentially, we are looking for a new coordinate system or a new basis under which **A** is transformed to a diagonal matrix.

More precisely, if T is a linear transformation and the usual coordinate axes are all invariant under T, then the matrix **A** corresponding to T with respect to the standard basis is a diagonal matrix. If the usual coordinate axes are not invariant under T, then we look for n distinct directions, if possible, which are invariant under T. Taking these directions as a new basis, the matrix associated with T in this new basis will be a diagonal matrix. When this happens, we say that T or **A** is diagonalizable. If it is not possible to get any set of n distinct directions invariant under T, then T will not be diagonalizable.

2.5.1   Diagonalizability and block diagonalizability

Suppose there is a non-zero vector $\mathbf{x} \in \mathbb{C}^n$ and $\lambda \in \mathbb{C}$ such that $\mathbf{Ax} = \lambda\mathbf{x}$, then we say λ is an *eigenvalue* of $\mathbf{A}$ with the corresponding *eigenvector* $\mathbf{x}$. Note that $\mathbf{x}, \lambda$, in general, are complex, even though $\mathbf{A}$ is real. The subspace $\{\mathbf{x} \in \mathbb{R}^n \; or \; \mathbb{C}^n : \mathbf{Ax} = \lambda\mathbf{x}\}$, where λ is an eigenvalue of $\mathbf{A}$, is called the *eigenspace* corresponding to λ. The matrix $\mathbf{A} = \begin{bmatrix} 0 & 1 \\ -1 & 0 \end{bmatrix}$ has no real eigenvalues, but the complex eigenvalues are given by $\lambda = \pm i$.

A sufficient condition for diagonalizability is as follows. Suppose $\mathbf{A}$ has eigenvalues $\lambda_1, \cdots, \lambda_n$, not necessarily distinct, with the corresponding eigenvectors $\mathbf{u}_1, \cdots, \mathbf{u}_n$. If $\mathbf{u}_1, \cdots, \mathbf{u}_n$ are real and linearly independent, put $\mathbf{Q} = [\mathbf{u}_1 \; \cdots \; \mathbf{u}_n]$. Then, $\mathbf{Q} \in M_n(\mathbb{R})$ and is non-singular. Thus, with $\mathbf{P} = \mathbf{Q}^{-1}$, we obtain $\mathbf{PAP}^{-1} = \mathbf{B} = \mathrm{diag}(\lambda_1, \cdots, \lambda_n)$ and $\mathbf{A}$ is therefore diagonalizable.

We next discuss block diagonalizability of a matrix, when it is not diagonalizable. Let $\mathbf{A} \in M_n(\mathbb{R})$. The eigenvalues of $\mathbf{A}$ are the roots of the *characteristic polynomial* $\det(\lambda\mathbf{I} - \mathbf{A})$ which is a real polynomial in λ of degree n. The roots may be real or complex. The set of all eigenvalues of $\mathbf{A}$ is known as the *spectrum* of $\mathbf{A}$ and is denoted by $\sigma(\mathbf{A})$. If $\mu \in \sigma(\mathbf{A})$ is real, then, there is a real eigenvector $\mathbf{x} \in \mathbb{R}^n$. Suppose $\mu \in \sigma(\mathbf{A})$ is non-real, that is, $\mu = a + ib, a, b \in \mathbb{R}, b \neq 0$ and $\mathbf{u} = \mathbf{x} + i\mathbf{y}$ is a corresponding eigenvector, where $\mathbf{x}, \mathbf{y} \in \mathbb{R}^n$, that is, $\mathbf{Au} = \mu\mathbf{u}$. Expanding and equating the real and imaginary parts, we get

$$\mathbf{Ax} = a\mathbf{x} - b\mathbf{y} \text{ and } \mathbf{Ay} = a\mathbf{y} + b\mathbf{x}. \tag{2.5.3}$$

It is straightforward to verify that these vectors $\mathbf{x}$ and $\mathbf{y}$ are linearly independent.

We now present some ideas on block diagonalization.

Invariant subspaces

Let M and N be two subspaces of $\mathbb{R}^n$ such that $M \cap N = \{\mathbf{0}\}$. We say M and N are disjoint subspaces though $\mathbf{0} \in M \cap N$ always. We say that $\mathbb{R}^n$ is a *direct sum* of M and N, if, by definition, for every $\mathbf{x} \in \mathbb{R}^n$, there exist unique $\mathbf{y} \in M, \mathbf{z} \in N$ such that $\mathbf{x} = \mathbf{y} + \mathbf{z}$. We denote the direct sum by $\mathbb{R}^n = M \oplus N$. For example, $\mathbb{R}^2 = \{(x,0) : x \in \mathbb{R}\} \oplus \{(0,y) : y \in \mathbb{R}\}$.

We can also introduce the direct sum of more than two subspaces $M_1, \cdots, M_k$ as $\mathbb{R}^n = M_1 \oplus \cdots \oplus M_k$, that is, each vector $\mathbf{x} \in \mathbb{R}^n$ has a unique representation $\mathbf{x} = \mathbf{u_1} + \cdots + \mathbf{u_k}$, where $\mathbf{u}_i \in M_i, i = 1, 2, \cdots, k$. For example,

$$\mathbb{R}^3 = \{(x,y,0) : x,y \in \mathbb{R}\} \oplus \{(0,0,z) : z \in \mathbb{R}\}$$

$$= \{(x,0,0) : x \in \mathbb{R}\} \oplus \{(0,y,0) : y \in \mathbb{R}\} \oplus \{(0,0,z) : z \in \mathbb{R}\}.$$
$$(2.5.4)$$

Definition 2.5.2

A subspace M of $\mathbb{R}^n$ is said to be *invariant* under a matrix $\mathbf{A}$ if $\mathbf{A}(M) \subset M$.

Assume $\mathbf{R}^n = M \oplus N$, where M and N are invariant subspaces under $\mathbf{A} \in M_n(\mathbb{R})$. Let $\dim M = k, \dim N = l$, so that $n = k + l$. Choose a basis $\{\mathbf{u}_1, \cdots, \mathbf{u}_k\}$ of M and $\{\mathbf{v}_1, \cdots, \mathbf{v}_l\}$ of N. Since M and N are disjoint subspaces, we see that $\{\mathbf{u}_1, \cdots, \mathbf{u}_k, \mathbf{v}_1, \cdots, \mathbf{v}_l\}$ is a linearly independent set and therefore, it is a basis of $\mathbb{R}^n$. Let $\mathbf{C}_1 = [\mathbf{u}_1 \quad \cdots \quad \mathbf{u}_k]$ and $\mathbf{C}_2 = [\mathbf{v}_1 \quad \cdots \quad \mathbf{v}_l]$ which are matrices of order $n \times k$ and $n \times l$, respectively and define $\mathbf{C} = [\mathbf{C}_1 \quad \mathbf{C}_2]$, which is an $n \times n$ invertible matrix as its columns form a basis.

It is not hard to see that if $\mathbf{A}(M) \subset M$ and $\mathbf{A}(N) \subset N$, then there exist matrices $\mathbf{A}_1$ of order $k \times k$ and $\mathbf{A}_2$ of order $l \times l$ such that

$$\mathbf{A}\mathbf{C} = \mathbf{C} \begin{bmatrix} \mathbf{A}_1 & \mathbf{O} \\ \mathbf{O} & \mathbf{A}_2 \end{bmatrix}$$

and hence,

$$\mathbf{C}^{-1}\mathbf{A}\mathbf{C} = \begin{bmatrix} \mathbf{A}_1 & \mathbf{O} \\ \mathbf{O} & \mathbf{A}_2 \end{bmatrix}, \quad \text{so that } \mathbf{A} \text{ is block diagonalizable.}$$

In a similar way, it can be shown that if $\mathbb{R}^n$ is a direct sum of the subspaces $M_1, \cdots, M_k$, each one of them invariant under $\mathbf{A}$, then, there exists a non-singular matrix $\mathbf{C}$ and square matrices $\mathbf{A}_1, \cdots, \mathbf{A}_k$ such that

$$\mathbf{C}^{-1}\mathbf{A}\mathbf{C} = \text{diag}(\mathbf{A}_1, \cdots, \mathbf{A}_k).$$

Further,

$$\mathbf{C}^{-1}e^{\mathbf{A}}\mathbf{C} = \mathrm{diag}(e^{\mathbf{A}_1}, \cdots, e^{\mathbf{A}_k})$$

The computation of $e^{\mathbf{A}_i}$ need not be easy in general. We shall next describe a procedure to find a suitable $\mathbf{C}$ so that $e^{\mathbf{A}_i}$ are easily computed.

2.5.2   Spectral analysis of A

Let $\mathbf{A} \in M_n(\mathbb{R})$ and $\sigma(\mathbf{A}) = \{\lambda_1, \cdots, \lambda_k, \mu_1, \cdots, \mu_l, \bar{\mu}_1, \cdots, \bar{\mu}_l\}$, where $\lambda_i, i = 1, \cdots, k$ are the real eigenvalues and $\mu_i, \bar{\mu}_i, i = 1, \cdots, l$ are complex eigenvalues with $Im(\mu_i) \neq 0$. Thus, $n = k + 2l$; it may happen that $l = 0$ or $k = 0$. If we get n linearly independent eigenvectors, then we see that $\mathbf{A}$ is diagonalizable. The problem arises if we do not have enough eigenvectors to form a basis of $\mathbb{R}^n$. In that case, we need to look for additional vectors to form a basis. This is done by introducing the concept of *generalized eigenvectors*.

Let $\lambda \in \sigma(\mathbf{A}) \cap \mathbb{R}$ and m be the algebraic multiplicity of λ, that is, its multiplicity as a root of the characteristic polynomial, $\det(\lambda \mathbf{I} - \mathbf{A})$, det denoting the determinant. Let $N_1 = \ker(\mathbf{A} - \lambda \mathbf{I})$ be the kernel or null space of $(\mathbf{A} - \lambda \mathbf{I})$, which is the *eigenspace* corresponding to λ. The number $\dim N_1$ is called the *geometric multiplicity*. This gives the number of linearly independent eigenvectors corresponding to the eigenvalue λ. Note that the geometric multiplicity is always less than or equal to the algebraic multiplicity. If they are not equal, there is a deficiency in the number of eigenvectors, namely, the difference between them. Define the generalized eigenspaces $N_j = \ker(\mathbf{A} - \lambda \mathbf{I})^j$, $j = 1, 2, 3, \cdots$. Clearly $N_1 \subset N_2 \subset \cdots$ and there exists a smallest integer $d \geq 1$ such that

$$N_1 \subset N_2 \subset \cdots\cdots \subset N_d = N_{d+1} = \cdots$$

Here, d is called the *index* of λ. Further, it is easy to see that N_js are invariant under $\mathbf{A}$ and a bit long to show that $\dim(N_d)$ equals algebraic multiplicity of λ. For $j > 1, N_j$s are called the *generalized eigenspaces*. If $d = 1$, then, geometric multiplicity = $\dim(N_1)$ = algebraic multiplicity and we have the required number of eigenvectors corresponding to λ. But if $d > 1$, then we may look for generalized eigenvectors from $N_2, N_3, \cdots, N_d$ to complete the deficient number of basis vectors, corresponding to λ.

Let μ be a non-real eigenvalue and let $N_j = \ker(\mathbf{A} - \mu \mathbf{I})^j$, $j = 1, 2, \cdots$ and consider the real and imaginary parts of the vectors of a basis of N_j.

It is not hard to see that these real vectors are linearly independent. In conclusion, we have the following theorem.

Theorem 2.5.3

Let $\mathbf{A} \in M_n(\mathbb{R})$.

Then, for each $\lambda \in \sigma(\mathbf{A})$ real or non-real, there exists an invariant subspace N_λ of $\mathbb{R}^n$ such that

$$\dim(N_\lambda) = \begin{cases} \text{algebraic multiplicity of } \lambda \text{ if } \lambda \text{ is real} \\ \text{twice the algebraic multiplicity of } \lambda \text{ if } \lambda \text{ is non-real} \end{cases}$$

Further, $N_\lambda \cap N_\mu = \{0\}$ if $\lambda \neq \mu$ and $\mathbb{R}^n$ can be decomposed as

$$\mathbb{R}^n = N_{\lambda_1} \oplus \cdots \oplus N_{\lambda_k} \oplus N_{\mu_1} \oplus \cdots \oplus N_{\mu_l},$$

where $\lambda_1, \cdots, \lambda_k$ are distinct real eigenvalues and $\mu_1, \cdots, \mu_l$ are distinct, non-real eigenvalues with positive imaginary parts. □

As we observed earlier, $\mathbf{A}$ can now be block diagonalized as

$$\mathbf{C}^{-1} \mathbf{A} \mathbf{C} = \text{diag}\left(\mathbf{A}_{\lambda_1}, \cdots, \mathbf{A}_{\lambda_k}, \mathbf{A}_{\mu_1}, \cdots, \mathbf{A}_{\mu_l}\right).$$

Our aim is to find suitable bases for N_{λ_i} and N_{μ_j} so that $\mathbf{A}_{\lambda_i}$ and $\mathbf{A}_{\mu_j}$ have simple structures. Hence, $\exp(\mathbf{A}_{\lambda_i})$ and $\exp(\mathbf{A}_{\mu_j})$ can be computed easily. Before proceeding further, we illustrate this with an example.

Example 2.5.4

Let λ be an eigenvalue with geometric multiplicity 2 and algebraic multiplicity 4.

Suppose that $N_1 \subset N_2 \subset N_3$ with $\dim(N_1) = 2$, $\dim(N_2) = 3$, $\dim(N_3) = 4$. Choose a vector, denoted by $\mathbf{u}_4 \in N_3$, but $\mathbf{u}_4 \notin N_2$. Define $\mathbf{u}_3 = (\mathbf{A} - \lambda\mathbf{I})\mathbf{u}_4, \mathbf{u}_2 = (\mathbf{A} - \lambda\mathbf{I})^2\mathbf{u}_4 = (\mathbf{A} - \lambda\mathbf{I})\mathbf{u}_3$. Then, $\mathbf{u}_3 \in N_2$ and $\mathbf{u}_2 \in N_1$. Further, $\{\mathbf{u}_2, \mathbf{u}_3, \mathbf{u}_4\}$ is linearly independent. Since $\mathbf{u}_2 \in N_1$, and $\dim(N_1) = 2$, we can choose $\mathbf{u}_1 \in N_1$ so that $\mathbf{u}_1, \mathbf{u}_2$ are linearly independent. Hence, $\{\mathbf{u}_1, \mathbf{u}_2, \mathbf{u}_3, \mathbf{u}_4\}$ is a linearly independent set. Moreover, we have

$$\mathbf{A}\mathbf{u}_1 = \lambda\mathbf{u}_1, \ \mathbf{A}\mathbf{u}_2 = \lambda\mathbf{u}_2, \ \mathbf{A}\mathbf{u}_3 = \mathbf{u}_2 + \lambda\mathbf{u}_3, \ \mathbf{A}\mathbf{u}_4 = \mathbf{u}_3 + \lambda\mathbf{u}_4.$$

With $\mathbf{C} = [\mathbf{u}_1, \mathbf{u}_2, \mathbf{u}_3, \mathbf{u}_4]$, a matrix with full rank, we obtain

$$\mathbf{AC} = \mathbf{C} \begin{bmatrix} \mathbf{A}_1 & \mathbf{O} \\ \mathbf{O} & \mathbf{A}_2 \end{bmatrix},$$

where $\mathbf{A}_1 = [\lambda]$ singleton matrix and $\mathbf{A}_2 = \begin{bmatrix} \lambda & 1 & 0 \\ 0 & \lambda & 1 \\ 0 & 0 & \lambda \end{bmatrix}$.

The submatrices $\mathbf{A}_1$ and $\mathbf{A}_2$ are called Jordan blocks. In general, the number of Jordan blocks corresponding to an eigenvalue coincides with its geometric multiplicity.

If $\mu = a + ib$, $b \neq 0$ is a non-real eigenvalue (complex eigenvalues appear in pairs as $\mathbf{A}$ is real), the Jordan blocks corresponding to μ are of the form

$$\begin{bmatrix} \mathbf{B}_2 & \mathbf{I}_2 & \mathbf{O} & \cdots & \mathbf{O} \\ \mathbf{O} & \mathbf{B}_2 & \mathbf{I}_2 & \cdots & \mathbf{O} \\ \cdots & \cdots & \cdots & \cdots & \\ \mathbf{O} & \mathbf{O} & \cdots & \cdots & \mathbf{B}_2 \end{bmatrix} \qquad (2.5.5)$$

where $\mathbf{B}_2 = \begin{bmatrix} a & b \\ -b & a \end{bmatrix}$, $\mathbf{I}_2 = \begin{bmatrix} 1 & 0 \\ 0 & 1 \end{bmatrix}$, $\mathbf{O} = \begin{bmatrix} 0 & 0 \\ 0 & 0 \end{bmatrix}$ are all 2×2 matrices.

This analysis can be worked out for every eigenvalue and we get the final decomposition known as Jordan decomposition theorem (JDT).

Theorem 2.5.5

[Jordan Decomposition Theorem] Given $\mathbf{A} \in M_n(\mathbb{R})$, there exists a non-singular matrix $\mathbf{C}$ such that

$$\mathbf{C}^{-1}\mathbf{AC} = \text{diagonal}(\mathbf{J}_1, \cdots \mathbf{J}_k), \qquad (2.5.6)$$

where each $\mathbf{J}_i$ is a Jordan block corresponding to an eigenvalue of $\mathbf{A}$. A typical Jordan block is a square matrix and has the form

$$\begin{bmatrix} \lambda & 1 & 0 & \cdots & 0 \\ 0 & \lambda & 1 & \cdots & 0 \\ \cdots & \cdots & \cdots & \cdots & \cdots \\ 0 & 0 & \cdots & \cdots & \lambda \end{bmatrix} \qquad\qquad (2.5.7)$$

if λ is a real eigenvalue or takes the form (2.5.5) if $\lambda = a + ib$, $b \neq 0$ is a non-real eigenvalue. We remark that for each λ, there may be several Jordan blocks depending on its geometric multiplicity as we have seen in Example 2.5.4.

2.5.3 Computation of $e^{\mathbf{J}}$ for a Jordan block $\mathbf{J}$

We begin with a simple observation. If $\mathbf{A}, \mathbf{B} \in M_n(\mathbb{R})$, which commute with each other, that is, $\mathbf{AB} = \mathbf{BA}$, then, the following binomial theorem holds:

$$(\mathbf{A} + \mathbf{B})^k = \sum_{j=0}^{k} \binom{k}{j} \mathbf{A}^j \, \mathbf{B}^{k-j}$$

for $k = 1, 2, \cdots$. Further, $e^{\mathbf{A}+\mathbf{B}} = e^{\mathbf{A}} \cdot e^{\mathbf{B}}$.

Assume $\mathbf{J}$ is a Jordan block of the block (2.5.7) of order $r \geq 2$ corresponding to a real eigenvalue λ of $\mathbf{A}$. Write, $\mathbf{J} = \lambda \mathbf{I} + \mathbf{N}$, where, $\mathbf{N}$ is a matrix such that the first upper diagonal have entries that are all 1:

$$\mathbf{N} = \begin{bmatrix} 0 & 1 & 0 & \cdots & 0 \\ 0 & 0 & 1 & \cdots & 0 \\ \cdots & \cdots & \cdots & \cdots & \cdots \\ 0 & 0 & \cdots & \cdots & 0 \end{bmatrix}.$$

Thus, since $\mathbf{I}$ and $\mathbf{N}$ commute with each other,

$$e^{\mathbf{J}} = e^{\lambda \mathbf{I}}.e^{\mathbf{N}} = e^{\lambda} \mathbf{I} e^{\mathbf{N}} = e^{\lambda} e^{\mathbf{N}}.$$

It is easy to see that $\mathbf{N}^r = \mathbf{N}^{r+1} = \cdots = \mathbf{O}$, the zero matrix. Hence,

$$e^{\mathbf{J}} = e^{\lambda} \left(\mathbf{I} + \mathbf{N} + \cdots + \frac{\mathbf{N}^{r-1}}{(r-1)!} \right)$$

The matrix $\mathbf{N}$ defined earlier is called a *nilpotent* matrix of order r. We leave it as an exercise to the reader that if $\mathbf{J}$ is a Jordan block of order $r = 2m$ of the form (2.5.5), then

$$e^{\mathbf{J}} = \mathrm{diag}\left(e^{\mathbf{B}_2}, \cdots, e^{\mathbf{B}_2}\right) \left(\mathbf{I} + \mathbf{D} + \cdots + \frac{\mathbf{D}^{2(m-1)}}{2(m-1)!} \right),$$

where

$$\mathbf{D} = \begin{bmatrix} \mathbf{O} & \mathbf{I}_2 & \mathbf{O} & \cdots & \mathbf{O} \\ \mathbf{O} & \mathbf{O} & \mathbf{I}_2 & \cdots & \mathbf{O} \\ \cdots & \cdots & \cdots & \cdots & \cdots \\ \mathbf{O} & \mathbf{O} & \mathbf{O} & \cdots & \mathbf{O} \end{bmatrix} \quad \text{and} \quad \mathbf{B}_2 = \begin{bmatrix} a & b \\ -b & a \end{bmatrix}.$$

Further, it is straightforward to see that $e^{\mathbf{B}_2} = e^a \begin{bmatrix} \cos b & \sin b \\ -\sin b & \cos b \end{bmatrix}$. From (2.5.6), it follows that

$$\mathbf{A} = \mathbf{C}\,\mathrm{diag}(\mathbf{J}_1, \cdots \mathbf{J}_k)\,\mathbf{C}^{-1},$$

$$e^{\mathbf{A}} = \mathbf{C}\,\mathrm{diag}(e^{\mathbf{J}_1}, \cdots e^{\mathbf{J}_k})\,\mathbf{C}^{-1} \tag{2.5.8}$$

and for any t

$$e^{t\mathbf{A}} = \mathbf{C}\,\mathrm{diag}(e^{t\mathbf{J}_1}, \cdots, e^{t\mathbf{J}_k})\,\mathbf{C}^{-1}.$$

Hence,

$$\left|e^{t\mathbf{A}}\right| \leq \left|\mathrm{diag}(e^{t\mathbf{J}_1}, \cdots, e^{t\mathbf{J}_k})\right| \leq \max_{1 \leq i \leq k} \left|e^{t\mathbf{J}_i}\right|.$$

With some more computation, one can prove the following theorem (using the representation of $e^{\mathbf{J}}$). See [CL72].

Theorem 2.5.6

Suppose $\sigma(\mathbf{A}) \subset \{\lambda \in \mathbb{C} : \mathrm{Re}\,\lambda < 0\}$. Then, there exist positive constants k, r such that

$$\left|e^{t\mathbf{A}}\right| \leq k e^{-rt}, \text{ for all } t \geq 0. \tag{2.5.9}$$

2.6   Linear Dependence and Independence of Functions

Recall the vector space V, introduced earlier, of all the real valued functions defined on a non-empty set I. Two vectors $u_1, u_2 \in V$ are linearly independent if $a_1 u_1 + a_2 u_2 = 0$ implies $a_1 = a_2 = 0$. Here, $a_1 u_1 + a_2 u_2 = 0$ means that $a_1 u_1(t) + a_2 u_2(t) = 0$ for all $t \in I$. Otherwise, u_1, u_2 are linearly dependent. Of course, the definition can be extended to a finite collection of functions.

We now discuss some sufficient conditions for two functions to be linearly independent. For any two points $t_1, t_2 \in I$, $t_1 \neq t_2$, if the matrix $\begin{bmatrix} u_1(t_1) & u_2(t_1) \\ u_1(t_2) & u_2(t_2) \end{bmatrix}$ is non-singular, then u_1 and u_2 are independent. To see this, let $a_1 u_1 + a_2 u_2 = 0$. Thus,

$$a_1\, u_1(t_1) + a_2\, u_2(t_1) \;=\; 0$$

$$a_1\, u_1(t_2) + a_2\, u_2(t_2) \;=\; 0.$$

The non-singularity of the matrix implies that $a_1 = 0 = a_2$. Since the class V is too large, we cannot make a statement about the converse. We now consider a special class from V. Let $C^1(I)$ be the class of continuously differentiable functions defined on I. Clearly $C^1(I) \subset V$, which again is a subspace of V. In the class $C^1(I)$, we get a simpler sufficient condition for linear independence. For $u_1, u_2 \in C^1(I)$, define the Wronskian of u_1, u_2, denoted by $W = W(t) = W(u_1, u_2)(t)$ by

$$W(u_1, u_2)(t) = u_1(t)\dot{u}_2(t) - \dot{u}_1(t)u_2(t),\ t \in I,$$

which is the determinant of the Wronskian matrix $\begin{bmatrix} u_1(t) & u_2(t) \\ \dot{u}_1(t) & \dot{u}_2(t) \end{bmatrix}$.

It is not hard to see the following. Suppose $u_1, u_2 \in C^1(I)$. If there is a point $t_0 \in I$ such that $W(t_0) \neq 0$, then u_1, u_2 are linearly independent.

The converse need not be true. The functions $u_1(t) = t^3, u_2(t) = |t|^3$, $t \in I = [-1, 1]$ are in $C^1(I)$ and are linearly independent, but $W(u_1, u_2)(t) = 0$ for all $t \in [-1, 1]$. This easy verification is left as an exercise for the reader.

It is interesting and important that this situation does not occur when we deal with functions which are solutions of linear second order ODE, as will be shown in Chapter 3.

2.7 Exercises

1. Consider $f_k : [0,1] \to \mathbb{R}$ defined by

$$f_k(x) = \begin{cases} k^2 x, & 0 \le x \le \dfrac{1}{k} \\[2mm] k^2\left(\dfrac{2}{k} - x\right), & \dfrac{1}{k} \le x \le \dfrac{2}{k} \\[2mm] 0, & \dfrac{2}{k} \le x \le 0. \end{cases}$$

Show that $f_k(x) \to f \equiv 0$, not uniformly and $\int_0^1 f_k(x) = 1$, $\int_0^1 f(t)\,dt = 0$.

2. Prove the following:

   (a) Show that $f(x) = |x|^{1/2}$ is not locally Lipschitz at 0, that is, f is not Lipschitz in any interval (a,b) containing the origin. But, it is Lipschitz in any interval (finite or infinite) away from the origin. More specifically, prove that it is Lipschitz in (a,b) if $a > 0$ and it is Lipschitz in (a,b) with $b < 0$. Is it Lipschitz in $(0,1)$? Justify your answer.

   (b) Write down 3 different solutions for $\dot{x} = |x|^{1/2}$ satisfying $x(0) = 0$.

3. Discuss the Lipschitz continuity of the following functions with respect to y.

   (a) $f(t,y) = y^{2/3}$

   (b) $f(t,y) = \sqrt{|y|}$

   (c) $f(t,y) = \begin{cases} \dfrac{4t^3 y}{t^4 + y^4} & (t,y) \ne (0,0) \\[2mm] 0 & (t,y) = (0,0) \end{cases}$

   (d) $f(t,y) = t|y|$ on $D : |t| \le a, |y| \le b$

   (e) $f(t,y) = t\sin y + y\cos t$ $D : |t| \le a, |y| \le b$

   (f) $f(t,y) = y + [t]$ on a bounded domain D in $\mathbb{R}^2$, where $[t]$ denotes the greatest integer less than or equal to t. Note that f is not continuous in t.

4. Show that the matrix $\mathbf{A} = \begin{bmatrix} 1 & 1 \\ 0 & 1 \end{bmatrix}$ is not diagonalizable by proving $\mathbf{A}$ has only one eigenvalue and the corresponding eigenspace is one dimensional. Thus, it will not be possible to obtain two linearly independent eigenvectors.

2.8 Notes

In this chapter, we have merely listed some results from analysis and linear algebra which are used throughout the book. For a comfortable understanding of the book, the reader is advised to get familiarized with these basics. A good course on basic analysis and linear algebra will be sufficient to follow the book. Quite often, the beauty and importance of many interesting notions like diagonalization, eigenvalues and eigenvectors are hidden in the abstraction. We have made an effort to introduce these notions in a very natural way and hence, the diagonalization of matrices is no longer unreachable to undergraduate students. There are many books for both linear algebra and analysis; for example, see [Apo11, BS05, Rud76] for analysis and [Apo11, HK97, Kum00, Str06] for linear algebra.

3

First and Second Order Linear Equations

3.1 First Order Equations

Recall from Chapter 1 that a general first order equation is written as

$$f(t,y,\dot{y}) = 0, \tag{3.1.1}$$

where $y = y(t)$, is the unknown function to be determined. Since this equation is extremely difficult to deal with, a slightly simpler, but quite general, possibly non-linear equation is given by

$$\dot{y} = f(t,y). \tag{3.1.2}$$

This is a *regular form* of first order equations. Equations with vanishing coefficient of $\dot{y}$ (highest derivative term for higher order equations) are classified as *singular equations*. These equations are difficult, but can have interesting features. We will see examples later.

In Chapter 1, we have given many interesting examples with explicit solutions for some of the problems. In this chapter, we plan to study a special class of ODE, known as linear ODE. In fact, we study in detail, the first and second order linear equations. It is an interesting fact that every equation of the nth order can be transformed into a system of n first order equations. A detailed study of linear systems will be carried out in Chapter 5. Before going to the linear equations, we briefly discuss initial value problems (IVP) and boundary value problems (BVP).

3.1.1   Initial and boundary value problems

There is a nice and beautiful theory regarding the existence and uniqueness of solutions to the differential equation or system of equations of the form (3.1.2) under suitable assumptions (like continuity and Lipschitz continuity etc.) on f which we will present in Chapter 4. We would like to remark that the uniqueness is proved for the initial value problem, that is, for the equation (3.1.2) together with a given initial condition of the form

$$y(t_0) = y_0. \tag{3.1.3}$$

This is relevant and essential in practical problems. Naturally, after the invention of differential calculus, solutions of first order ODE became part of the theory of integral calculus. That is, given a function $f = f(t)$, does there exist an antiderivative? In other words, given a function $f = f(t)$, does there exist a function $y = y(t)$ such that

$$\dot{y}(t) = f(t). \tag{3.1.4}$$

Then, we may write $y(t) = \int^t f(\tau)\, d\tau$, that is, y is an antiderivative of f. We use the notation $\int^t f(\tau)d\tau$ or $\int^t f(t)\, dt$ to denote an antiderivative of f, throughout the rest of this book. Using the area concept and continuity assumption on f, we indeed prove that such a function y exists. In fact, all the solutions are given by $y(t) = \int^t f(\tau)\, d\tau + C$, where C is a constant. This really is the content of the *fundamental theorem of calculus*. Thus, if we know the value of y at some point, say at t_0, that is, $y(t_0) = y_0$, then C can be determined uniquely, as $C = y_0$ and the solution is

$$y(t) = y_0 + \int_{t_0}^{t} f(\tau)\, d\tau \tag{3.1.5}$$

In many situations, even for 'smooth' functions f, we may not be able to integrate f explicitly to obtain y, which shows the difficulty even at this stage of dealing with the simplest differential equation (DE) (3.1.4). In this case, we can only say that y has a representation. For example, the function $f(t) = e^{t^2}$ cannot be integrated explicitly. Now, physically, the solution of IVP (3.1.4), (3.1.3) can be viewed as a function describing the motion of a particle starting from the point y_0 at time t_0 with the velocity of the particle at time $t > t_0$ is $f(t)$. More specifically, we may ask the question that knowing the instantaneous velocity $(f(t))$ of a vehicle moving on a straight road, is it possible to determine its position $(y(t))$ at any instance of time (t)? The answer is affirmative if the starting point y_0 at an initial time $t = t_0$ is known and f satisfies a continuity condition. If the DE is non-linear, that is, $f = f(t,y)$ depends nonlinearly on the unknown y as well, it means that the velocity at time t not only depends on time t, but the position $y = y(t)$ at time t also. In any case, this results in the following general IVP :

$$\left.\begin{array}{l} \dot{y} = f(t,y) \\[2mm] y(t_0) = y_0 \end{array}\right\} \tag{3.1.6}$$

If we have a second order equation in regular form, namely $\ddot{y} = f(t,y,\dot{y})$, then by putting $y_1 = y$ and $y_2 = \dot{y} = \dot{y}_1$, the second order equation can be written as a system of two first order equations for the two unknowns y_1 and y_2 as

$$\dot{y}_1 = y_2, \dot{y}_2 = f(t,y_1,y_2).$$

Thus, we have two first order ODEs to be integrated (though they are coupled) and hence, we require initial conditions for each variable y_1 and y_2; $y_1(t_0) = y(t_0) = y^0$, $y_2(t_0) = \dot{y}(t_0) = y^1$. Thus, one form of the initial value problem for the second order equation is given by

$$\left.\begin{array}{l} \ddot{y} = f(t,y,\dot{y}) \\[2mm] y(t_0) = y^0, \ \dot{y}(t_0) = y^1 \end{array}\right\} \tag{3.1.7}$$

It is possible to formulate other types of initial value problems which are physically relevant.

We now briefly discuss boundary value problems. There are many interesting physical problems described by second order equations defined on an interval $[a,b]$. The examples include Bessel's, Legendre, Hermite, Mathieu equations and physical conditions are given at the end points, namely at the boundary of the interval $[a,b]$, in contrast with (3.1.7), where conditions are prescribed only at one point. Such systems are generally classified as regular or singular Sturm–Liouville systems. Hence, a typical boundary value problem (BVP) for a second order equation in regular form can be stated as

$$\left.\begin{array}{l} \ddot{y} = f(t,y,\dot{y}) \text{ for } t \in (a,b) \\[2mm] \alpha_1 y(a) + \beta_1 \dot{y}(a) = \gamma_1, \; \alpha_2 y(b) + \beta_2 \dot{y}(b) = \gamma_2 \end{array}\right\}\qquad (3.1.8)$$

We remark that boundary value problems are generally more difficult than initial value problems. We will discuss some of these issues in Chapter 7 and Chapter 9.

3.1.2 Concept of a solution

In the integral calculus problem, we have seen that it may not always be possible to integrate the given function to get the solution in an explicit form. In fact, the solution may also be given implicitly as an algebraic relation between t and y as we have seen in the nonlinear model for the atomic waste disposal problem in Chapter 1. Thus, the *concept of a solution* itself has to be viewed in a very general sense, like: may exist, but may not have an explicit/implicit formula; may have an implicit relation; possible to represent y in terms of t (explicitly); may be represented as a power series. It is unfortunate that in most practical problems, we may not be able to obtain a solution to the DE in the implicit/explicit form. This necessitates the importance of studying differential equations (DEs) from the point of view of mathematical analysis and numerical computations. Thus, we may ask various questions regarding the solutions of differential equations like methods to solve DEs; existence, uniqueness, and continuous dependence on the initial data; numerical methods and computation; qualitative analysis like stability, large time behavior $\left(\lim\limits_{t \to +\infty} y(t)\right)$ and so on. Existence, uniqueness, continuous dependence on the initial data, continuation of solutions etc. will be the topics of discussion in Chapter 4, whereas qualitative analysis for systems of linear and nonlinear equations, respectively, will be studied in Chapter 5 and Chapter 8.

> **Definition 3.1.1**

[Solution] Let f be defined in a rectangle $R := (a,b) \times (c,d)$ containing the initial data (t_0, y_0). A solution to the IVP (3.1.6) is a function $y : (\bar{a}, \bar{b}) \to \mathbb{R}$ which is differentiable and satisfies the DE (3.1.6) together with the initial condition $y(t_0) = y_0$. $\square$

This means that for each $t \in (\bar{a},\bar{b})$, $y(t) \in (c,d)$ and $\dot{y}(t) = f(t,y(t))$ and $y(t_0) = y_0$. The interval $(\bar{a},\bar{b})$ is referred to as an interval of existence of the solution. Here, $t_0 \in (\bar{a},\bar{b}) \subset (a,b)$ for some interval $(\bar{a},\bar{b})$ and $y(t) \in (c,d)$ for all $t \in (\bar{a},\bar{b})$. If $(\bar{a},\bar{b}) = (a,b)$, then we say y is a *global solution* to the IVP; otherwise, it is known as a *local solution*. If the function f is continuous, then y is continuously differentiable, that is $y \in C^1(\bar{a},\bar{b})$. It is also possible to define a weaker notion of the solution concept. Throughout this book, we will assume that f is continuous and hence, we seek a solution in $C^1(\bar{a},\bar{b})$.

A similar concept of a solution may be extended to a system of first order equations. Let $\mathbf{f} : (a,b) \times \Omega \to \mathbb{R}^n$ be a vector valued continuous function so that $\mathbf{f} = (f_1, \cdots, f_n)$ and each f_i is a real valued continuous function, where Ω is an open domain in $\mathbb{R}^n$. For a given initial value $\mathbf{y}_0 \in \Omega$, the IVP is given by

$$
\left.
\begin{aligned}
\dot{\mathbf{y}} &= \mathbf{f}(t,\mathbf{y}) \\[2mm]
\mathbf{y}(t_0) &= \mathbf{y}_0
\end{aligned}
\right\}
\tag{3.1.9}
$$

Thus, a local solution to the system (3.1.9) is a smooth function $\mathbf{y} \in C^1((\bar{a},\bar{b}),\Omega)$ satisfying the aforementioned system. The definition of a solution can also be extended to higher order equations of order k, in a similar fashion by demanding that the solution is k times differentiable in an appropriate interval.

3.1.3 First order linear equations

We have already classified differential equations by their orders. In addition to the order classification, we can also classify them as *linear* and *nonlinear* ODEs. In this chapter, we will study first order and second order linear equations. There is a rich theory regarding linear ODEs and systems of linear ODE, developed invoking the machinery from linear algebra and analysis. Recall that the general first order ODE can also be written in the form $f(t,y,\dot{y}) = h(t)$, where h consists of all the terms that does not involve the unknown y and its derivative $\dot{y}$. Now treat f as a function of $y,\dot{y}$. The equation is said to be *linear* if $f(t,\alpha(z_1,w_1) + \beta(z_2,w_2)) = \alpha f(t,z_1,w_1) + \beta f(t,z_2,w_2)$ for all real numbers $\alpha, \beta, z_1, w_1, z_2, w_2$ and all t. Note that we do not demand linearity with respect to t. If the ODE is linear, it is an easy exercise from linear algebra that f takes the form $f(t,y,\dot{y}) = p_0(t)\dot{y} + p_1(t)y$.

Thus, the first order general linear, homogeneous ODE is given by

$$Ly \equiv \left(p_0(t)\frac{d}{dt} + p_1(t)\right)y = p_0(t)\dot{y} + p_1(t)y = 0. \tag{3.1.10}$$

The corresponding linear, first order, non-homogeneous ODE can be written as

$$Ly = q(t), \tag{3.1.11}$$

where p_0, p_1 and q are given functions of t. More precisely, it is linear in the sense that the differential operator L defined by $Ly = p_0(t)\dot{y} + p_1(t)y$ is linear in the class of functions where it is defined. That is, $L(y_1 + y_2) = Ly_1 + Ly_2$ and $L(\alpha y) = \alpha L(y)$, where α is any scalar. The set of all solutions of $Ly = 0$ has a linear structure. The importance of the linear structure lies in its superposition principle; that is, if y_1, y_2 are solutions of the homogeneous equation $Ly = 0$, then $\alpha y_1 + \beta y_2$ is also a solution for any constants α, β.

When the coefficient of the highest order term, namely p_0, vanishes at one or more points, the equation can bring surprises and is much more difficult to handle. Such equations are classified as *singular equations*. A few examples of second order singular equations that occur in applications are Bessel's equation, Legendre equation, Lagrange equation. We will not pursue such equations in this book; however, the reader can refer to [BR03]. We only present here two simple examples. Later, in Chapter 6, we deal with some more examples.

Example 3.1.2

Consider the equation $t\dot{y} - 2y = 0$.

Equivalently, $\dot{y} = \dfrac{2y}{t}$ when $t \neq 0$. Separating the variables and integrating, we obtain the general solution as $y(t) = Ct^2$, where C is a constant. For any fixed C, the solution y, therefore represents a parabola in the $t - y$ plane passing through the origin. Thus, if we consider the IVP for this equation with the initial value $y(0) = 0$, there are infinitely many solutions satisfying the initial condition, but no solution if the initial value is $y(0) = y_0 \neq 0$.

Example 3.1.3

Consider the equation $\dot{y} = -\dfrac{t}{y}$.

We can easily see that y and t satisfy the implicit equation $y^2 + t^2 = C^2$, where C is a constant, which implies $y = \pm\sqrt{C^2 - t^2}$ and $|t| \le |C|$. Therefore, for $-|C| \le t \le |C|$, there exist solutions. The solution is not defined for $|t| > |C|$.

A general *regular* (that is, the coefficient of highest order term is never zero) first order linear ODE can be written as

$$Ly := \dot{y} + p(t)y = q(t), \tag{3.1.12}$$

where, p and q are functions of t. We assume that p and q are continuous functions of t. For the basic equation, namely the integral calculus problem, $\dot{y} = f(t)$, the general solution is given by

$$y(t) = \int^t f(\tau)d\tau + C.$$

Now recall the linear equation (3.1.12) and consider the corresponding homogeneous equation $Ly = 0$; that is, $\dot{y} + p(t)y = 0$ or $\dot{y} = -p(t)y$. Writing this formally as: $\dfrac{\dot{y}}{y} = -p(t)$, an integration gives

$$\frac{d}{dt}\log|y(t)| = -p(t)$$

and therefore

$$|y(t)| = C\exp\left(-\int^t p(\tau)d\tau\right), \text{ that is, } \left|y(t)\exp\left(\int^t p(\tau)d\tau\right)\right| = C,$$

where C is a positive constant. Since $y(t)\exp\left(\int^t p(\tau)d\tau\right)$ is a continuous function, we get

$$y(t)\exp\left(\int^t p(\tau)d\tau\right) = C \quad \text{or} \quad y(t)\exp\left(\int^t p(\tau)d\tau\right) = -C.$$

In either case, we can write

$$y(t) = \tilde{C}\exp\left(-\int^t p(\tau)d\tau\right) \tag{3.1.13}$$

for some arbitrary constant $\tilde{C}$. The reader should verify that if f is a continuous function defined in an interval in $\mathbb{R}$ whose modulus is a constant, then f itself is a constant. It is also easy to directly verify that y given by (3.1.13) indeed satisfies $Ly = 0$.

Remark 3.1.4

From this discussion, we see that a solution of the IVP

$$\dot{y} + p(t)y = 0, \; y(t_0) = y_0, \tag{3.1.14}$$

is given by $y(t) = y_0 \exp\left(-\int_{t_0}^{t} p(\tau)d\tau \right)$. The uniqueness of this solution follows from the general theory explained in Chapter 4.   □

However, in this simple situation, we can prove the uniqueness directly without appealing to the general theory. Suppose z is another solution to the homogeneous linear equation $Ly = 0$ and consider the function $x(t) = z(t)\exp\left(\int_{t_0}^{t} p(\tau)d\tau \right)$. Then, it is easy to verify that $\dot{x} = 0$ which implies x is a constant. Thus, $z(t) = C\exp\left(-\int_{t_0}^{t} p(\tau)d\tau \right)$. Since z satisfies the same initial condition $z(t_0) = y_0$, we get $z = y$, proving the uniqueness of the solution. (see also the exercise section.)

We now consider the non-homogeneous equation

$$\dot{y} + p(t)y = q(t). \tag{3.1.15}$$

Note that we obtained a solution to the first order linear homogeneous equation by reducing it to an integral calculus problem, namely, $\dot{h}(t) = -p(t)$, where $h(t) = \log|y(t)|$.

In a similar fashion, for the non-homogeneous problem, if it is possible to find a function $h(t)$ such that

$$Ly = \dot{y} + p(t)y = \dot{h},$$

then solving (3.1.15) is reduced to an integral calculus problem. That is, we just need to integrate $\dot{h} = q(t)$ to obtain a solution. In this situation, the equation (3.1.15) is referred to as an *exact differential equation (EDE)*. However, this may not be possible in general. If $\mu(t)$ is any differentiable function, then $\frac{d}{dt}(\mu y) = \mu\dot{y} + \dot{\mu}y$. A comparison of this with the expression on the left side of (3.1.15), suggests that if we multiply (3.1.15) by $\mu(t)$, the resulting equation may become an EDE. This is the idea behind the concept of an *integrating factor*. Thus, we consider the equation

$$\mu(t)\dot{y}(t) + \mu(t)p(t)y(t) = \mu(t)q(t). \tag{3.1.16}$$

If μ is positive, then any solution of (3.1.15) is a solution of (3.1.16) and vice versa. The term on the left hand side of (3.1.16) can be written as $\frac{d}{dt}(\mu y)$, provided μ satisfies $\dot{\mu}(t) - p(t)\mu(t) = 0$. Thus, (3.1.16) is exact. Note that the equation satisfied by μ is a homogeneous linear DE in μ and hence, $\mu(t) = \exp\left(\int^t p(\tau)d\tau\right)$ is a solution and it is positive. Thus, (3.1.16) becomes

$$\frac{d}{dt}\left(\exp\left(\int^t p(\tau)d\tau\right)y(t)\right) = \exp\left(\int^t p(\tau)d\tau\right)q(t)$$

which gives

$$\exp\left(\int^t p(\tau)d\tau\right)y(t) = \int^t \exp\left(\int^s p(\tau)d\tau\right)q(s)ds + C$$

and we have the general solution as

$$y(t) = \exp\left(-\int^t p(\tau)d\tau\right)\left[\int^t q(s)\exp\left(\int^s p(\tau)d\tau\right) + C\right].$$

So, we have basically solved a homogeneous equation to get the solution of the non-homogeneous equation. The function $\mu(t)$ is called an *integrating factor* (I.F) associated with the homogeneous part.

Example 3.1.5

Consider the differential equation $\dot{y} + 2ty = t$. Here, the I.F. is $e^{\int 2t dt} = e^{t^2}$. Thus, $e^{t^2}(\dot{y} + 2ty) = te^{t^2}$, which implies

$$\frac{d}{dt}(ye^{t^2}) = te^{t^2} \Rightarrow ye^{t^2} = \int te^{t^2} + C$$

or $y(t) = \frac{1}{2} + Ce^{-t^2}$

Another class of equations which can be explicitly integrated are the *variable separable* equations.

3.1.4   Variable separable equations

Recall the general regular ODE $\dot{y} = f(t,y)$. If $f(t,y)$ has the form $f(t,y) = h(t)g(y)$, where h is a function of t alone and g is a function of y alone, then the equation is called *variable separable* and can be solved as

$$\int \frac{dy}{g(y)} = \int h(t)dt + C. \tag{3.1.17}$$

Equation (3.1.17) can be interpreted using the change of variable formula for integration as follows: Since y is function of t, the change of variable formula gives

$$\int \frac{dy}{g(y)} = \int \frac{1}{g(y(t))}\dot{y}(t)dt = \int h(t)dt + C,$$

using the given ODE. This is precisely (3.1.17), which is usually (symbolically) written as $\dfrac{dy}{g(y)} = h(t)dt$.

> **Example 3.1.6**

Consider the equation $\dot{y} = ty$.

Separating the variables and integrating using the change of variable formula, we get $\log|y| = \frac{t^2}{2} + C_1$, that is, $|y| = Ce^{t^2/2}, C \geq 0$. We may conclude that $y = C_1 e^{t^2/2}$, for an arbitrary constant C_1.

> **Example 3.1.7**

Consider the second order equation $\ddot{y} = t^2 y$.

Putting $\dot{y} = v$, we reduce the given equation to a first order equation for $v : \dot{v} = t^2 v$. Solving the latter equation, we obtain $v = C_1 e^{\frac{t^3}{3}}$ for some constant C_1. Therefore, $\dot{y} = C_1 e^{\frac{t^3}{3}}$ and thus, $y = C_1 \int^t e^{\frac{t^3}{3}} dt + C_2$, with C_2 another constant.

Example 3.1.8

Consider another second order equation

$$\ddot{y}\dot{y} = t(1+t), \, y(0) = 1, \, \dot{y}(0) = 2.$$

Putting $\dot{y} = v$, we have $v\dot{v} = t(1+t)$ with $v(0) = 2$. This can be written as $\dfrac{d}{dt}\left(\dfrac{v^2}{2}\right) = t(1+t)$. Integrating and using the initial condition, we get $\dfrac{v^2}{2} = \dfrac{t^2}{2} + \dfrac{t^3}{3} + 2$. That is, $v^2 = t^2 + \frac{2}{3}t^3 + 4$. Therefore, we have

$$\dot{y} = \pm\sqrt{\dfrac{3t^2 + 2t^3 + 12}{3}} \quad \text{and} \quad \text{solving} \quad \text{it} \quad \text{gives} \quad \text{us}$$

$y(t) = \pm\dfrac{1}{\sqrt{3}}\displaystyle\int_0^t \sqrt{12 + 3s^2 + 2s^3}\,ds + C$. Using the initial values, we have $y(t) = 1 \pm \dfrac{1}{\sqrt{3}}\displaystyle\int_0^t \sqrt{12 + 3s^2 + 2s^3}\,ds$.

We remark that more generally, a second order equation of the form $\ddot{y} = f(t,\dot{y})$, where f does not contain y explicitly, may always be reduced to a first order equation for $\dot{y}$. Once $\dot{y}$ is obtained, finding y reduces to an integral calculus problem.

3.2 Exact Differential Equations

We next consider a class of equations termed as *exact differential equations*. We have already encountered such an equation while attempting to solve a first order linear non-homogeneous equation. If the equation $\dot{y} = f(t,y)$ could be written as $\dfrac{d}{dt}\varphi(t,y) = 0$ for some two variable function φ in a domain in the (t,y) plane, then we can represent the solution implicitly as $\varphi(t,y) \equiv C$, a constant.

Definition 3.2.1

If the differential equation $\dot{y} = f(t,y)$ can be written as $\dfrac{d}{dt}\varphi(t,y(t)) = 0$ for a two variable function φ in a domain in the (t,y) plane, then the differential equation is said to be an *exact differential equation* (EDE).

$\square$

Example 3.2.2

The equation $1+\cos(t+y)+\cos(t+y)\dot{y}=0$ can be written as $\frac{d}{dt}[t+\sin(t+y)]=0$ and hence is exact. The solution is implicitly given by $t+\sin(t+y)=$ constant.

We will consider a first order equation in a little more general form:

$$M(t,y)+N(t,y)\dot{y}=0 \qquad (3.2.1)$$

Of course, $M=-f, N=1$ will give $\dot{y}=f(t,y)$. Similarly, taking $N\equiv 1$ and $M(t,y)=p(t)y$, we obtain a linear homogeneous equation. But the present form has an advantage that the equation $\dot{y}=f(t,y)$ can be written in the form (3.2.1) for different choices of M and N out of which certain representations may be exact and others not. For example, f can be written as $\frac{f}{1}$; it can also be written as $\frac{f\mu}{\mu}$ for any non-zero $\mu=\mu(t,y)$. This is essentially the process of making an equation exact by multiplying by suitable functions. We will also write (3.2.1) in a more traditional (symbolic) way as $Mdt+Ndy=0$. We now ask the question that under what conditions on the given functions M and N, the equation (3.2.1) is exact? The equation (3.2.1) is *exact* if and only if there exists a function $\varphi=\varphi(t,y)$ such that

$$M(t,y)+N(t,y)\dot{y}=\frac{d}{dt}\varphi(t,y)=\frac{\partial\varphi}{\partial t}+\frac{\partial\varphi}{\partial y}\dot{y}.$$

Therefore, if there exists $\varphi=\varphi(t,y)$ satisfying

$$M(t,y)=\frac{\partial\varphi}{\partial t}, N(t,y)=\frac{\partial\varphi}{\partial y},$$

then, (3.2.1) is exact.

The aforementioned conditions involve not only the given functions M and N, but also the unknown function φ. If we assume that φ is twice differentiable with respect to both the variables, we immediately arrive at the necessary condition $\frac{\partial M}{\partial y}=\frac{\partial N}{\partial t}$, by using the equality of the mixed partial derivatives: $\frac{\partial^2\varphi}{\partial y\partial t}=\frac{\partial^2\varphi}{\partial t\partial y}$.

Question: Given two functions M, N, does there exist a function φ such that $M = \dfrac{\partial \varphi}{\partial t}$, $N = \dfrac{\partial \varphi}{\partial y}$? The answer is given in the following theorem.

Theorem 3.2.3

Assume M, N are defined on a rectangle $D = (a, b) \times (c, d)$ and $M, N \in C^1(D)$. Then, there exists a function φ defined in D, such that $M = \dfrac{\partial \varphi}{\partial t}$ and $N = \dfrac{\partial \varphi}{\partial y}$ if and only if $\dfrac{\partial M}{\partial y} = \dfrac{\partial N}{\partial t}$.

Proof: If there exists a φ satisfying the conditions in the theorem, then the first relation $M = \dfrac{\partial \varphi}{\partial t}$, suggests that φ must be of the form $\varphi(t, y) = \int M(t, y) dt + h(y)$, for some h, which is a function of y alone. To satisfy the second relation, we must have

$$N = \frac{\partial \varphi}{\partial y} = \int \frac{\partial M}{\partial y}(t, y) dt + \frac{dh}{dy}. \tag{3.2.2}$$

Therefore, we have a first order ODE for h as

$$\frac{dh}{dy}(y) = N(t, y) - \int \frac{\partial M}{\partial y}(t, y) dt. \tag{3.2.3}$$

Note that on the left hand side is a function of y alone and hence, so should there be on the right side. Therefore, the right side is independent of t and thus, we get

$$\frac{\partial}{\partial t} \left[N(t, y) - \int \frac{\partial M}{\partial y}(t, y) dt \right] = 0.$$

Note that the integration in the second term is with respect to t and hence, on differentiation with respect to t, we arrive at

$$\frac{\partial N}{\partial t} = \frac{\partial M}{\partial y},$$

which is the necessary condition.

On the other hand, if M and N satisfy the condition $\dfrac{\partial N}{\partial t} = \dfrac{\partial M}{\partial y}$, we show how to obtain φ as follows. Let $(t_0, y_0) \in D$ be a fixed point. Define

$$\varphi(t,y) = \int_{t_0}^{t} M(s,y)\,ds + h(y),$$

where h is to be determined. Therefore,

$$\frac{\partial \varphi}{\partial y} = \int_{t_0}^{t} \frac{\partial M}{\partial y}(s,y)\,ds + \frac{dh}{dy}$$

$$= \int_{t_0}^{t} \frac{\partial N}{\partial t}(s,y)\,ds + \frac{dh}{dy}, \text{ using the given relation}$$

$$= N(t,y) - N(t_0,y) + \frac{dh}{dy}.$$

Therefore, the second relation, namely $\frac{\partial \varphi}{\partial y} = N$, is satisfied if we choose h such that $\frac{dh}{dy} = N(t_0,y)$. But, this is an integral calculus problem for h and we obtain $h(y) = \int_{y_0}^{y} N(t_0,\xi)\,d\xi$. Thus, the required function is given by

$$\varphi(t,y) = \int_{t_0}^{t} M(s,y)\,ds + \int_{y_0}^{y} N(t_0,\xi)\,d\xi.$$

$\square$

We remark that φ is determined only up to a constant. Thus, if we change t_0, y_0 in this equation, only the constant term is going to change. Therefore, the role of t_0, y_0 is minimal and one can discard all the constants in the expression for φ. We will observe this in the following examples. First, make the following definition.

Definition 3.2.4

The DE, $M(t,y) + N(t,y)\dot{y} = 0$ is said to be exact if $\dfrac{\partial M}{\partial y} = \dfrac{\partial N}{\partial t}$.

Example 3.2.5

The DE, $3y + e^t + (3t + \cos y)\dfrac{dy}{dt} = 0$ is exact.

We have

$$M = 3y + e^t, \ N = 3t + \cos y, \text{ and thus } \frac{\partial M}{\partial y} = 3 = \frac{\partial N}{\partial t}.$$

Therefore, $M = \dfrac{\partial \varphi}{\partial t}$, that is $\dfrac{\partial \varphi}{\partial t} = 3y + e^t$ which gives $\varphi(t,y) = 3yt + e^t + h(y)$. Differentiating with respect to y, we get $N = \dfrac{\partial \varphi}{\partial y} = 3t + \dfrac{dh}{dy}$.

Thus, $\dfrac{dh}{dy} = \cos y$ or $h(y) = \sin y$. We may take the constant of integration as 0. Hence, $\varphi(t,y) = 3yt + e^t + \sin y$. Therefore, the given DE can be written as $\dfrac{\partial}{\partial t}\varphi(t,y) = 0$. The solution is given by $\varphi(t,y) = 3yt + e^t + \sin y = $ constant.

We now discuss the notion of an **integrating factor**. If the DE (3.2.1) is not exact, we may possibly make it exact by multiplying it with a suitable function, which is called an *integrating factor* (I.F.). Multiplying (3.2.1) by $\mu(t,y)$, we get

$$\mu(t,y)M(t,y) + \mu(t,y)N(t,y)\dot{y} = 0. \tag{3.2.4}$$

Note that if the function $\mu > 0$, then any solution y of (3.2.4) is also a solution of (3.2.1) and vice versa. Equation (3.2.4) is exact if and only if $\dfrac{\partial}{\partial y}(\mu M) = \dfrac{\partial}{\partial t}(\mu N)$, which implies

$$\frac{\partial \mu}{\partial y}M + \mu\frac{\partial M}{\partial y} = \frac{\partial \mu}{\partial t}N + \mu\frac{\partial N}{\partial t}. \tag{3.2.5}$$

If this equation has a solution μ, then (3.2.4) is exact and μ is an I.F. of the original equation. As (3.2.5) is a PDE for μ, it is more difficult to solve and goes beyond the realm of ODE! However, since we have some freedom in choosing μ, we will try to choose it as simple as possible, say μ is a function of only t or only y. Fortunately, such an assumption works in many situations.

Consider a special case $\mu = \mu(t)$ is a function of t alone. Then, (3.2.5) becomes $\mu(t)\left[\dfrac{\partial M}{\partial y} - \dfrac{\partial N}{\partial t}\right] = \dot{\mu}(t)N$ and hence,

$$\frac{\dot{\mu}(t)}{\mu(t)} = \frac{1}{N}\left(\frac{\partial M}{\partial y} - \frac{\partial N}{\partial t}\right).$$

As the expression on the left is a function of t alone, this equation makes sense only when the expression on the right side is also a function of t alone, say $R(t)$; then one can find an I.F. $\mu(t) = \exp\left(\int^t R(t)dt\right)$.

Similarly, if the expression $\dfrac{1}{M}\left(\dfrac{\partial N}{\partial t}-\dfrac{\partial M}{\partial y}\right)$ is a function of y alone, then we can choose μ as function of only y.

Example 3.2.6

Consider the linear equation $\dot{y}+p(t)y=q(t)$.

Here,

$$M=p(t)y-q(t), N=1.$$

Now

$$\frac{1}{N}\left(\frac{\partial M}{\partial y}-\frac{\partial N}{\partial t}\right)=p(t).$$

Hence, $\mu(t)=\exp\left(\displaystyle\int^{t}p(t)dt\right)$ is an I.F. as we have already seen Section 1.2.

Example 3.2.7

Consider the equation

$$2t\sin y+y^{3}e^{t}+(t^{2}\cos y+3y^{2}e^{t})\frac{dy}{dt}=0.$$

Here, we have $M=2t\sin y+y^{3}e^{t}$ and $N=t^{2}\cos y+3y^{2}e^{t}$. Thus,

$$\frac{\partial M}{\partial y}-\frac{\partial N}{\partial t}=(2t\cos y+3y^{2}e^{t})-(2t\cos y+3y^{2}e^{t})=0,$$

which shows that the given equation is exact. So we have

$$\varphi(t,y)=\int^{t}(2t\sin y+y^{3}e^{t})dt+h(y)=t^{2}\sin y+y^{3}e^{t}+h(y).$$

Now

$$N=t^{2}\cos y+3y^{2}e^{t}=\frac{\partial\varphi}{\partial y}=t^{2}\cos y+3y^{2}e^{t}+\frac{dh}{dy}.$$

Therefore, $\dfrac{dh}{dy}(y) = 0$, $h(y) = $ constant. We can take $\varphi(t,y) = t^2 \sin y + y^3 e^t$. The equation becomes $\dfrac{d}{dt}(t^2 \sin y + y^3 e^t) = 0$ which implies $t^2 \sin y + y^3 e^t = k$, a constant.

Example 3.2.8

Consider the DE $t\dot{y} - 2y = 0$.

For $t > 0$, we can see that the function $\mu(t) = \dfrac{1}{t^3}$ is an integrating factor since $\dfrac{d}{dt}\left(\dfrac{y}{t^2}\right) = \dfrac{1}{t^2}\dot{y} - \dfrac{2}{t^3}y = 0$. Thus, $y = ct^2$ is a solution for any constant c.

3.3 Second Order Linear Equations

In Chapter 1, we have seen interesting examples of the spring–mass–dashpot system and the LCR circuit leading to second order linear equations with constant coefficients. Newton's second law of motion, describing the motion of a body under the influence of a given force, is another example. We can also write the second order equation as a system of two first order equations. In fact any n^{th} order equation can be written as a system of n first order equations. This will be seen in Chapter 5 dealing with linear systems.

We have introduced in this chapter the IVPs for the second order ODE together with two initial conditions. Now, we will study the following IVP for general, regular, second order linear equations:

$$\ddot{y} + p(t)\dot{y} + q(t)y = r(t), \; t \in I,$$

$$y(t_0) = y_0, \; \dot{y}(t_0) = y_1. \tag{3.3.1}$$

Here p, q, r are real valued continuous functions in a given interval I containing t_0 and y_0, y_1 are given real numbers. We denote by $I(t_0)$, a general interval containing t_0. As introduced earlier, a general second order equation can be written in the form $f(t, y, \dot{y}, \ddot{y}) = 0$ and a *second order linear differential equation* (SLDE) is given by

$$Ly \equiv \ddot{y} + p(t)\dot{y} + q(t)y = r(t). \tag{3.3.2}$$

If $r(t) \equiv 0$, the equation is called a homogeneous SLDE (HSLDE). The operator $Ly = L(t,y,\dot{y},\ddot{y})$ is multi-linear, that is, L is linear in $(y,\dot{y},\ddot{y})$. More precisely, if we define $L(t,y,z,w)$ by $L(t,y,z,w) = w + p(t)z + q(t)y$, then it is linear in (y,z,w). That is, $L(\alpha(y_1,z_1,w_1) + \beta(y_2,z_2,w_2)) = \alpha L(y_1,z_1,w_1) + \beta L(y_2,z_2,w_2)$ for all real numbers α and β. We can also see this in another way. We view L as the differential operator given by $Ly = \left(\dfrac{d^2}{dt^2} + p(t)\dfrac{d}{dt} + q(t) \right) y$. Then L is linear in y, that is, $L(a_1y_1 + a_2y_2) = a_1Ly_1 + a_2Ly_2$, for any C^2 functions y_1,y_2 and scalars a_1,a_2. Further, if y and $\tilde{y}$ are two solutions to the homogeneous equation, then $\alpha y + \beta \tilde{y}$ is also a solution of the homogeneous system for all constants α and β. This is known as the *superposition principle* and it is an important and useful property of linear equations.

Unlike the case of a first order equation, it is not possible to solve the equation (3.3.2), in general, to obtain a solution of (3.3.1) in explicit form. This necessitates us to do analysis for second and, more generally, higher order equations like existence, uniqueness, etc., which are discussed in Chapter 4. The proof of the existence theorem is similar to the one for the general first order equations using Picard's iterations. In fact, (3.3.1)) can be converted to a first order system and then Picard's iteration can be employed to obtain the existence of a solution.

Theorem 3.3.1

([Sim91]) Let p,q,r be continuous functions defined in a compact interval $I(t_0)$ and y_0, y_1 be any two real numbers. Then, the IVP (3.3.1) has a unique solution y defined in $I(t_0)$ satisfying $y(t_0) = y_0$, $\dot{y}(t_0) = y_1$.

We remark that the compactness assumption on $I(t_0)$ may be relaxed, if we assume that the functions p,q and r are bounded in $I(t_0)$. See Chapter 4.

As we do not have a general procedure to obtain a solution for second order equations like in the case of first order equations, we need to have methods, if possible, to obtain solutions in addition to the existence and uniqueness of a solution. We now try to understand the structure of the set of solutions via linear algebra.

3.3.1   Homogeneous SLDE (HSLDE)

By taking $r = 0$ in (3.3.1), we will first study the homogeneous equation

$$Ly := \ddot{y} + p(t)\dot{y} + q(t)y = 0. \tag{3.3.3}$$

Proposition 3.3.2

Let z, w be two solutions of (3.3.3) in some interval. Then, for any $\alpha, \beta \in \mathbb{R}$, the function $y = \alpha z + \beta w$ is also a solution to (3.3.3). Further, if z and w are two linearly independent solutions of HSLDE (3.3.3) in an interval $I(t_0)$, then every solution of HSLDE (3.3.3) is given by $y = \alpha z + \beta w$, where α and β are suitable constants.

A *general solution* of HSLDE therefore is a linear combination of any two linearly independent solutions of HSLDE.

Proof: The first part of the proposition is trivial to verify. Now, let y be any solution of (3.3.3) with $y_0 - y(t_0)$ and $y_1 = \dot{y}(t_0)$. We now show that there are constants α and β such that $y(t) = \alpha z(t) + \beta w(t)$, for all $t \in I(t_0)$. In particular, taking $t = t_0$, we see that α and β should satisfy the 2×2 matrix system

$$\begin{cases} z(t_0)\alpha + w(t_0)\beta = y_0, \\ \dot{z}(t_0)\alpha + \dot{w}(t_0)\beta = y_1. \end{cases} \tag{3.3.4}$$

This equation is solvable uniquely for α and β if and only if $\det \begin{bmatrix} z(t_0) & w(t_0) \\ \dot{z}(t_0) & \dot{w}(t_0) \end{bmatrix} \neq 0$. Define *Wronskian W* of z, w by

$$W(t) = \det \begin{bmatrix} z(t) & w(t) \\ \dot{z}(t) & \dot{w}(t) \end{bmatrix} = z(t)\dot{w}(t) - \dot{z}(t)w(t). \tag{3.3.5}$$

Thus, the system (3.3.4) is uniquely solvable for α, β if $W(t_0) \neq 0$. In fact, if z, w are linearly independent, then we have $W(t) \neq 0$ for all $t \in I(t_0)$. To see this, consider

$$\frac{d}{dt}W(t) = z\ddot{w} - \ddot{z}w = -p(z\dot{w} - \dot{z}w) = pW$$

z, w satisfies (3.3.3). Thus, W is given by $W(t) = C \exp\left(\int^t p(t)\,dt\right)$, for some constant C. Hence, $W \equiv 0$ if $C = 0$. If $C \neq 0$, then $W(t) \neq 0$, for all t.

The following proposition will complete the proof of Proposition 3.3.2.
$\square$

Proposition 3.3.3

$W \equiv 0$ if and only if z and w are dependent.

Proof: If z and w are dependent, that is, either $z = kw$ or $w = kz$ for some constant k, then clearly $W \equiv 0$. On the other hand, assume $W \equiv 0$. If one of z, w is identically zero, then z, w are linearly dependent. So, assume $z \not\equiv 0$ and $w \not\equiv 0$. Since z is continuous, there is an interval $[c,d] \subset I(t_0)$ such that $z(t) \neq 0$, for all $t \in [c,d]$. Thus, for all $t \in [c,d]$

$$0 = \frac{W}{z^2} = \frac{z\dot{w} - \dot{z}w}{z^2} = \frac{d}{dt}\left(\frac{w}{z}\right).$$

Thus, we get $w = kz$ in $[c,d]$. Observe that both w and kz are solutions to (3.3.3) in $I(t_0)$ with the same values in $[c,d]$ (in particular, they can be treated as initial conditions). Thus, by uniqueness, $w = kz$ in $I(t_0)$. Hence, z, w are linearly dependent.

This also completes the proof of the theorem since now there exist unique α, β satisfying (3.3.4). It follows then that $y = \alpha z + \beta w$. $\square$

Let S be the set of all the solutions to the homogeneous equation $Ly = 0$. By the superposition principle, it is easily verified that S is a linear space and it is a subspace of the infinite dimensional space $C^2(I(t_0))$, the space of twice continuously differentiable functions. Proposition 3.3.2 shows the amazing fact that $\dim S \leq 2$. In fact, we have the following theorem.

Theorem 3.3.4.

$\dim(S) = 2$.

Proof: Let z and w be the solutions to the homogeneous equation $Ly = 0$ with the initial conditions $z(t_0) = 1, \dot{z}(t_0) = 0$ and $w(t_0) = 0, \dot{w}(t_0) = 1$, respectively. Existence and uniqueness of z and w follow from Theorem 3.3.1. Then, the Wronskian $W(t)$ of z and w satisfies $W(t_0) = 1 \neq 0$.

Hence, the Wronskian is non-zero for all t and by Proposition 3.3.3, z and w are linearly independent. It now follows from Proposition 3.3.2 that any solution of (3.3.3) can be written as a linear combination of z and w. □

We remark that the aforementioned proposition holds true for an nth order linear equation as well. Further, the existence of a unique solution for the aforementioned IVP is guaranteed under the assumption that the functions p, q are continuous in a compact interval $I(t_0)$. But, in general, even for second order equations, it is difficult to find independent solutions to $Ly = 0$ in explicit form. We present two methods describing the possibility of obtaining linearly independent solutions. When applicable, these methods generate two linearly independent solutions.

Method 1:   The idea is to remove the term involving the first order derivative $\dot{y}$ via an integrating factor. We look for a solution of the form $y = uv$, where u and v are to be properly chosen. In this case, $\dot{y} = u\dot{v} + \dot{u}v$ and $\ddot{y} = u\ddot{v} + 2\dot{u}\dot{v} + \ddot{u}v$. Substituting in $Ly = 0$, we get

$$(u\ddot{v} + 2\dot{u}\dot{v} + \ddot{u}v) + p(t)(u\dot{v} + \dot{u}v) + q(t)uv = 0. \qquad (3.3.6)$$

Rearranging the terms in (3.3.6), we obtain

$$u\ddot{v} + (2\dot{u} + p(t)u)\dot{v} + (\ddot{u} + p(t)\dot{u})v + q(t)uv = 0.$$

Now choose u so that the coefficient of $\dot{v}$ in this equation vanishes. That is, choose u satisfying

$$2\dot{u} + p(t)u = 0, \qquad (3.3.7)$$

which can be easily solved for u. Note that u never vanishes, if it is not zero initially. The equation satisfied by v now becomes

$$\ddot{v} + q(t)v = -(\ddot{u} + p(t)\dot{u})/u. \qquad (3.3.8)$$

Since $\dot{v}$ term is absent and u is known in (3.3.8), it may be possible to solve this equation for v, at least in some situations.

Example 3.3.5

Solve $\ddot{y} + 2t\dot{y} + (1 + t^2)y = 0$.

Here $p(t) = 2t$, $q(t) = 1 + t^2$, $u(t) = e^{-\frac{1}{2}t^2}$. It is easy to see that v satisfies $\ddot{v} = 0$. Thus, $v(t) = C_1 t + C_2$ and the solution is given by

$$y(t) = u(t)v(t) = e^{-\frac{1}{2}t^2}(C_1 t + C_2).$$

Method 2 (Reduction of order): This method describes a way to find a second (linearly independent) solution, given a non-trivial solution of (3.3.3). This method is particularly useful when we are able to obtain a solution of a given equation by simple inspection. Assume $y_1(t)$ is a known non-trivial solution to (3.3.3). Then, any constant multiple of $y_1(t)$ is also a solution. If we wish to produce another independent solution, the idea is to take a variable multiple of $y_1(t)$. Therefore, we seek a second solution in the form $y_2(t) = C(t)y_1(t)$ which will be independent of y_1. That is, $\dfrac{y_2}{y_1}$ must be a non-constant function for y_2 to be linearly independent of y_1. Using the fact that y_1 is a solution and assuming y_2 is another solution, we obtain

$$\ddot{C}(t)y_1 + \dot{C}(t)(2\dot{y}_1 + py_1) = 0. \tag{3.3.9}$$

Though this is a second order equation for C, it is only a first order linear equation for $\dot{C}$. Hence, we get $\dot{C}(t) = \dfrac{1}{y_1^2}\exp\left(-\int^t p\,dt\right)$ (reduction of order) and solving for C reduces it to an integral calculus problem. One more integration will give

$$C(t) = \int^t \frac{1}{y_1^2}\exp\left(-\int^s p\,ds\right)dt. \tag{3.3.10}$$

It is easy to compute the Wronskian of y_1 and y_2 and conclude that the two solutions are indeed linearly independent.

3.3.2 Linear equation with constant coefficients

In this section, we consider the equation

$$Ly := a\ddot{y} + b\dot{y} + cy = 0, \tag{3.3.11}$$

where a, b, c are constants with $a \neq 0$. Observe the interesting fact: If $\dot{y} \propto y$, that is, $\dot{y} = ry$ for some constant r, then $\ddot{y} = r\dot{y} = r^2 y$, that is $\ddot{y} \propto y$.

Substituting these relations into the equation, we get $(ar^2 + br + c)y = 0$. For non-trivial solution y, r must, therefore, satisfy

$$ar^2 + br + c = 0. \tag{3.3.12}$$

This motivates us to look for a solution of $\dot{y} = ry$, that is, a solution of the form $y(t) = e^{rt}$ for (3.3.11). Then, r satisfies (3.3.12), which is called the *characteristic equation* of (3.3.11). The two roots of (3.3.12) are given by

$$r_1 = \frac{1}{2a}\left(-b + \sqrt{b^2 - 4ac}\right) \text{ and } r_2 = \frac{1}{2a}\left(-b - \sqrt{b^2 - 4ac}\right).$$
$$\tag{3.3.13}$$

We now analyse various cases, depending on the nature of the discriminant of the quadratic equation (3.3.12).

Case (i) $b^2 - 4ac > 0$:   In this case, the roots r_1 and r_2 of (3.3.12) are real and distinct and we get two linearly independent solutions $y_1(t) = e^{r_1 t}$ and $y_2(t) = e^{r_2 t}$. Hence, the general solution can be written as

$$y(t) = Ae^{r_1 t} + Be^{r_2 t}, \tag{3.3.14}$$

where, A and B are arbitrary constants.

Case (ii) $b^2 - 4ac = 0$:   Here, we have a double real root $r_1 = r_2 = r = \frac{-b}{2a}$ and we get one solution $y_1(t) = e^{-\frac{b}{2a}t}$. Now, using the reduction of order method, the reader can construct a second linearly independent solution as $y_2(t) = ty_1(t) = te^{-\frac{b}{2a}t}$. Thus, the general solution is given by

$$y(t) = (A + Bt)e^{-\frac{b}{2a}t}, \tag{3.3.15}$$

where, A and B are arbitrary constants.

Case (iii) $b^2 - 4ac < 0$:   Then, the roots r_1 and r_2 are complex and $e^{r_1 t}$ and $e^{r_2 t}$ are complex valued solutions. Clearly, if $y(t) = u(t) + iv(t)$ is a complex valued solution, then u and v are real valued solutions. Thus, if $r_1 = \alpha + i\beta$ and $r_2 = \alpha - i\beta$, $\beta \neq 0$, the two independent solutions are given by $y_1(t) = e^{\alpha t}\cos\beta t$ and $y_2(t) = e^{\alpha t}\sin\beta t$, where $\alpha = -\frac{b}{2a}$ and $\beta = \frac{\sqrt{4ac - b^2}}{2a}$. Thus, the general solution is given by

$$y = e^{\alpha t}(A\cos\beta t + B\sin\beta t),$$

for arbitrary constants A and B.

The reader should directly verify the linear independence of solutions in all the aforementioned cases.

3.3.3 Non-homogeneous equation

Now, we consider the non-homogeneous case

$$Ly := \ddot{y} + p(t)\dot{y} + q(t)y = r(t). \tag{3.3.16}$$

Solution Structure: Let $\tilde{S}$ be the set of all solutions to the non-homogeneous equation (3.3.16). Any solution $y = y_p \in \tilde{S}$ of (3.3.16) is called a particular solution. If z and w are solutions to the homogeneous equation (3.3.3), then $y = \alpha z + \beta w + y_p$ is also in $\tilde{S}$ for any constants α and β. In fact, the converse is also true; that is, any $y \in \tilde{S}$ is given by the form $y = \alpha z + \beta w + y_p$, where z and w are linearly independent solutions to the homogeneous equation (3.3.3), y_p is any particular solution and α, β are constants. To see this, let $y \in \tilde{S}$ be arbitrary; then, the function $y - y_p$ satisfies the homogeneous equation $L(y - y_p) = 0$. In other words, $y - y_p \in S$, the two dimensional space of solutions of the homogeneous equation. Hence, the result follows from Proposition 3.3.2.

Theorem 3.3.6

Assume that the functions p, q, r in (3.3.16) are continuous. Let S and $\tilde{S}$ be, respectively, the solution set of homogeneous equation (3.3.3) and non-homogeneous equation (3.3.16). Then, S is a linear space of dimension 2 and $\tilde{S}$ is an affine space given by $\tilde{S} = S + y_p$ for any particular solution y_p of (3.3.16).

Example 3.3.7

Consider $\ddot{y} + y = t$. The characteristic equation is given by $r^2 + 1 = 0$ whose roots $r = \pm i$. Thus, $\cos t$, $\sin t$ are linearly independent solutions of $\ddot{y} + y = 0$ and $y_p(t) = t$ is a particular solution. The general solution is $y(t) = C_1 \cos t + C_2 \sin t + t$.

We now provide two methods to obtain a particular solution of the non-homogeneous equation.

Method of Variation of Parameters: In this method, given two linearly independent solutions y_1 and y_2 of HSLDE (3.3.3), we seek a solution of the non-homogeneous equation (3.3.16) of the form

$$y(t) = C_1(t)y_1(t) + C_2(t)y_2(t), \tag{3.3.17}$$

where, the functions C_1 and C_2 are to be suitably found. Note that if C_1, C_2 are constants, then y can never be a solution of (3.3.16), if $r \neq 0$. A simple computation leads to

$$r = Ly = \left[2(\dot{C}_1\dot{y}_1 + \dot{C}_2\dot{y}_2) + \ddot{C}_1 y_1 + \ddot{C}_2 y_2\right] + p(t)(\dot{C}_1 y_1 + \dot{C}_2 y_2). \tag{3.3.18}$$

We need two equations to determine $\dot{C}_1$ and $\dot{C}_2$. We choose the first equation as

$$\dot{C}_1 y_1 + \dot{C}_2 y_2 = 0. \tag{3.3.19}$$

Using this in (3.3.18), we obtain

$$Ly = \dot{C}_1\dot{y}_1 + \dot{C}_2\dot{y}_2 = r(t). \tag{3.3.20}$$

Thus, the equations (3.3.19), (3.3.20) give the system of equations for $\dot{C}_1, \dot{C}_2$ which is uniquely solvable for $\dot{C}_1, \dot{C}_2$, since the Wronskian W of y_1 and y_2 is non-zero. Hence, by Cramer's rule, we have

$$\dot{C}_1(t) = \frac{r(t)y_2}{W} \quad \text{and} \quad \dot{C}_2(t) = -\frac{r(t)y_1}{W}. \tag{3.3.21}$$

By integrating, we get $C_1(t)$ and $C_2(t)$. This gives a particular solution:

$$y_p(t) = y_1(t) \int \frac{r(t)y_2(t)}{W} dt - y_2(t) \int \frac{r(t)y_1(t)}{W} dt.$$

Example 3.3.8

Consider $L(y) = r(t)$, where the homogeneous equation $L(y) = 0$ has two linearly independent solutions $\cos t$ and $\sin t$. Let $r(t) = \tan t$. Therefore, a particular solution can be obtained as

$$y_p(t) = \sin t \int^t \cos t \tan t \, dt - \cos t \int^t \sin t \tan t \, dt = \cos t \log|\sec t + \tan t|.$$

Example 3.3.9

Now suppose that the homogeneous equation $L(y) = 0$ has two linearly independent solutions e^{-t} and e^{3t} and let $r(t) = te^{-t}$. Now, the Wronskian of these linearly independent solutions is $W = 4e^{2t}$. A particular solution is then given by

$$y_p(t) = e^{-t} \int^t \frac{te^{-t}}{4e^{2t}} e^{3t}\, dt - e^{3t} \int^t \frac{te^{-t}}{4e^{2t}} e^{-t}\, dt = \frac{1}{16} t(2t+1)e^{-t}.$$

Example 3.3.10

Consider the linear equation

$$\ddot{y} - \frac{1+t}{t}\dot{y} + \frac{1}{t}y = te^{2t}.$$

By inspection, we observe that $y_1(t) = e^t$ is a solution of the homogeneous equation. By the method of reduction of order, it can be shown that $y_2(t) = 1+t$ is another (linearly independent) solution. Therefore, a particular solution of the given non-homogeneous equation is given by

$$y_p(t) = e^t \int^t \frac{te^{2t}}{te^t}(1+t)\, dt - (1+t) \int^t \frac{te^{2t}}{te^t} e^{-t}\, dt = \frac{1}{2}(t+1)e^{2t}.$$

Method of Undetermined Coefficients (Special Non-homogeneous Terms)

1) Consider the equation

$$Ly := \ddot{y} + p\dot{y} + qy = e^{at}, \tag{3.3.22}$$

where p, q and a are constants. Suggested by the exponential function on the right side, whose derivatives are also multiples of the same function, we look for a solution of the form $y_p(t) = Ae^{at}$. This gives

$$A(a^2 + pa + q) = 1.$$

Thus, if $a^2 + pa + q \neq 0$, that is, a is not a root of the characteristic equation (3.3.12), we get a particular solution as

$$y_p(t) = \frac{e^{at}}{a^2 + pa + q}.$$

If a is root of (3.3.12), we now look for a solution of the form $y_p = Ate^{at}$. In fact, one can use the reduction order method to get the coefficient as At. A computation will lead to $A(2a + p) = 1$. Again, we get a particular solution by choosing $A = \dfrac{1}{2a + p}$ if $2a + p \neq 0$ as

$$y_p(t) = \frac{te^{at}}{2a + p}.$$

Now, note that $2a + p = \dfrac{d}{da}(a^2 + ap + q)$. Thus, $2a + p = 0$ is equivalent to a being a double root. If a is a double root (that is, $a^2 + ap + q = 0$ and $2a + p = 0$), then look for a solution of the form $y_p(t) = At^2 e^{at}$ which will give us $A = \frac{1}{2}$. In summary, we have

$$y_p(t) = \begin{cases} \dfrac{e^{at}}{a^2 + pa + q} & \text{if } a \text{ is not a root of (3.3.12),} \\[2ex] \dfrac{te^{at}}{2a + p} & \text{if } a \text{ is a simple root of (3.3.12),} \\[2ex] \dfrac{1}{2}t^2 e^{at} & \text{if } a \text{ is a double root of (3.3.12).} \end{cases}$$

$$(3.3.23)$$

2) Here, we consider a similar equation as $Ly := \ddot{y} + p\dot{y} + qy = \sin bt$ or $\cos bt$. Look for a particular solution of the form

$$y_p(t) = A \sin bt + B \cos bt.$$

In a similar fashion, we can determine A and B, unless y_p itself is a solution to the homogeneous system. If y_p happens to be a solution of the homogeneous equation, then, we may try $y_p(t) = t(A \sin bt + B \cos bt)$ and proceed as earlier.

3) Consider $Ly := \ddot{y} + p\dot{y} + qy = a_0 + a_1 t + \cdots + a_n t^n$. One can try for a particular solution, an expression of the form

$$y_p(t) = A_0 + A_1 t + \cdots + A_n t^n.$$

By plugging in this expression into the ODE, we get

$$(2A_2 + 3.2A_3t + \cdots + n(n-1)A_nt^{n-2}) + p(A_1 + 2A_2t + \cdots + nA_nt^{n-1})$$

$$+ q(A_0 + A_1t + \cdots + A_nt^n) = a_0 + a_1t + \cdots + a_nt^n.$$

Comparing the coefficients of equal powers, we obtain

$$qA_n = a_n, \ pnA_n + qA_{n-1} = a_{n-1}, \ \cdots \ 2A_2 + pA_1 + qA_0 = a_0.$$

Therefore, we get a solution if $q \neq 0$. If $q = 0$, we may try a solution of the form $y_p(t) = A_0t + \cdots + A_nt^{n+1}$.

Example 3.3.11

Consider $\ddot{y} + y = \sin t$.

Since $\sin t, \cos t$ are solutions to $\ddot{y} + y = 0$, we may try $y_p(t) = t(A\cos t + B\sin t)$ and arrive at $A = -\frac{1}{2}$, $B = 0$. Hence, the general solution is $y(t) = C_1 \cos t + C_2 \cos bt - \frac{1}{2}t \cos t$, with arbitrary constants C_1 and C_2.

We now do a complete analysis of the spring–mass–dashpot system introduced in Chapter 1, using the solutions obtained for equations with constant coefficients. We can do a similar analysis for the LCR circuit problem and other systems.

Example 3.3.12

[Spring–Mass–Dashpot System] Let us recall, the SMD system;

$$m\ddot{y} + c\dot{y} + ky = F(t),$$

where $m > 0$, $c \geq 0$, $k > 0$, are, respectively, the mass of the body attached to the spring, damping parameter and spring constant, and F is the applied external force. This is a second order differential equation with constant coefficients. We analyse various cases using the solution of the homogeneous equation.

Case (i) (Free undamped vibrations: $F = 0, c = 0$):   In this case, the general solution is oscillatory which is given by

$$y(t) = a\cos \omega_0 t + b\sin \omega_0 t,$$

where $\omega_0 = \sqrt{\dfrac{k}{m}}$, called the *natural frequency* of the system. This can also be written as

$$y(t) = R\cos(\omega_0 t - \delta).$$

Here, $R = \sqrt{a^2 + b^2}$, $\delta = \tan^{-1}\left(\dfrac{b}{a}\right)$ are, respectively, the *amplitude* and *phase angle*. Further, $T_0 = \dfrac{2\pi}{\omega_0}$ is the period of the motion and the motion is periodically oscillating between $-R$ and R (see Fig. 3.1). Note that the term involving c is the damping term. Indeed, Newton's law is justified as the motion never stops.

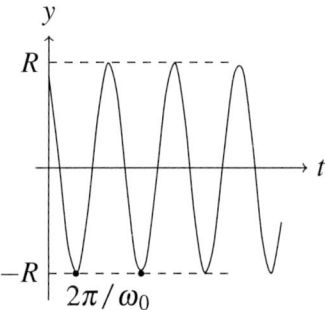

Fig. 3.1   Free undamped vibrations

Case (ii) (Damped, free motion: $F = 0, c > 0$): If r_1, r_2 are the roots of the characteristic equation $mr^2 + cr + k = 0$, we can write the general solution as

$$y(t) = \begin{cases} ae^{r_1 t} + be^{r_2 t} & \text{if } c^2 - 4mk > 0 \\[2mm] (a + bt)e^{-\frac{c}{2m}t} & \text{if } c^2 - 4mk = 0 \\[2mm] e^{-\frac{ct}{2m}}[a\cos\mu t + b\sin\mu t] & \text{if } c^2 - 4mk < 0, \end{cases} \qquad (3.3.24)$$

where $\mu = \dfrac{\sqrt{4mk - c^2}}{2m}$. Further, it is easy to see that r_1, r_2 are negative real numbers or have negative real parts. Hence, in the first two cases, $y(t) \to 0$ as $t \to \infty$, $y(t)$ creeps back to the equilibrium position and there are no oscillations at all. These are referred to as *over-damped* or *critically damped systems*, respectively. The interesting third case is known as *under damped motion* which occurs quite often in mechanical vibrations. Thus, it is a case of damped vibrations. Now, rewrite the solution as

$$y(t) = Re^{-ct/2m}\cos(\mu t - \delta). \tag{3.3.25}$$

The displacement $y(t)$ oscillates between the curves $y = \pm Re^{-ct/2m}$ and $y(t) \to 0$ as $t \to \infty$, in this case as well. See Fig. 3.2.

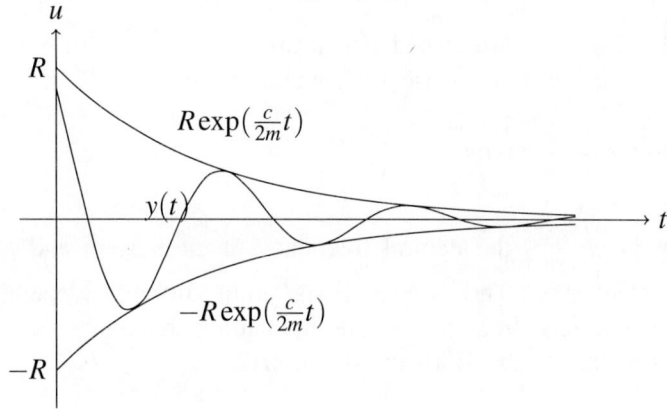

Fig. 3.2   Free damped vibration

As remarked earlier, the motion dies out if there is damping in the system. In other words, the initial disturbance is dissipated by damping. That is why it is very useful in practical mechanical systems. It can eliminate undesirable vibrations like shocks transmitted in an automobile, the principle behind shock absorbers, gun barrel etc.

Case (iii): (Forced, damped vibrations: $F \neq 0$, $c > 0$): Of course, this will depend on the applied force. Consider the case of a periodic applied force $F(t) = F_0 \cos \omega t$, $F_0 \neq 0$. A particular solution can be obtained as

$$y_p(t) = A\cos \omega t + B\sin \omega t, \tag{3.3.26}$$

where A, B are given by

$$A = F_0 \frac{m(\omega_0^2 - \omega^2)}{m^2(\omega_0^2 - \omega^2)^2 + c^2\omega^2}, \quad \text{and } B = F_0 \frac{c\omega}{m^2(\omega_0^2 - \omega^2)^2 + c^2\omega^2}$$

and $\omega_0 = \sqrt{\dfrac{k}{m}}$. The general solution can be written as

$$y(t) = \varphi(t) + y_p(t), \qquad\qquad (3.3.27)$$

where φ is the general solution to the homogeneous equation and as observed earlier, $\varphi(t) \to 0$ as $t \to \infty$. Thus, for large time, $y(t)$ behaves like $y_p(t)$. The solution y_p is called the *steady state* part of $y(t)$ and $\varphi(t)$ is called the *transient part*.

Case (iv): (Forced, undamped vibrations: $c = 0$, $F \neq 0$):   Again, we take $F = F_0 \cos \omega t$. In this case, the equation becomes

$$\ddot{y} + \omega_0^2 y = \dfrac{F_0}{m} \cos \omega t. \qquad\qquad (3.3.28)$$

Here, $\omega_0 = \sqrt{\dfrac{k}{m}}$ is the natural frequency of the system and ω is the external frequency. The behaviour is quite different depending on whether the applied frequency ω and the natural frequency are the same or not. This leads to the following situations.

Case (iv a) (Without resonance):   This is the case with $c = 0$, $F_0 \neq 0$, $\omega \neq \omega_0$: If $\omega \neq \omega_0$, that is, the applied frequency ω is different from the natural frequency ω_0, then, the case is similar to an earlier case and the solution is the sum of oscillatory functions in the form

$$y(t) = c_1 \cos \omega_0 t + c_2 \sin \omega_0 t + \dfrac{F_0}{m(\omega_0^2 - \omega^2)} \cos \omega t.$$

Case (iv b) (With resonance): $c = 0$, $F_0 \neq 0$, $\omega = \omega_0$: Consider the interesting case of $\omega = \omega_0$. Here, the external force and the given system have the same frequency ω_0, the natural frequency. We have

$$\ddot{y} + \omega_0^2 y = \dfrac{F_0}{m} \cos \omega_0 t.$$

More generally, consider the equation

$$\ddot{y} + \omega_0^2 y = \dfrac{F_0}{m} e^{i\omega_0 t}.$$

Here $e^{i\omega_0 t}$ is a solution of the homogeneous equation. Taking the expression $Ate^{i\omega_0 t}$ for a particular solution, we get $A = \dfrac{-iF_0}{2m\omega_0}$. Therefore, a particular solution is $\dfrac{F_0 t}{2m\omega_0}\sin\omega_0 t - i\dfrac{F_0 t}{2m\omega_0}\cos\omega_0 t$. Thus, we get the general solution as

$$y(t) = C_1\cos\omega_0 t + C_2\sin\omega_0 t + \frac{F_0}{2m\omega_0}t\sin\omega_0 t.$$

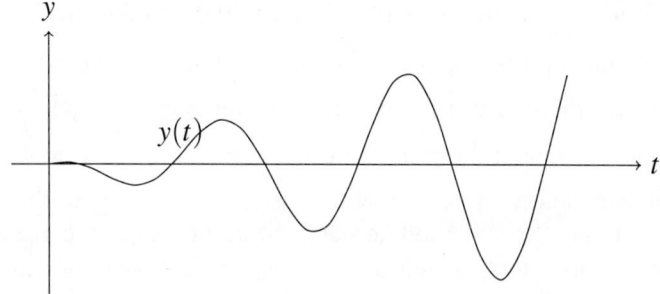

Fig. 3.3   Forced undamped vibrations with resonance

The first two terms are periodic functions of time. The last term is oscillatory and its amplitude keeps increasing due the presence of t. Thus, if the forcing term $F_0\cos\omega_0 t$ is in **resonance** with the natural frequency of the system, then it will cause unbounded oscillations, leading to mechanical catastrophes.

A criterion, similar to the lack of sufficient damping, was the reason for the collapse of the Tacoma bridge on November 7, 1940 at 11.00 am. This is also the cause for the collapse of the Broughton suspension bridge near Manchester. This occurred when a column of soldiers marched in cadence over the bridge, thereby setting up a periodic force with a rather large amplitude. The frequency was almost equal to the natural frequency of the bridge, and thus, large oscillations were induced and the bridge collapsed. It is for this reason that soldiers are ordered to break cadence when crossing a bridge.

Among many similarities with mechanical vibrations, electrical circuits also have the property of resonance. Unlike mechanical systems, resonance is put to good use here like, the tuning knob of radio or television is used to vary the capacitance in such a manner that the resonant frequency is changed until it agrees with the frequency of the external signal, that is, from a radio or television station; the amplitude of the current produced by this signal will be much greater than that of other signals so that we get the desired sound quality or picture quality or both.

3.3.4   Green's functions

We end this section by introducing an important concept in DE, namely Green's functions, which help us to represent the solutions of non-homogeneous equations with an arbitrary non-homogeneous term via a kernel which is defined independent of the non-homogeneous term; it solves a homogeneous DE with special initial conditions. Recall the integral calculus problem $\dot{y} = f(t)$; then y is given by $y(t) = \int^t f(t)\,dt$. In other words, the solution is represented by an integral which is in some sense the *inverse* operation to the differential operator $\dfrac{d}{dt}$. Can we have a similar representation for general equations $Ly = r$ leading to the integral representation via the so-called kernels? More precisely, we are looking for a solution operator S given by $y = Sr$, where S symbolically is the inverse of the given differential operator L. Hence, the motivation is to look for the integral representation.

First of all observe that the solution operator, namely $S : r \to y$, where y solves (3.3.16) is linear. In other words, define $S(r) = y$, that is, S acts on the function space where the non-homogeneous term r belongs to, for example, space of continuous functions and maps to the corresponding solution. Then the operator S is linear. In general, a class of important linear operators are given by integral operators defined via a kernel. In fact, history tells us that the study of linear operators was indeed motivated by integral operators. Thus, we would like to know whether the solution operator S is an integral operator. In other words, does there exist a two variable function $G(t,\xi)$ so that

$$y(t) = (Sr)(t) = \int_a^t G(t,\xi)r(\xi)\,d\xi. \tag{3.3.29}$$

In fact, it is quite reasonable to expect that the solution operator is an integral operator because formally S is the inverse of a differential operator L. Assume such a G exists. Then, differentiating y twice, we get

$$\dot{y}(t) = \int_a^t G_t(t,\xi)r(\xi)\,d\xi + G(t,t)r(t)$$

and

$$\ddot{y}(t) = \int_a^t G_{tt}(t,\xi)r(\xi)d\xi + G_t(t,t)r(t) + \frac{d}{dt}(G(t,t)r(t)).$$

Now substituting in (3.3.16), we get

$$Ly(t) = \int_a^t LG(\cdot,\xi)r(\xi)d\xi + G_t(t,t)r(t) + p(t)G(t,t)r(t)$$

$$+ \frac{d}{dt}(G(t,t)r(t)).$$

This motivates us to define G as a solution to the following homogeneous problem: For fixed $\xi \geq a$ as a parameter (in fact initial point), define G, for $t \geq \xi$ to be the solution of

$$LG(.,\xi) = 0, G(\xi,\xi) = 0, G_t(\xi,\xi) = 1.$$

Further, define $G(t,\xi) = 0$ for $t \in [a,\xi]$. Then, y given by (3.3.29) will satisfy the non-homogeneous equation $Ly = r$ satisfying the initial conditions $y(0) = \dot{y}(0) = 0$. The kernel G is called *Green's function* associated with the problem (3.3.16). For a class of problems associated with second order equations, the process of obtaining G is done in detail in Chapter 9.

3.4   Partial Differential Equations and ODE

In this section, we give two examples from PDE whose solution can be obtained via ODE. In the first example, we consider a first order linear equation with constant coefficients. The solution to the IVP for the PDE can be derived by solving two ODEs. The interesting fact is that the so-called *method of characteristics* can be applied to any general first order PDE. We will study the general case in great detail in Chapter 10. The second example is the *heat equation* which is a second order linear PDE. A precise study of this equation needs more sophisticated tools like semigroup theory, distribution theory and Sobolev spaces which are not within the scope of this book. The purpose of presenting this example is that we can reduce the PDE problem to that of an IVP for ODE, though in an infinite dimensional Hilbert space. Hence, it will not come under the purview of this book as it involves the modern tools indicated earlier.

Example 3.4.1

Consider the simplest PDE, namely the transport equation in two independent variables t, x:

$$u_t(x,t) + c u_x(x,t) = 0, \qquad (3.4.1)$$

for $t > 0$, $x \in \mathbb{R}$, where $c > 0$ is a given constant. This is a linear, first order PDE in two dimensions. Introduce the ODE $\dot{x} = c$ in the upper half (x,t) plane, $t > 0$, whose solutions are given by $x - ct = \xi_0$. These are straight lines with slope c (see Fig. 3.4) and we denote these curves by

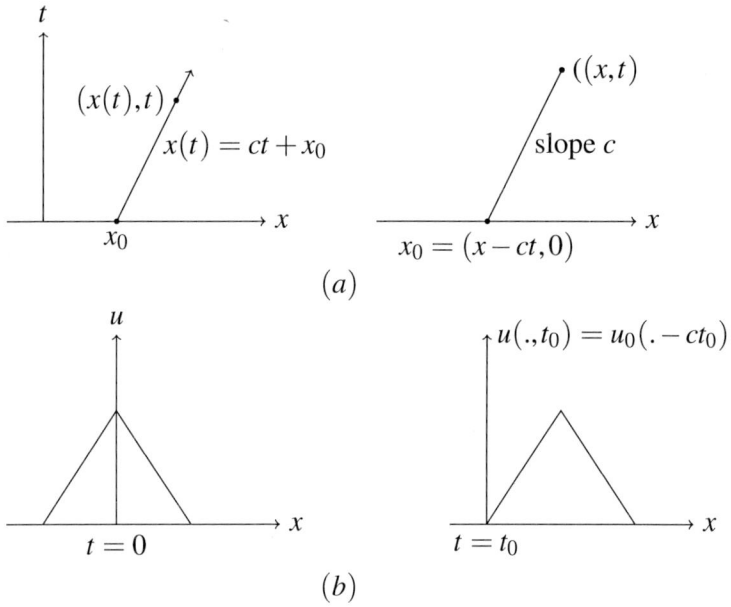

(a)

(b)

Fig. 3.4   (a) Characteristic curves, (b) Solution curves

$x(t)$, that is, $x(t) = ct + \xi_0$. These curves, straight lines in this case, are called the *characteristic curves* of (3.4.1). In general, when c is a function of t and x, these characteristic curves need not be straight lines. Fix $\xi = \xi_0$ and restrict u along this curve (line). Consider the function of one variable $g(t) = u(x(t),t)$. Now, it is easy to see by the chain rule that

$$\frac{d}{dt}g(t) = \frac{d}{dt}u(x(t),t) = u_x \dot{x}(t) + u_t . 1 = u_x . c + u_t = 0. \qquad (3.4.2)$$

Solving this ODE, we observe that u is constant along the curve $x - ct = \xi_0$. In other words, u takes the same value along the fixed line constructed earlier. This observation can be used to solve the IVP for the PDE (3.4.1) as follows: Suppose the initial values of u are given on the x-axis, that is, $u(x,0) = u_0(x)$ is given and u_0 is differentiable. Now, for any point (x,t) in the upper half plane, $t > 0$, draw the straight line with slope c which will meet the x-axis at the pint $(x - ct, 0)$. By this observation, we get $u(x,t) = u(x - ct, 0) = u_0(x - ct)$.

Verify that this indeed is a solution to the IVP for PDE. Thus, we have obtained the solution of (3.4.1) which satisfies the prescribed initial values.

We observe that the solution of the PDE was obtained by solving two ODEs. The initial curve, where initial value u_0 is assigned, namely the x-axis in this case, can be general. But, at the same time, one cannot choose the initial curve arbitrarily. For example, one can choose a curve Γ which intersects all the characteristics, (see Fig. 3.5 (a)), but at the same time, it should not coincide with a characteristic curve as in Fig. 3.5 (b), because this will contradict the fact that u is a constant along the characteristics. So the initial curve will have to satisfy certain *transversality conditions*, which we will discuss in Chapter 10. We will also see in Chapter 10 that the solvability of any first order PDE can be reduced to the solvability of systems of ODE by the method of characteristics, but the ODE system may be more complicated.

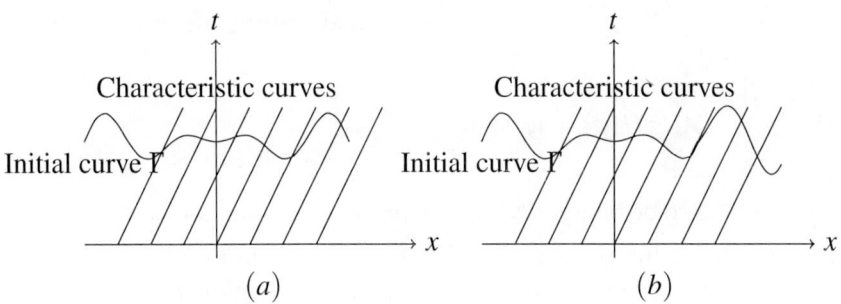

Fig. 3.5   Characteristic and initial curves

We now discuss a second order linear PDE, namely, the heat equation. Our only purpose here is to indicate that this equation may be viewed as an ODE, though in an infinite dimensional (Hilbert) space. As such, many of the following terminology are not explained in a rigorous manner.

An inquisitive reader may explore this and similar topics after gaining sufficient knowledge in functional analysis and related topics.

Example 3.4.2

[Heat Equation] Consider the IVP for the heat equation

$$u_t - u_{xx} = 0, \ u(x,0) = u_0(x), \tag{3.4.3}$$

for $t > 0$, $x \in \mathbb{R}$. Again $u = u(x,t)$ is an unknown function to be determined. We transform this problem to an ODE problem in a Hilbert space. Let X be the space of all compactly supported smooth (twice differentiable) functions $v = v(x)$ in $\mathbb{R}$ such that both v and $\dfrac{dv}{dx}$ are square integrable over $\mathbb{R}$. We equip X with the norm given by

$$\|v\|^2 \equiv \int_{\mathbb{R}} |v(x)|^2 + \int_{\mathbb{R}} \left|\frac{dv}{dx}\right|^2.$$

Unfortunately, this space is not complete. Let H be the completion of X. One of the major difficulties is the characterization of H, which can be done only with modern tools like distributions and Sobolev spaces.

We will be looking for a solution $u(x,t)$ such that for each t, the function $U(t) \equiv u(\cdot,t) \in H$. Of course, we need certain smoothness in the t variable as well. We define an operator A in an appropriate subspace of H by $Av \equiv v_{xx}$. With these notations, we may rewrite the problem (3.4.3) as

$$\frac{dU}{dt} = AU, \ U(0) = u_0.$$

Here $\dfrac{dU}{dt}$ has to be interpreted in an appropriate infinite dimensional space. The essence is that we have an IVP for ODE in an infinite dimensional space as well. The operator A is linear, but it is *unbounded*. Again, we need a new tool such as the *semi group theory* from functional analysis, to study the existence and uniqueness of solutions to the aforementioned ODE. When A is a bounded operator in a finite dimensional space, it is the standard linear system which we will study in Chapter 5.

3.5   Exercises

1. Prove that every separable equation is exact.

2. Find the unique solution to the IVP $\dot{y} = \frac{2y}{t}, y(t_0) = y_0$, where $t_0 \neq 0$. Also find the interval of existence and plot the solution for different values of y_0.

3. Show that the solution of $\dot{y} + (\sin t)y = 0$, $y(0) = \frac{3}{2}$ is given by $y(t) = \frac{3}{2}e^{\cos t - 1}$.

4. The solution of $\dfrac{dy}{dt} + e^{t^2}y = 0$, $y(1) = 2$ can be represented as $y(t) =$
$$2\exp\left(-\int_1^t e^{s^2} ds\right).$$

5. Classify the following into linear or non-linear:

   (a) $\dot{y} = ay - by^2$, $\dot{y} = -t/y$, $\dot{y} = -y/t$, $\dot{y}(t) = \sin(t)$,
   $\sin y + x\cos(\dot{y}) = 0$ $\dot{y} = |y|$, $y\dot{y} = y$,
   $\dot{y} = \sin y$, $y\dot{y} = \frac{g}{W}(W - B - cy)$.

   (b) (Duffing equation): $\ddot{y} + \delta\dot{y} + \alpha y + \beta x^3 = 0$

   (c) (van der Pol equation): $\ddot{y} - \mu(y^2 - 1)\dot{y} + y = 0$

   (d) (Prey–predator system): $\dot{x} = ax - bxy$, $\dot{y} = -cy + dxy$

   (e) (Epidemiology): $\dot{S} = -\beta SI$, $\dot{I} = \beta SI - \gamma I$

   (f) (Bernoulli equation): $\dot{y} + \phi(t)y = \psi(t)y^n$

   (g) (Reduced Bernoulli equation): $\dot{y} + (1-n)\phi(t)y = (1-n)\psi(t)$

   (h) (Generalized Riccati equation): $\dot{y} + \psi(t)y^2 + \phi(t)y + \chi(t) = 0$

6. Consider the Bernoulli equation

   $$\dot{x} + \phi x = \psi x^n,$$

   where ϕ, ψ are continuous functions. For $n \neq 1$, it is non-linear; show that it can be reduced to a linear equation by the substitution $y = x^{1-n}$. Then, solve the equation.

7. Find the general solution of (i) $\dot{x} + e^t x = e^t x^2$ (ii) $\dot{x} + t^n x = x^n$.

8. Consider the Jacobi equation

   $$(a_1 + b_1 t + c_1 x)(t\,dx - x\,dt) - (a_3 + b_3 t + c_3 x)dx$$

   $$+(a_3 + b_3 t + c_3 x)dt = 0,$$

where a_i, b_i, c_is are constants. Transform the variables $(t,x) \to (\tau,y)$, where $t = \tau + \alpha$ and $x = y + \beta$. Choose α, β appropriately to get the equation

$$\tau dy - y d\tau + \phi(y/\tau)dy + \psi(y/\tau)d\tau = 0.$$

Now make the substitution $y = \tau u$ to bring it to Bernoulli's form

$$\frac{d\tau}{du} + h(u)\tau + g(u)\tau^2 = 0.$$

9. Consider the generalized Riccati equation

$$\dot{y} + \psi(t)y^2 + \phi(t)y + \chi(t) = 0,$$

where ψ, χ, ϕ are functions of t. In general, we do not have solutions in explicit form. Assume $x = x_1$ is one known solution, let x be any other solution. Write $x = x_1 + y$. Show that y satisfies the Bernoulli equation

$$\dot{y} + (2x_1\psi + \phi)y + \psi y^2.$$

10. Find the general solution of the equation $\dot{y} + y^2 + y - (1 + t + t^2) = 0$.

11. Assume $\psi(t) \neq 0$ for all t in the Riccati equation $\dot{y} + \psi(t)y^2 + \phi(t)y + \chi(t) = 0$. Show that the transformation $x = \frac{1}{\psi}\frac{\dot{z}}{z}$, reduces the Riccati equation to a linear second order equation for z.

12. Reduce the original Riccati equation, $\dot{y} + ay^2 = bt^m$, where a, b are constants, to a second order linear equation $\ddot{z} - abt^m z = 0$ using the transformation in Exercise 11.

13. Find the general solution of the following Riccati equations:

   (a) $\dot{y} = (1 - 2t - t^3) + 2(1 + t^2)y - ty^2$, given that $y(t) = t$ is a solution of this equation.

   (b) $\dot{y} = (t + \frac{1}{t})y - \frac{1}{2}(t^2 + y^2)$, given that this equation has a solution given by $y(t) = t$.

14. a) Consider the linear problem $\dot{y} + py = q$. Show that, if $q \geq 0$, then, $y \geq 0$ if it is initially so, that is, if $y(0) \geq 0$. Now consider the equation $\dot{x} + px = q_1$ and $\dot{y} + py = q_2$, then compare the solutions when $q_1 \geq q_2$.

b) Consider $\dot{x} + p_1 x = q$ and $\dot{y} + p_2 y = q$. Show that, if $p_2 \geq p_1$, $x(0) \geq y(0)$ and $y \geq 0$, then $x \geq y$.

c) Consider the inequality $\dot{y} + py \leq q$. Derive the inequality

$$y(t) \leq \exp\left(-\int_0^t p(s)ds\right)\left[y(0) + \int_0^t q(s)\exp\left(\int_0^s p(z)dz\right)ds\right].$$

d) Derive Gronwall's inequality. Assume f and g are continuous real valued functions defined on the interval $[a,b]$ and $g \geq 0$. Assume

$$f(t) \leq c + k\int_{t_0}^t f(s)g(s)ds,$$

where c,k are constants, $k \geq 0$. Then

$$f(t) \leq c\exp\left(k\int_{t_0}^t g(s)ds\right), \; t_0 \in [a,b].$$

e) (Uniqueness) Let p,q be continuous functions on $[a,b]$. Show that the linear initial value problem $\dot{x} + p(t)x = q(t), x(t_0) = x_0$ has at most one solution.

15. Verify linear independence of solutions in Subsection 3.3.2.

16. Find the general solution of the following equations

(a) $\dot{y} - a\dfrac{y}{t} = \dfrac{t+1}{t}$

(b) $(t - t^3)\dot{y} + 2(t^2 - 1)y - at^3 = 0$

(c) $\dot{y} + ty = t^3 y^3$

(d) $y - \dot{y}\cos t = y^2 \cos t(1 - \sin t)$

(e) $\left(\dfrac{1}{t^2} + \dfrac{3y^2}{t^4}\right)dt = \dfrac{2y}{t^3}dy$

(f) $\dfrac{t^2 dy - y^2 dt}{(t - y)^2} = 0$

17. Find the general solution of the following equations

(a) $t^2\ddot{y} + t\dot{y} - y = 0$.

(b) $t\ddot{y} - (t+n)\dot{y} + ny = 0.$

(c) $\ddot{y} - f(t)\dot{y} + (f(t) - 1)y = 0.$

(d) $t\dfrac{d^3y}{dt^3} = 2$ and $\ddot{y} = \dfrac{a}{y^3}.$

18. Three solutions of a certain second order non-homogeneous linear equation in $\mathbb{R}$ are

$$\varphi_1(t) = t^2, \ \varphi_2(t) = t^2 + e^{2t}, \ \varphi_3(t) = 1 + t^2 + 2e^{2t}.$$

Find the general solution of this equation.

19. Three solutions of a certain second order non-homogeneous linear equation $\mathscr{L}y = g$ in $\mathbb{R}$ are

$$\psi_1(t) = t^2, \ \psi_2(t) = t^2 + e^{2t}, \ \psi_3(t) = 1 + t^2 + 2e^{2t}.$$

Here g is a continuous function in $\mathbb{R}$. Find the solution y of $\mathscr{L}y = g$ satisfying $y(0) = 1, \ y'(0) = 2.$

20. Assume the unique existence of a solution to the nth order IVP

$$Ly := y^{(n)} + p_1(t)y^{(n-1)} + \cdots + p_{n-1}(t)y^{(1)} + p_n(t)y = 0$$

$$y(t_0) = y_0, \dot{y}(t_0) = y_1, \cdots y^{(n-1)}(t_0) = y_{n-1},$$

$$(3.5.1)$$

where $p_1, \cdots, p_n$ are continuous functions. Let S be the set of all solutions to $Ly = 0$. Show that S is a linear space of dimension n.

21. Let $f(t, y, \dot{y}) = h(t)$ be the general form of the first order equation, where $h = h(t)$ is all the combined non-homogeneous terms. Consider $L(r, s) = f(t, r, s)$, where t is fixed. If L is linear in r and s, show that there exists functions $p_0 = p_0(t)$ and $p_1 = p_1(t)$ so that f takes the form,

$$f(t, y, \dot{y}) = p_0(t)\dot{y}(t) + p_1(t)y(t).$$

More generally, an nth order linear equation has the general form,

$$p_0(t)y^{(n)}(t) + p_1(t)y^{(n-1)}(t) + \cdots + p_n((t)y(t) = h(t).$$

22. **Escape Velocity Problem:** Determine the smallest velocity with which a body must be thrown vertically upwards so that it will not return to the earth; air resistance may be neglected.

Let M denote the mass of the earth and m, the mass of the body to be thrown. By Newton's law of gravitation, the force of attraction f acting on the body is $f = k\dfrac{Mm}{r^2}$, where r is the distance between the center of the earth and the center of gravity of the body and k is the gravitational constant. The second law of motion, then, implies that the motion of the body is governed by the differential equations

$$m\frac{d^2 r}{dt^2} = -k\frac{Mm}{r^2}; \text{ that is, } \frac{d^r}{dt^2} = -k\frac{M}{r^2}.$$

The minus sign indicates that the acceleration is negative; air resistance has been omitted. Solve this equation with initial conditions $r + R$, $\dfrac{dr}{dt} = v_0$ at time $t = 0$, where R is the radius of the earth and further, show that

$$\frac{r^2}{2} \equiv \frac{M}{r} + \left(\frac{v_0^2}{2} - \frac{kM}{r} \right),$$

where $v = \dfrac{dr}{dt}$. Conclude that the escape velocity v_0 satisfies $v_0^2 \geq \dfrac{2kM}{R}$. In the usual CGS units, $k = 6.66 \times 10^{-8} \text{cm}^3/\text{g. sec}^2$ and $R = 63 \times 10^7 \text{cm}$. At the earth's surface $r = R$, the acceleration due to gravity is $g = 981 \text{cm}/\text{sec}^2$. Therefore, $g = \dfrac{kM}{R^2}$. Hence, we require, $v_0^2 \geq 2gR$, which gives an approximate value of v_0 as $11.2 \times 10^5 \text{cm}/\text{sec}^2$.

3.6 Notes

The discussion on linear first and second order equations is available in every basic book on ODE. In addition to the linear equations, we have also introduced a section on exact differential equations. On one hand, we have shown how every first order linear equation can be reduced to an integral calculus problem, namely $\dot{y}(t) = h(t)$ with the introduction of an integrating factor (I.F.). This also makes it clear why an IVP for a first order equation requires just one condition as integral calculus problems require only one condition to find the constant of integration. By transforming the second order equation into a system of two first order equations, we see that two (initial) conditions are needed for the corresponding IVP. We have made an attempt to clarify these issues regarding the number of (initial) conditions. We have also presented exact differential equations in a natural way.

Though first order linear equations can be reduced to an integral calculus problem, such a scheme is not available for a second order linear equation. A complete solution structure is presented by showing that the solution space is a two-dimensional vector space. This analysis may not be available in many books as most books deal with second order linear equations with constant coefficients. However, here we have emphasized that there is no general method to find two linearly independent solutions. We have also presented some methods in this chapter to find the general solution. An analysis of the spring–mass–dashpot system has been carried out to further exhibit the importance of analysis even when an explicit solution is available. As far as references are concerned, there are many; see [Bra78, Bra75, BR03, CL72, Sim91, SK07, Tay11, MU78].

4

General Theory of Initial Value Problems

4.1 Introduction

4.1.1 Well-posed problems

In this chapter, we address the problem of the existence and uniqueness of solutions of initial value problems (IVP). For this purpose, our first task is to ensure that the given differential equation has a solution. A mathematical model originating from a real life system may exhibit more than one solution starting from the same initial condition, though a unique solution is expected. This may be due to rough approximations and assumptions made while making a mathematical model of the physical system. On the other hand, a mathematical model may not have a solution at all. Similarly, it is also important to study the behaviour of the solution with respect to the initial data as the initial data is usually measured by using some devices and is bound to have some small errors. Continuous dependence of solutions on initial data guarantees that a small error in the initial data does not cause a drastic change in the solution of the system. According to the French mathematician Jacques Hadamard, if an initial value problem arising from a physical phenomenon qualifies the above mentioned tests, namely, a solution exists (*existence problem*), the solution is unique (*uniqueness problem*) and the solution depends continuously on the initial conditions (*stability problem*) in appropriate norms, then the IVP is said to be *well-posed*. Otherwise, the problem is *ill-posed*. In this chapter, we will address these issues and prove results which ensure the well-posedness of an IVP, under suitable assumptions. We consider the following IVP

$$\dot{y} = f(t,y), \quad y(t_0) = y_0, \tag{4.1.1}$$

where, $f : D \subset \mathbb{R}^2 \to \mathbb{R}$ is a function not necessarily linear and is assumed to be continuous in an open connected set D of $\mathbb{R}^2$ containing the initial point (t_0, y_0). By a solution of the initial value problem (4.1.1), we mean a continuously differentiable function $y \in C^1(I)$, where $t_0 \in I$ is an interval in $\mathbb{R}$, satisfying (4.1.1). This means that $y(t_0) = y_0$ and for each $t \in I$, the point $(t, y(t)) \in D$ and $\dot{y}(t) = f(t, y(t))$. In Chapter 3, we have already seen some examples including an ill-posed linear problem!

In general, the solvability property of (4.1.1) will fall under one of the following three cases: The IVP (4.1.1) has

(i) a unique solution.

(ii) no solution.

(iii) infinitely many solutions.

When the function $f(t,y)$, (called the vector field) is not continuous at a point in the (t,y)-plane, then there may be a possibility of non-existence of a solution at that point. Similarly, if the vector field is not differentiable at a point, then there may be a possibility of non-uniqueness of solutions through that point. The initial condition also plays a crucial role in the existence and behaviour of the solution of an IVP.

Before proceeding further, we will consider some examples which exhibit one or more phenomena discussed here. Also see Examples 3.1.2, 3.1.3.

4.1.2   Examples

Example 4.1.1

Consider the linear differential equation $\dot{y} = 2y$.

It has been shown in Chapter 3 that the function $y(t) = ce^{2t}$ is a solution of the equation for every arbitrary constant c. Thus, the differential equation has infinitely many solutions. Geometrically, this is a one-parameter family of curves.

Let us remark that a differential equation always comes with some associated physically meaningful conditions such as initial conditions and boundary conditions. So when we talk about well-posedness, it is for the differential equation together with the associated conditions provided.

In this example, if the solution is required to satisfy an initial condition, for example, $y(0) = 1$ then, the unique solution is given by $y(t) = e^{2t}$.

Example 4.1.2

We consider the problem $\dot{y} = \dfrac{3}{t}y$,

with various initial conditions to exhibit multiplicity of solutions, non-existence and a unique existence. Note that the vector field is not defined at $t = 0$. By separable variable method, we get

$$y = ct^3$$

as the solution for the arbitrary constant c. Now consider the initial condition $y(0) = 0$, we see that $y = ct^3$ is a solution to the initial value problem for any value of c. Thus, the IVP has infinitely many solutions with the initial condition $y(0) = 0$. All the solution curves pass through $(0,0)$ in the (t,y)-plane.

On the other hand, if we take the same differential equation but with the initial condition $y(0) = 2$, then the IVP has no solution as the general solution is $y = ct^3$. The trouble is due to the singularity at $t = 0$. It is also easy to see that the IVP with $y(t_0) = y_0$, with $t_0 \neq 0$ has a unique solution $y(t) = \dfrac{y_0}{t_0^3}t^3$.

We now give an example of an initial value problem in which f is continuous but not linear and exhibit infinitely many solutions.

Example 4.1.3

Consider the nonlinear IVP $\dot{y} = 3y^{2/3}, \quad y(0) = 0.$

Clearly $y \equiv 0$ is a solution known as the trivial solution. Separating the variables and integrating, we get

$$y(t) = (t+c)^3.$$

Using the initial condition $y(0) = 0$, we have $y(t) = t^3$ as a solution. Thus, we get two solutions. In fact, this IVP has infinitely many solutions. It is left as an easy exercise to verify that for any fixed $a > 0$, the function $y(t)$ defined by

$$y(t) = \begin{cases} 0, & 0 \leq t \leq a \\ (t-a)^3, & t > a \end{cases}$$

is a solution to the given IVP.

Example 4.1.3 shows that even continuity is not enough for uniqueness of solution to an IVP. We shall see later that if f has a bounded derivative, then the solution of the IVP is unique. We will also prove the uniqueness under the assumption of Lipschitz continuity. But, even when a unique solution exists for an IVP and the vector field is very smooth, the solution may not exist for all values of t. This fact is illustrated in the following example.

Example 4.1.4

Consider the nonlinear initial value problem:

$$\dot{y} = y^2, \quad y(0) = y_0 > 0.$$

By an integration and using $y(0) = y_0$, we get a solution $y(t) = \dfrac{y_0}{1 - y_0 t}$. It is the only solution to the problem. We will see this fact in the next section. Note that $y(t)$ is defined only for $t < \dfrac{1}{y_0}$, despite the fact that the vector field $f(t,y) = y^2$ is very smooth on the entire real line. If y_0 is large, then we have solutions only on a very small interval to the right of $t = 0$; however, they exist for all $t < 0$.

We now see an example of a nonlinear IVP which does not have a solution.

Example 4.1.5

Consider the nonlinear initial value problem $\dot{y} = -sgn(y)$, where the vector field $sgn(y)$ is defined by

$$sgn(y) = \begin{cases} 1, & y \geq 0 \\ -1, & y < 0. \end{cases}$$

Let $y(0) = 0$ be the initial condition. The vector field $-sgn(y)$ is discontinuous at $y = 0$. If there is a solution to the IVP, then, the solution must be decreasing initially as $\dot{y}(0) = -1$. Therefore, $y(t) < 0$ for some $t > 0$. However, for all negative values of y, the solution must be increasing. Since these two statements contradict each other, there is no solution to this IVP.

4.2   Sufficient Condition for Uniqueness of Solution

We first address the question of uniqueness of a solution to IVP. In fact, if the function $f(t,y)$ is continuous in a domain D in the (t,y)-plane containing the initial point (t_0, y_0), then existence of a local solution is guaranteed in some interval $(t_0 - \alpha, t_0 + \alpha)$, $\alpha > 0$, being a constant. This is known as the *Cauchy–Peano theorem*, which will be proved later. To ensure the uniqueness of solution of the IVP (4.1.1), we need a stronger assumption, namely, the Lipschitz continuity of $f(t,y)$ with respect to y (see Chapter 2).

4.2.1   A basic lemma

We now convert the IVP into an equivalent integral equation, which will be used to establish existence and uniqueness theorems for the IVP. The IVP (4.1.1) can be reduced to a nonlinear Volterra integral equation as we see in the following basic lemma.

Lemma 4.2.1

[Basic Lemma]

Let D be an open connected set in $\mathbb{R}^2$ and $(t_0, y_0) \in D$. Assume that f is continuous in D. If y is a solution to the IVP (4.1.1) in an interval (a,b) containing t_0, then y satisfies the integral equation

$$y(t) = y_0 + \int_{t_0}^{t} f(\tau, y(\tau)) d\tau, \ t \in (a,b). \tag{4.2.1}$$

Conversely, if y is a continuous function defined on (a,b), which satisfies the integral equation (4.2.1) for all $t \in (a,b)$, then $y \in C^1(a,b)$ and satisfies the IVP (4.1.1).

Proof: Let y be a function satisfying the IVP (4.1.1). Now integrating (4.1.1) over (t_0, t), we get

$$\int_{t_0}^t \dot{y}(\tau)d\tau = \int_{t_0}^t f(\tau,y(\tau))d\tau,$$

that is,

$$y(t) = y_0 + \int_{t_0}^t f(\tau,y(\tau))d\tau.$$

Therefore, y satisfies the integral equation (4.2.1). Conversely, since f and y are continuous, the composite function $t \mapsto f(t,y(t)), t \in (a,b)$ is continuous and hence integrable. By the fundamental theorem of integral calculus, we then infer that the function defined by the integral on the right side in (4.2.1) is a differentiable and hence, y is also differentiable. Differentiating (4.2.1) with respect to t, we obtain

$$\dot{y}(t) = f(t,y(t)).$$

Further, $y(t_0) = y_0$ proving that y satisfies (4.1.1).   □

Remark 4.2.2

One can analyse the solution of (4.2.1) without the continuity assumption of f but with certain integrability assumptions. This will only give continuity of y and hence, we will not be able to derive (4.1.1) from the integral formulation (4.2.1). Such solutions are called *mild solutions* or *weak solutions* of the initial value problem. Mild solutions are important in applications, for example, in control theory, where the control function (continuous or discontinuous) may appear through the source term f and one has to interpret the solutions in the form of mild solutions.

Before proceeding to prove the existence, we will prove the uniqueness which is quite easy. This is done via Gronwall's inequality which is given as an exercise in Chapter 1. However, we now state and prove Gronwall's inequality.

Lemma 4.2.3

[Gronwall's Inequality] Suppose that p and q are continuous real valued functions defined on $[a,b]$ with $q(t) \geq 0$ on $[a,b]$. Assume p,q satisfy

$$p(t) \le C + k \int_{t_0}^t q(s)p(s)ds,$$

for all $t \in [a,b]$, where $t_0 \in [a,b]$ is fixed and C, k are constants with $k \ge 0$. Then,

$$p(t) \le C \exp\left(k \int_{t_0}^t q(s)ds\right) \tag{4.2.2}$$

for all $t \in [a,b]$.

Proof: The proof is quite easy. Put $F(t) = C + k \int_{t_0}^t q(s)p(s)ds$. Then, by hypothesis, we have

$$p(t) \le F(t). \tag{4.2.3}$$

Using this inequality, we get

$$\dot{F}(t) = kq(t)p(t) \le kq(t)F(t),$$

as k and q are non-negative. Now, using the integrating factor $\exp\left(-k \int_{t_0}^t q(s)ds\right)$ for the aforementioned differential inequality, we get

$$\frac{d}{dt}\left(\exp\left(-k \int_{t_0}^t q(s)ds\right)F(t)\right) \le 0.$$

An integration over $[t_0, t]$ gives

$$\exp\left(-k \int_{t_0}^t q(s)ds\right)F(t) \le F(t_0) = C.$$

Now using (4.2.3), we get the required Gronwall's inequality (4.2.2).   □

Remark 4.2.4

If $C \equiv 0$, that is, p satisfies

$$p(t) \le k \int_{t_0}^t q(s)p(s)ds$$

with k,q non-negative, then $p \leq 0$ by Gronwall's inequality. In particular, if $p \geq 0$, then $p \equiv 0$.   □

4.2.2   Uniqueness theorem

We now prove the uniqueness theorem using Gronwall's inequality, the basic lemma and the Lipschitz continuity of $f(t,y)$ with respect to y. Here, we wish to clarify that uniqueness means that it does not have more than one solution. We do not interpret that it has a solution.

Theorem 4.2.5

[Uniqueness] Let a,b be positive real numbers. Suppose that $f = f(t,y)$ is continuous on the rectangle $\mathscr{R} = \{(t,y) \in \mathbb{R}^2 : |t - t_0| \leq a, |y - y_0| \leq b\}$ and Lipschitz continuous with respect to y in $\mathscr{R}$. Then, the IVP (4.1.1) has at most one solution.

Proof: Suppose that y and z are two solutions of the IVP (4.1.1) defined on an interval $[c,d]$ contained in the interval $[t_0 - a, t_0 + a]$ and $t_0 \in [c,d]$. Thus by the basic lemma, we have

$$y(t) = y_0 + \int_{t_0}^{t} f(\tau, y(\tau)) d\tau \text{ and } z(t) = y_0 + \int_{t_0}^{t} f(\tau, z(\tau)) d\tau$$

for all $t \in [c,d]$. Subtracting, we get

$$y(t) - z(t) = \int_{t_0}^{t} [f(\tau, y(\tau)) - f(\tau, z(\tau))] d\tau,$$

Using the Lipschitz continuity of f, we have

$$|y(t) - z(t)| \leq \int_{t_0}^{t} \alpha |y(\tau) - z(\tau)| d\tau,$$

where α is the Lipschitz constant of f. Applying Gronwall's inequality with $C = 0, k = \alpha, q \equiv 1$ and $p = |y - z|$, we get

$$|y(t) - z(t)| = 0$$

for all $t \in [c,d]$ and hence the solution is unique.   □

4.3   Sufficient Condition for Existence of Solution

We now provide sufficient conditions to ensure the existence of a solution to IVP (4.1.1). We discuss three theorems on the existence namely, Picard's theorem, Cauchy–Peano theorem and fixed point theorem. We assume Lipschitz continuity of the function f in Picard's and fixed point iterations whereas, we prove Peano's existence theorem with only the continuity assumption on f. We first prove the theorem due to Picard, where Lipschitz condition is also assumed and hence uniqueness is guaranteed in the light of the uniqueness theorem we just proved.

Theorem 4.3.1

[Picard's Existence Theorem]

Let D be an open connected set in $\mathbb{R}^2$. Assume $f : D \to \mathbb{R}$, satisfies the following conditions:

(i) $f(t,y)$ is continuous on D.

(ii) $f(t,y)$ is Lipschitz continuous with respect to y on D with Lipschitz constant $\alpha > 0$.

Let $(t_0, y_0) \in D$ and a and b be positive constants such that the rectangle $\mathscr{R}$ defined by

$$\mathscr{R} = \{(t,y) : |t - t_0| \le a, \ |y - y_0| \le b\}$$

is a subset of D. Let $M = \max\limits_{(t,y) \in \mathscr{R}} |f(t,y)|$ and $h = \min\left(a, \dfrac{b}{M}\right)$. Then, IVP (4.1.1) has a unique solution in the interval $|t - t_0| \le h$.

Proof: Since $\mathscr{R}$ is a closed rectangle inside D, f satisfies all the properties assumed in the theorem inside the rectangle $\mathscr{R}$ as well. Let $\mathscr{R}_1$ be the rectangle defined by:

$$\mathscr{R}_1 = \{(t,y) : |t - t_0| \le h, \ |y - y_0| \le b\}.$$

If $a \le \dfrac{b}{M}$, then $h = a$ and $\mathscr{R}_1 = \mathscr{R}$, and if $\dfrac{b}{M} \le a$, then $h = \dfrac{b}{M}$ and $\mathscr{R}_1 \subset \mathscr{R}$ as shown in Fig. 4.1.

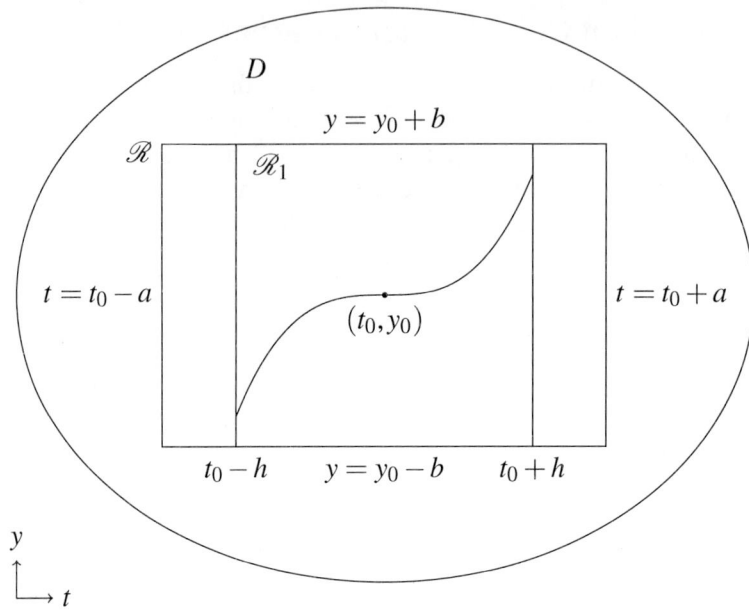

Fig. 4.1   Picard's theorem

Consider the interval $[t_0, t_0 + h]$. Similar arguments hold for the interval $[t_0 - h, t_0]$. The proof will be established by the construction of successive approximations, called Picard's iterates $\{y_n\}$, $n = 0, 1, 2, \cdots$ and showing that $\{y_n\}$ converges uniformly to some y defined on $[t_0, t_0 + h]$, a solution of the integral equation (4.2.1). The basic lemma (Lemma 4.2.1), then gives the existence of a solution to (4.1.1). The proof will be accomplished through the following three steps.

Step 1: Here we define Picard's iterates. For $t \in [t_0, t_0 + h]$, let

$$y_0(t) = y_0$$

and define successively,

$$y_n(t) = y_0 + \int_{t_0}^{t} f(\tau, y_{n-1}(\tau)) \, d\tau \tag{4.3.1}$$

for $n = 1, 2, \cdots$. Our first task is to show that the y_ns are well-defined on $[t_0, t_0 + h]$ and are continuously differentiable. We do this by using an induction argument. Note that y_n is well-defined only if $(t, y_{n-1}(t)) \in \mathcal{R}_1$ for all $t \in [t_0, t_0 + h]$. This holds trivially for y_0. Assume, for $n \geq 1$,

$(t, y_{n-1}(t)) \in \mathcal{R}_1$ and $|y_{n-1}(t) - y_0| \leq b$ holds, for all $t \in [t_0, t_0 + h]$. We show that the same statements are true when y_{n-1} is replaced by y_n and that completes the induction argument. Since $\mathcal{R}_1 \subset \mathcal{R}$, we have

$$|f(t, y_{n-1}(t))| \leq M \quad \text{on } [t_0, t_0 + h].$$

Now consider $y_n(t) = y_0 + \int_{t_0}^t f(\tau, y_{n-1}(\tau)) \, d\tau$. The aforementioned induction assumption implies that the definition of $y_n(t)$ makes sense and y_n is continuously differentiable on $[t_0, t_0 + h]$. Now,

$$|y_n(t) - y_0| = \left| \int_{t_0}^t f(\tau, y_{n-1}(\tau)) \, d\tau \right| \leq \int_{t_0}^t |f(\tau, y_{n-1}(\tau))| \, d\tau$$

$$\leq M(t - t_0) \leq Mh \leq b.$$

Thus, $(t, y_n(t))$ lies in the rectangle $\mathcal{R}_1$ and hence, $f(t, y_n(t))$ is defined and continuous on $[t_0, t_0 + h]$. Hence, the said properties hold also for y_n and induction is complete.

Step 2:   In this step, we show that the sequence of functions $\{y_n\}$, constructed in Step 1 satisfy the inequality

$$|y_n(t) - y_{n-1}(t)| \leq \frac{M}{\alpha} \frac{(\alpha(t - t_0))^n}{n!} \quad \text{on } [t_0, t_0 + h], \tag{4.3.2}$$

also using mathematical induction. Assume that, for $n \geq 2$,

$$|y_{n-1}(t) - y_{n-2}(t)| \leq \frac{M\alpha^{n-2}}{(n-1)!} (t - t_0)^{n-1} \quad \text{on } [t_0, t_0 + h]. \tag{4.3.3}$$

Then,

$$|y_n(t) - y_{n-1}(t)| = \left| \int_{t_0}^t f(\tau, y_{n-1}(\tau)) - f(\tau, y_{n-2}(\tau)) \, d\tau \right|$$

$$\leq \int_{t_0}^t |f(\tau, y_{n-1}(\tau)) - f(\tau, y_{n-2}(\tau))| \, d\tau.$$

Since $|y_n(t) - y_0| \leq b$ for all n and $t \in [t_0, t_0 + h]$, we get $(t, y_{n-1}(t)), (t, y_{n-2}(t))$ in $\mathcal{R}_1$ for $t \in [t_0, t_0 + h]$. By applying the Lipschitz continuity of f, we have

$$|y_n(t) - y_{n-1}(t)| \leq \alpha \int_{t_0}^{t} |y_{n-1}(\tau) - y_{n-2}(\tau)| \, d\tau$$

$$\leq \alpha \int_{t_0}^{t} \frac{M\alpha^{n-2}}{(n-1)!} (\tau - t_0)^{n-1} \, d\tau, \text{ by } (4.3.3)$$

$$= \frac{M\alpha^{n-1}}{(n-1)!} \left[\frac{(\tau - t_0)^n}{n} \right]_{t_0}^{t}$$

$$= \frac{M\alpha^{n-1}}{n!} (t - t_0)^n.$$

Therefore, the inequality is true for n. For the case $n = 1$, we have

$$|y_1(t) - y_0| \leq \int_{t_0}^{t} |f(\tau, y_0)| \, d\tau \leq M(t - t_0).$$

Thus, the inequality (4.3.2) is true for $n = 1$, and hence, it is also true for any $n \geq 1$ as shown earlier using mathematical induction.

Step 3: In this final step, we show the sequence of functions $\{y_n\}$, defined in Step 1, converges uniformly to a function y on $[t_0, t_0 + h]$. Now, consider the series

$$\frac{M}{\alpha} \sum_{n=1}^{\infty} \frac{(\alpha h)^n}{n!} = \frac{M}{\alpha} \left(\frac{\alpha h}{1!} + \frac{(\alpha h)^2}{2!} + \frac{(\alpha h)^3}{3!} + \cdots \right) \qquad (4.3.4)$$

formed by the constants on the right side in the inequality (4.3.2). This series converges to $\frac{M}{\alpha}(e^{\alpha h} - 1)$. Now consider the infinite series

$$\sum_{n=1}^{\infty} |y_n(t) - y_{n-1}(t)|$$

which is dominated by the series (4.3.4). Hence, by Weierstrass M-test, the series $y_0 + \sum_{n=1}^{\infty} [y_n(t) - y_{n-1}(t)]$ converges uniformly on $[t_0, t_0 + h]$ to a limit function, say y. Therefore, y is continuous on $[t_0, t_0 + h]$. Observe that the sequence of partial sums $S_n(t) = y_0 + \sum_{i=1}^{n} [y_i(t) - y_{i-1}(t)] = y_n(t)$.

Thus, the sequence $\{y_n\}$ converges uniformly to y on $[t_0, t_0 + h]$.

Since each y_n satisfies $|y_n(t) - y_0| \leq b$ on $[t_0, t_0 + h]$, we get $|y(t) - y_0| \leq b$ on $[t_0, t_0 + h]$. This implies that $f(t, y(t))$ is defined on $[t_0, t_0 + h]$. Further, by the Lipschitz continuity of f, we have

$$|f(t, y_n(t)) - f(t, y(t))| \leq \alpha \, |y_n(t) - y(t)|$$

which shows that $f(t, y_n(t)) \to f(t, y(t))$ uniformly on $[t_0, t_0 + h]$. This follows from the uniform convergence of $\{y_n\}$ to y on $[t_0, t_0 + h]$.

Thus, by Theorem 2.2.4, (on interchangeability of limit and integral), we get

$$\begin{aligned} y(t) &= \lim_{n \to \infty} y_{n+1}(t) = y_0 + \lim_{n \to \infty} \int_{t_0}^t f(\tau, y_n(\tau)) \, d\tau \\[2mm] &= y_0 + \int_{t_0}^t \lim_{n \to \infty} f(\tau, y_n(\tau)) \, d\tau \\[2mm] &= y_0 + \int_{t_0}^t f(\tau, y(\tau)) \, d\tau \quad \text{on } [t_0, t_0 + h]. \end{aligned}$$

Therefore, y satisfies (4.2.1). Thus, by the basic lemma, the limit function $y(t)$ satisfies IVP (4.1.1) on $[t_0, t_0 + h]$. Using similar arguments, one can show the existence of a solution on the interval $[t_0 - h, t_0]$. Thus, Picard's iterates converge uniformly to the unique solution of IVP (4.1.1). This completes the proof. $\qquad \square$

Remark 4.3.2

Although the Lipschitz continuity was used in Theorem 4.3.1 to establish the existence result, it is possible to establish the existence theorem just by assuming continuity of f. However, to establish the uniqueness of a solution, we need to use conditions like the Lipschitz continuity of f with respect to y, or a weaker version of the Lipschitz type condition given by

$$|f(t, y_1) - f(t, y_2)| \leq \alpha |y_1 - y_2| \log \frac{1}{|y_1 - y_2|}$$

for all $(t, y_1), (t, y_2) \in \mathscr{R}_1, y_1 \neq y_2$. $\qquad \square$

4.3.1 Cauchy–Peano existence theorem

We now state and prove the Cauchy–Peano existence theorem for the IVP, where the function $f(t,y)$ is only assumed to be continuous on a domain $D \subseteq \mathbb{R}^2$.

To prove the existence theorem, we first define an ε-approximate solution to the IVP and prove that continuity on f is sufficient to guarantee the existence of an ε-approximate solution for the IVP. Subsequently, we show that the sequence of ε-approximate solutions of the IVP is equi-continuous and uniformly bounded. Then, by invoking the Arzela–Ascoli theorem, we show that there exists a convergent subsequence which converges to a solution of the IVP.

> **Definition 4.3.3**
>
> [ε-approximate solution] Consider IVP (4.1.1), where $f(t,y)$ is a real valued continuous function defined on a domain $D \subseteq \mathbb{R}^2$. Let $\varepsilon > 0$. An ε-approximate solution of IVP (4.1.1) on an interval $I = [t_0 - a, t_0 + a]$, $a > 0$, is a function y defined on I such that
>
> (i) $(t, y(t)) \in D$, for all $t \in I$.
>
> (ii) $y \in C^1$ on I except possibly for a finite set $S \subset I$, that is, $y \in C^1(I \setminus S)$ and $\dot{y}(t)$ may have simple discontinuities, that is, jump type discontinuities, on I.
>
> (iii) $|\dot{y}(t) - f(t, y(t))| \leq \varepsilon$ for $t \in I \setminus S$.

A function $y \in C(I)$ satisfying (ii) is said to have piecewise-continuous derivative on I and this is indicated by writing $y \in C_p^1(I)$.

> **Theorem 4.3.4**
>
> [Existence of ε-Approximate Solution] Suppose that $f(t,y)$ in IVP (4.1.1) is continuous on the rectangle
>
> $$\mathscr{R} = \{(t,y) \,:\, |t - t_0| \leq a, \ |y - y_0| \leq b\},$$
>
> where, a, b are positive real numbers. Let
>
> $$M = \max_{(t,y) \in \mathscr{R}} |f(t,y)| \quad \text{and} \quad h = \min\left(a, \frac{b}{M}\right).$$

Then, for given $\varepsilon > 0$, there exists an ε-approximate solution y for the IVP (4.1.1) on $|t - t_0| \le h$. Note that h does not depend on ε.

Proof: Let $\varepsilon > 0$ be given. An ε-approximate solution will be constructed for the interval $[t_0, t_0 + h]$; a similar construction will define the solution in the interval $[t_0 - h, t_0]$. The idea is to divide the interval $[t_0, t_0 + h]$ into subintervals by points $t_0, t_1, \cdots, t_n$ such that $t_0 < t_1 < \cdots < t_n = t_0 + h$ and in each interval $[t_k, t_{k+1}]$, we give a linear approximation as follows:

Starting with the initial interval $[t_0, t_1]$, approximate the given differential equation (4.1.1) with $\dot{z} = f(t_0, y_0)$. Note that the latter ODE has a constant slope $f(t_0, y_0)$. If t_1 is close to t_0, then $f(t, y_0)$ will be close to $f(t, y(t))$ in $[t_0, t_1]$ and we expect z to be close to y in the small interval $[t_0, t_1]$. Solving we get, $z(t) = y_0 + f(t_0, y_0)(t - t_0)$. With the new initial value $y_1 = z(t_1)$, we approximate IVP (4.1.1) by the IVP $\dot{z} = f(t_1, y_1)$, $z(t_1) = y_1$ in the interval $[t_1, t_2]$ and continue this process till we reach t_n. Effectively in each small interval, we are actually solving an integral calculus problem. This idea is made precise as follows:

Since f is uniformly continuous on $\mathscr{R}$, there exists $\delta = \delta(\varepsilon) > 0$, such that

$$|f(t,y) - f(\tilde{t},\tilde{y})| < \varepsilon \quad \text{whenever} \quad |t - \tilde{t}| \le \delta, \ |y - \tilde{y}| \le \delta.$$

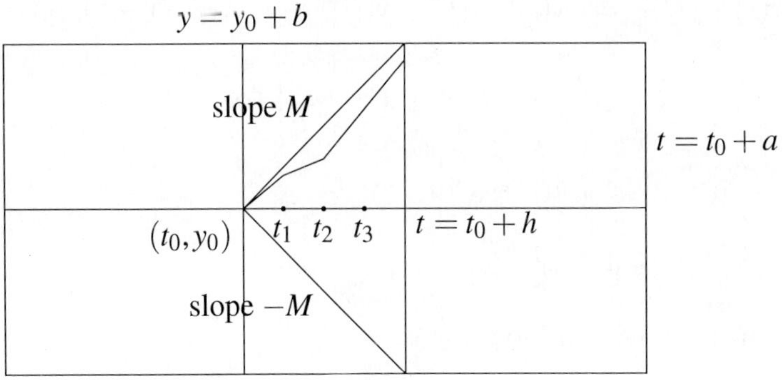

Fig. 4.2   Cauchy–Peano existence

Let $\delta_1 = \min\left(\delta, \frac{\delta}{M}\right)$. Now divide the interval $[t_0, t_0 + h]$ into n parts $t_0 < t_1 < t_2 < t_3 < \cdots < t_n = t_0 + h$ such that

$$\max_{1 \le k \le n} |t_k - t_{k-1}| \le \delta_1.$$

Then, in each interval $[t_{k-1}, t_k]$, we solve the IVP

$$\dot{z} = f(t_{k-1}, z_{k-1}), \quad z(t_{k-1}) = z_{k-1},$$

where z_{k-1} is the value of z at t_{k-1} obtained from the previous interval. From (t_0, y_0) construct a straight line with slope $f(t_0, y_0)$ proceeding to the right of t_0 until it intersects the line $t = t_1$ at some point (t_1, z_1). Now, repeat this process starting at (t_1, y_1) with slope $f(t_1, z_1)$. See Fig. 4.2.

This line segment lies inside the triangular region bounded by the lines starting from (t_0, y_0) with slope M and $-M$. This follows from the definition of h and the fact that $|f(t, y)| \leq M$. In fact, in each interval $[t_{k-1}, t_k]$, z is given by

$$z(t) = z(t_{k-1}) + f(t_{k-1}, z(t_{k-1}))(t - t_{k-1}).$$

Now define y on $[t_0, t_0 + h]$ by

$$y(t) = z(t_{k-1}) + f(t_{k-1}, z(t_{k-1}))(t - t_{k-1}) \tag{4.3.5}$$

for $t \in [t_{k-1}, t_k]$, $k = 1, 2, 3, \cdots, n$. Thus, y is piecewise linear and may fail to be differentiable only at $t = t_k$, $k = 0, 1, 2, \cdots, n$. Therefore, $y \in C_p^1[t_0, t_0 + h]$. For any t and $\tilde{t} \in [t_0, t_0 + h]$, say $\tilde{t} \in [t_{i-1}, t_i]$, $t \in [t_{j-1}, t_j]$, $i \geq j$, we have

$$|y(t) - y(\tilde{t})| = |y(t) - y(t_j) + y(t_j) - \cdots - y(t_{i-1}) + y(t_{i-1}) - y(\tilde{t})|$$

$$\leq M|t - t_j| + M|t_j - t_{j+1}| + \cdots + M|t_{i-1} - \tilde{t}|$$

$$\leq M|t - t_j + t_j - \cdots t_{i-1} - \tilde{t}|$$

$$= M|t - \tilde{t}|. \tag{4.3.6}$$

In particular, with $\tilde{t} = t_0$, we get

$$|y(t) - y(t_0)| = |y(t) - y_0| \leq M|t - t_0| \leq Mh \leq M.\frac{b}{M} = b.$$

Thus, $|y(t) - y_0| \leq b$, which implies that $(t, y(t)) \in \mathscr{R}$ and hence, $f(t, y(t))$ is well-defined. Now, for $t \in (t_{k-1}, t_k)$, we get from (4.3.5) that

$$|y(t) - y(t_{k-1})| \leq M|t - t_{k-1}| \leq M\frac{\delta}{M} = \delta$$

and by uniform continuity of f, it follows that

$$|\dot{y}(t) - f(t,y(t))| = |f(t_{k-1}, y(t_{k-1})) - f(t,y(t))| \le \varepsilon.$$

This, in turn, implies that $y(t)$ is an ε-approximate solution of IVP (4.1.1).

$\square$

Theorem 4.3.5

[Cauchy–Peano Existence Theorem] Let $f(t,y)$ be continuous on the rectangle $\mathscr{R} = \{(t,y) : |t - t_0| \le a, |y - y_0| \le b\}$. Then, there exists a solution to IVP (4.1.1) in the interval $|t - t_0| \le h$, where $h = \min\left(a, \dfrac{b}{M}\right)$, $M = \max_{\mathscr{R}} |f|$.

Proof: Choose $\varepsilon_n = \dfrac{1}{n}$, $n = 1, 2, \cdots$. From Theorem 4.3.4, we have for each ε_n, there exists an ε_n-approximate solution, which we denote by $y_n(t)$, defined on $|t - t_0| \le h$. This implies that

$$|y_n(t) - y_0| \le b$$

and hence, $|y_n(t)| \le |y_0| + b$. Thus, the family of approximate solutions $\{y_n\}$ is uniformly bounded. Again, from (4.3.6), we have

$$|y_n(t) - y_n(\tilde{t})| \le M|t - \tilde{t}|, \text{ for all } t, \tilde{t} \in [t_0, t_0 + h].$$

Therefore, $\{y_n\}$ is an equicontinuous family of functions; see Chapter 2. Thus, by the Arzela–Ascoli theorem (Theorem 2.2.8), there exists a subsequence $\{y_{n_k}\}$ of $\{y_n\}$ such that $y_{n_k} \to y$ uniformly on $[t_0 - h, t_0 + h]$ as $n_k \to \infty$. This implies that y is continuous and $|y(t) - y(\tilde{t})| \le M|t - \tilde{t}|$.

We, now prove that this limit function y is a required solution to IVP (4.1.1). Consider the error defined by

$$\Delta_{n_k}(t) = \begin{cases} \dot{y}_{n_k}(t) - f(t, y_{n_k}(t)) & \text{if } \dot{y}_{n_k} \text{ exists} \\ 0 & \text{otherwise.} \end{cases}$$

That is, except possibly at finite number of points, we have

$$\dot{y}_{n_k} = f(t, y_{n_k}(t)) + \Delta_{n_k}(t).$$

Integrating over (t_0, t), we get

$$y_{n_k}(t) \; = \; y_0 + \int_{t_0}^{t} \left(f(\tau, y_{n_k}(\tau)) + \Delta_{n_k}(\tau) \right) d\tau. \tag{4.3.7}$$

Now, $y_{n_k} \to y$ uniformly implies $f(t, y_{n_k}(t)) \to f(t, y(t))$ uniformly on $[t_0 - h, t_0 + h]$. This implies that

$$\int_{t_0}^{t} f(\tau, y_{n_k}(\tau)) \, d\tau \to \int_{t_0}^{t} f(\tau, y(\tau)) \, d\tau.$$

Furthermore, we have $|\Delta_{n_k}| \leq \varepsilon_{n_k} = \frac{1}{n_k}$. Therefore, we can pass to the limit in (4.3.7), to get

$$y(t) = y_0 + \int_{t_0}^{t} f(\tau, y(\tau)) \, d\tau.$$

Obviously $y(t_0) \; = \; y_0$. This completes the proof, using the basic lemma 4.2.1. □

Remark 4.3.6

In general, the entire sequence $\{y_n\}$ may not converge to a solution. In fact, there may exist some paths from y_n which need not converge in a neighbourhood of (t_0, y_0). But, if we know a priori that the solution is unique when it exists, then the entire sequence must converge on $|t - t_0| \leq h$. For, any subsequence of polygonal paths, by the aforementioned result, there exists a further subsequence which converges to a solution. As the solution is unique, all these limits are the same. Hence, the entire sequence converges to the solution. □

4.3.2   Existence and uniqueness by fixed point theorem

We now establish the existence and uniqueness of a solution to IVP (4.1.1), by invoking a fixed point theorem, namely the generalized Banach contraction principle. We arrive at the same conclusions as in Theorem 4.3.1, under the same hypotheses therein. The proof is as follows.

From the basic lemma, we have seen that the solvability of the initial value problem follows from that of the nonlinear integral equation

$$y(t) = y_0 + \int_{t_0}^{t} f(\tau, y(\tau)) d\tau.$$

Define $X = \{y \in C[t_0, t_0 + h] : |y(t) - y_0| \leq b$, for all $t \in [t_0, t_0 + h]\}$ which is a closed ball in the Banach space $C[t_0, t_0 + h]$ with the sup norm

$$\|y\| = \sup_{t \in [t_0, t_0 + h]} |y(t)|.$$

Thus, X is a complete metric space. Define an operator $T : X \to X$ by

$$(Ty)(t) = y_0 + \int_{t_0}^{t} f(\tau, y(\tau))d\tau.$$

Note that for $y \in X$, the function Ty is also in X by the choice of X and h. Clearly, the fixed points of T are solutions of the integral equation. We now prove that T^n is a contraction for sufficiently large $n \geq 1$. Let $y_1, y_2 \in X$; then, using the Lipschitz continuity of f, we have

$$
\begin{aligned}
|(Ty_1)(t) - (Ty_2)(t)| &= \left| \int_{t_0}^{t} f(\tau, y_1(\tau)) - f(\tau, y_2(\tau))d\tau \right| \\
&\leq \alpha \int_{t_0}^{t} |y_1(\tau) - y_2(\tau)|d\tau \\
&\leq \alpha(t - t_0)\|y_1 - y_2\|.
\end{aligned}
$$

Successively applying the first and second inequalities, we get

$$
\begin{aligned}
\left|(T^2 y_1)(t) - (T^2 y_2)(t)\right| &\leq \alpha \int_{t_0}^{t} |(Ty_1)(\tau) - (Ty_2)(\tau)|d\tau \\
&\leq \alpha^2 \int_{t_0}^{t} (\tau - t_0)\|y_1 - y_2\|d\tau \\
&= \alpha^2 \frac{(t - t_0)^2}{2} \|y_1 - y_2\|.
\end{aligned}
$$

Hence,

$$\left\|T^2 y_1 - T^2 y_2\right\| \leq \frac{\alpha^2 h^2}{2} \|y_1 - y_2\|.$$

An induction argument now gives that, for any $n \geq 1$,

$$\|T^n y_1 - T^n y_2\| \leq \frac{\alpha^n h^n}{n!} \|y_1 - y_2\|.$$

By choosing large n, the quantity $\dfrac{\alpha^n h^n}{n!}$ can be made less than 1, and hence, T^n is a contraction. Thus, by the generalized Banach contraction principle (Theorem 2.3.2 and its corollary 2.3.3), T has a unique fixed point. This completes the proof.   □

Note that the fixed point of T here is approximated by the iterates

$$y_{n+1} = Ty_n = y_0 + \int_{t_0}^{t} f(\tau, y_n(\tau)) d\tau,$$

with $y_0(t) = y_0$. These are nothing but Picard's iterates defined earlier.

Remark 4.3.7

Suppose y_{n0} and f_n be sequences of initial data such that $y_{n0} \to y_0$ in $\mathbb{R}$ and $f_n \to f$ in $C(\mathscr{R})$. Then $y_n \to y$, where y_n and y are, respectively, the solutions corresponding to the initial data y_{n0}, f_n and y_0, f.   □

Example 4.3.8

Apply Picard's iterates to solve the IVP

$$\frac{dy}{dt} = y, \quad y(0) = 1.$$

Let $y_0(t) = y_0 = 1$ and consider Picard's iterates

$$y_{n+1}(t) = y_0 + \int_0^t y_n(\tau) d\tau.$$

We successively get

$$y_1(t) = 1 + \int_0^t y_0(\tau) d\tau = 1 + \int_0^t 1 d\tau = 1 + t$$

$$y_2(t) = 1 + \int_0^t y_1(\tau) d\tau = 1 + \int_0^t (1 + \tau) d\tau = 1 + t + \frac{t^2}{2}$$

$$y_3(t) = 1 + \int_0^t y_2(\tau) d\tau = 1 + t + \frac{t^2}{2!} + \frac{t^3}{3!}$$

... .

$$y_n(t) = \sum_{m=0}^{n} \frac{t^m}{m!} \to e^t \text{ as } n \to \infty.$$

But we know that the solution to the IVP is indeed given by $y(t) = e^t$.

4.4 Continuous Dependence of the Solution on Initial Data and Dynamics

The initial data includes the given initial value y_0 and the dynamics f. In practical applications, it is important to know how small errors in the initial data affects the solution. In other words, we would like to know that if the initial data is close to another initial data in appropriate norm, then the corresponding solutions are also close to each other. This is known as the continuous dependence of the solution on the initial condition and dynamics.

Theorem 4.4.1

Let $\mathscr{R}$ be as in Theorem 4.3.1. Suppose $f, \tilde{f} \in C(\mathscr{R})$ and be Lipschitz continuous with respect to y on $\mathscr{R}$ with Lipschitz constants α, $\tilde{\alpha}$, respectively. Let y and $\tilde{y}$ be, respectively, the solutions of the IVP $\dot{y} = f(t,y)$, $y(t_0) = y_0$ and $\dot{\tilde{y}} = \tilde{f}(t,\tilde{y})$, $\tilde{y}(\tilde{t}_0) = \tilde{y}_0$ in some closed intervals I_1, I_2 containing t_0 and $\tilde{t}_0$. For small $|t_0 - \tilde{t}_0|$, let I any finite interval containing t_0 and $\tilde{t}_0$, where both y and $\tilde{y}$ are defined. Then,

$$\max_{t \in I} |y(t) - \tilde{y}(t)| \le \left(|y_0 - \tilde{y}_0| + |I| \max_{\mathscr{R}} |f(t,y) - \tilde{f}(t,y)| + M|t_0 - \tilde{t}_0| \right) e^{\alpha_0 |I|},$$

where $|I|$ is the length of the interval I, $M = \max\left(\max_{\mathscr{R}} f, \max_{\mathscr{R}} \tilde{f} \right)$ and $\alpha_0 = \min(\alpha, \tilde{\alpha})$.

Proof: We give a proof when $t_0 = \tilde{t}_0$. From the basic lemma, the solutions y and $\tilde{y}$ satisfy the following integral equations:

$$y(t) = y_0 + \int_{t_0}^{t} f(\tau, y(\tau))\, dt, \quad \tilde{y}(t) = \tilde{y}_0 + \int_{t_0}^{t} \tilde{f}(\tau, \tilde{y}(\tau))\, d\tau,$$

for all $t \in I$. Subtracting the second equation from the first, we get

$$y(t) - \tilde{y}(t) = y_0 - \tilde{y}_0 + \int_{t_0}^{t} \left(f(\tau, y(\tau)) - \tilde{f}(\tau, \tilde{y}(\tau)) \right) d\tau.$$

When $t_0 \neq \tilde{t}_0$, an extra integral appears in this equation, which can be estimated easily. Now add and subtract the term $f(\tau, \tilde{y}(\tau))$ in the aforementioned integral. We, then obtain

$$|y(t) - \tilde{y}(t)| \leq |y_0 - \tilde{y}_0| + \int_{t_0}^{t} |f(\tau, y(\tau)) - f(\tau, \tilde{y}(\tau))|$$

$$+ \int_{t_0}^{t} |f(\tau, \tilde{y}(\tau)) - \tilde{f}(\tau, \tilde{y}(\tau))| \, d\tau.$$

Using the Lipschitz continuity of f with respect to y, this inequality can be estimated as

$$|y(t) - \tilde{y}(t) \leq |y_0 - \tilde{y}_0| + \alpha \int_{t_0}^{t} |y(\tau) - \tilde{y}(\tau)| \, dt + |I| \max_{\mathscr{R}} |f(t,y) - \tilde{f}(t,y)|.$$

Applying Grownwall's inequality, we get

$$|y(t) - \tilde{y}(t)| \leq \left(|y_0 - \tilde{y}_0| + |I| \max_{\mathscr{R}} |f(t,y) - \tilde{f}(t,y)| \right) \exp(\alpha|I|).$$

The same inequality is true when α is replaced by $\tilde{\alpha}$ as well if we add and subtract $\tilde{f}(\tau, y(\tau))$ instead of $f(\tau, \tilde{y}(\tau))$. This completes the proof.   □

4.5   Continuation of a Solution into Larger Intervals and Maximal Interval of Existence

Observe that the existence results discussed in Section 4.3 are local in nature, that is, we could only claim the existence of a solution in a small interval containing the initial time t_0. However, when the differential equation is linear, that is, $f(t,y) = a(t)y + b(t)$, where $a(t)$ and $b(t)$ are continuous functions defined on $[t_0 - a, t_0 + a]$, then f is defined on $\mathscr{R} = [t_0 - a, t_0 + a] \times \mathbb{R}$. In this case, the solution is defined on the entire interval $[t_0 - a, t_0 + a]$, that is, h can be taken as a itself. But, this is not true when f is not linear, that is, the solution may not exist in the interval where f is defined. At the same time, it is possible that $[t_0 - h, t_0 + h]$ may not be the largest possible interval of existence. This leads to the following question. Can we enlarge the domain of the solution y further? More generally, what is the largest possible interval of existence?

Example 4.5.1

Recall the ODE in Example 4.1.4, namely $\dot{y} = y^2$, $y(1) = -1$.

Let $\mathscr{R} = \{(t,y) : |t-1| \le 1, \ |y+1| \le 1\}$. Here, $f(t,y) = y^2$ satisfies continuity and Lipschitz continuity assumptions on $\mathscr{R}$, $t_0 = 1, y_0 = -1$ and let $|t-1| \le h$, be the interval on which existence is guaranteed (by theorems), which can be computed as $h = \dfrac{1}{4}$. Hence by Theorem 4.3.1, the IVP has a solution on the interval $[\frac{3}{4}, \frac{5}{4}]$.

Now by the method of separation of variables, we can integrate the differential equation and use the initial condition to obtain the solution as $y(t) = -\dfrac{1}{t}$. Thus, the solution exists for $0 < t < \infty$. In other words, we can continue the solution outside the interval $[\frac{3}{4}, \frac{5}{4}]$. At the same time, since $|y(t)| \to \infty$ as $t \to 0+$, it ceases to exist to the left of 0, even though f is a differentiable function defined everywhere. This is the typical nonlinear phenomena and f is not Lipschitz in the entire $\mathbb{R}$, but is Lipschitz in any bounded interval.

4.5.1   Continuation of the solution outside the interval $|t-t_0| \le h$

The existence theorem (Section 4.3) guarantees that IVP (4.1.1) has a solution ϕ_0 on the interval $[t_0 - h, t_0 + h]$. Consider the right end point of $[t_0 - h, t_0 + h]$. Let $t_1 = t_0 + h$, $y_1 = \phi_0(t_1)$. The point (t_1, y_1) is inside $\mathscr{R}$, which is inside D. Now consider the IVP $\dot{y} = f(t,y)$, $y(t_1) = \phi_0(t_1) = y_1$. Reapplying the existence theorem with this new initial condition to get a solution ϕ_1 on the interval $t_1 - h_1 \le t \le t_1 + h_1$ for some $h_1 > 0$. Define

$$y(t) = \begin{cases} \phi_0(t), & t_0 - h \le t \le t_0 + h = t_1 \\[2mm] \phi_1(t), & t_1 \le t \le t_1 + h_1. \end{cases}$$

Then,

$$\phi_0(t) \ = \ y_0 + \int_{t_0}^{t} f(\tau, \phi_0(\tau))\, d\tau, \quad \text{for } t_0 - h \le t \le t_1$$

$$\phi_1(t) = \phi_0(t_1) + \int_{t_1}^{t} f(\tau, \phi_1(\tau)) \, d\tau, \quad \text{for } t_1 \leq t \leq t_1 + h_1.$$

Thus, we have

$$y(t) = y_0 + \int_{t_0}^{t} f(\tau, y(\tau)) \, d\tau$$

for $t \in [t_0 - h, t_1 + h_1] = [t_0 - h, t_0 + h + h_1]$. It is easy to see that y is differentiable at $t = t_1$ also, from the aforementioned two expressions and verify that y indeed satisfies the DE in question.

This solution $y(t)$ is called a *continuation* of the solution ϕ_0 to the interval $[t_0 - h, t_1 + h_1]$. Now repeating this process at the new end point $t_1 + h_1$, we obtain a solution on $[t_0 - h, t_1 + h_1 + h_2]$. In this manner, we may get longer intervals $[t_0 - h, t_n + h_n]$. Unfortunately, this still may not lead to the maximum interval of existence.

In Example 4.1.4, the function f is not Lipschitz in $\mathbb{R}$. When f is not globally Lipschitz, the bounds on Picard's iterates may become larger and larger, thus reducing the interval of existence. This is really due to *bad* non-linearity even though the function is very smooth as in Example 4.6.1. If f is Lipschitz globally, then we get the existence in the entire interval of definition as in the following theorem. The proof will follow along the same lines as in Picard's existence theorem and we leave the details as an exercise to the reader. In fact, because f is defined on the entire real line in the second variable, one need not check the validity of Picard's iterates as they are always defined.

Theorem 4.5.2

Let $f(t, y)$ be a bounded continuous function defined in the unbounded domain $\mathscr{R} = \{(t, y) : a < t < b, \ -\infty < y < \infty\}$. Let f be Lipschitz in y on $\mathscr{R}$. Then, a solution y of $\dot{y} = f(t, y)$, $y(t_0) = y_0, t_0 \in (a, b)$ is defined on the entire open interval $a < t < b$. In particular, if $a = -\infty$, and $b = +\infty$, then y is defined for all t in $\mathbb{R}$.

Theorem 4.5.3

Suppose $f : D \to \mathbb{R}$ is bounded and y is the solution of IVP (4.1.1) in some interval (a, b) containing t_0. If $b < \infty$, then, $y(b-) = \lim_{t \to b-} y(t)$ exists. If $(b, y(b-)) \in D$, then, the solution y may be continued to an interval $(a, \bar{b}]$ with $\bar{b} > b$. Similar statements hold at the left end point a.

Proof: Let M be the bound on $|f|$ on D and $t_0 < t_1 < t_2 < b$. Then,

$$|y(t_2) - y(t_1)| \leq \int_{t_1}^{t_2} |f(s, y(s))| ds \leq M|t_2 - t_1|.$$

Thus, if $t_1, t_2 \to b$ from the left, it follows that

$$|y(t_2) - y(t_1)| \to 0$$

which is the Cauchy criterion for the existence of the above mentioned limit. Since $(b, y(b-)) \in D$, we can now consider the IVP with the initial condition at b as $y(b-)$ and the solution can be continued beyond b as asserted in the theorem. □

4.5.2   Maximal interval of existence

Consider IVP (4.1.1). Assume the existence of a unique solution in a neighbourhood t_0. Call such a neighbourhood, an interval of existence. Suppose I_1 and I_2 are intervals of existence containing t_0, then their union is an interval of existence (why?).

Definition 4.5.4

Let J be the union of all possible intervals of unique existence of (4.1.1). Then, J is an interval and, is called the maximal interval of existence. More precisely, let $\{I_\alpha\}$ be the collection of all intervals of unique existence, containing t_0. This collection is non-empty and $J = \bigcup_\alpha I_\alpha$. □

Indeed, we can a define a unique solution y in J as follows: for any $t \in J$, t is in some interval I of existence and $y(t)$ is the value given by the solution in I. This is well defined by the uniqueness.

Proposition 4.5.5

The maximal interval J of existence is an open interval (α, β), where α can be $-\infty$ and β can be $+\infty$.

Proof: If not, suppose $J = (\alpha, \beta]$. In this case, $\beta < \infty$. Then, one can consider the IVP for the same ODE with initial condition at β, to get a solution in $[\beta, \beta + h]$ for some $h > 0$. This will produce a solution in $(\alpha, \beta + h]$ contradicting the maximality of J. Similar contradiction can be arrived at if α is a point in J. $\square$.

Theorem 4.5.6

Let $J = (\alpha, \beta)$ be the maximal interval of existence and y be the solution to the IVP in J. Assume $\beta < \infty$. If K is any compact subset of $\mathbb{R}$ such that $[t_0, \beta] \times K \subset D$, then there exists a $t_1 \in (\alpha, \beta)$ such that $y(t_1)$ does not belong to K. Similar statements hold at the end point α.

We infer the following from the conclusion of the theorem. Only one of the following statements is true:

- If $y(\beta-) = \lim\limits_{t \to \beta-} y(t)$ exists, then $(\beta, y(\beta-)) \in \bar{D} \setminus D$, the boundary of D.

- The solution y becomes unbounded near β, that is, given any large positive number C, there exists $t_1 < \beta$ such that $|y(t_1)| \geq C$.

  The following examples illustrate both these situations.

The proof of the theorem follows immediately from Theorem 4.5.3.

Example 4.5.7

Consider the equation $\dot{y} = \dfrac{1}{ty}$.

Here, the function $f(t,y) = \dfrac{1}{ty}$ is defined in the entire (t,y)-plane, except the t-axis and y-axis. We will consider the IVP in the first quadrant in the (t,y)-plane: $y(t_0) = y_0$ where both t_0, y_0 are positive. The solution is given by $y = [2\log(t/t_0) + y_0^2]^{1/2}$. Therefore, the maximal interval of existence is (α, ∞), where $\alpha = t_0 e^{-y_0^2/2}$ and as $t \to \alpha-$, $y(t) \to 0$ with $(\alpha, 0)$ belonging to the boundary of the domain in question.

Example 4.5.8

Consider the equation $\dot{y} = \dfrac{1}{t+y}$.

In this case, we take the domain as $\{(t,y) : t+y > 0\}$ and impose the initial condition as $y(t_0) = y_0$ with $t_0 + y_0 > 0$. By introducing a new variable $u(t) = t + y(t)$, we see that the solution is implicitly given by $\dfrac{e^{u(t)}}{1+u(t)} = \dfrac{e^{y_0}}{1+t_0+y_0} e^t$. We notice that the maximal interval of existence in this case is given by (α, ∞), where $\alpha = t_0 + \log(1+t_0+y_0) - (t_0+y_0) < t_0$. Again, it is not hard to see that as $t \to \alpha-$, $(t, y(t))$ approaches the boundary of the domain in question, that is, $t + y(t) \to 0$.

Example 4.5.9

Consider the IVP: $\dot{y} = \dfrac{\pi}{2}(1+y^2)$, $y(0) = 0$.

We now see that the solution $y(t) = \tan(\frac{\pi}{2}t)$ cannot be extended beyond the interval $(-1,1)$. Note that if we take any rectangle $\{(t,y) : |t| \leq a, |y| \leq b\}$ around the origin $(0,0)$, then as in the local existence theorem, we get the existence of a unique solution in an interval $[-h,h]$, where $h = \min(a, \frac{2}{\pi}\frac{b}{1+b^2})$, which is always less than $\frac{1}{\pi}$.

4.6 Existence and Uniqueness of a System of Equations

In this section, we study the existence and uniqueness of the solution of a system of differential equations of the form:

$$\dot{x}_1 = f_1(t, x_1, x_2, \cdots, x_n)$$

$$\dot{x}_2 = f_2(t, x_1, x_2, \cdots, x_n)$$

$$\cdots\cdots\cdots\cdots$$

$$\dot{x}_n = f_n(t, x_1, x_2, \cdots, x_n)$$

with initial conditions given by:

$$x_1(t_0) = x_{01}; \ x_2(t_0) = x_{02},\ldots,x_n(t_0) = x_{0n},$$

where, for $i = 1,2,\ldots,n$, $f_i \ : \ \mathbb{R} \times \mathbb{R}^n \to \mathbb{R}^n$ is possibly a nonlinear function which is assumed to be continuous on $\mathbb{R}^{n+1}$. Put

$$\mathbf{x}(t) = [x_1(t), x_2(t), \cdots, x_n(t)]^T,$$

$$\mathbf{f}(t, \mathbf{x}) = [f_1(t, x_1(t), \cdots, x_n(t)), f_2(t, x_1(t), \cdots, x_n(t)), \cdots, f_n(t, x_1(t),$$

$$\cdots, x_n(t))]^T$$

and

$$\mathbf{x}_0 = [x_{01}, x_{02}, \cdots, x_{0n}]^T.$$

Here, the superscript T denotes the transpose of a matrix/vector. Using these notations, the aforementioned system of differential equations can be written in the following compact form

$$\dot{\mathbf{x}} = \mathbf{f}(t, \mathbf{x}), \ \ \mathbf{x}(t_0) = \mathbf{x}_0. \tag{4.6.1}$$

An nth order scalar differential equation can be reduced into a system of n first order differential equations of this form. Such a representation is known as the state-space representation of the system.

Example 4.6.1

Consider the differential equation

$$y^{(n)}(t) = g\left(t, y(t), y^{(1)}(t), \cdots, y^{(n-1)}(t)\right)$$

with initial conditions

$$y(t_0) = x_{01}, \ y^{(1)}(t_0) = x_{02}, \cdots, y^{(n-1)}(t_0) = x_{0n}.$$

Define $x_1, x_2, \cdots, x_n$ as follows:

$$x_1(t) = y(t),$$

$$x_2(t) = y^{(1)}(t) = \dot{x}_1(t)$$

$$x_3(t) = y^{(2)}(t) = \dot{x}_2(t)$$

$$\ldots\ldots\ldots\ldots$$

$$x_n(t) = y^{(n-1)}(t) = \dot{x}_{n-1}(t).$$

Then, the given n^{th} order differential equation can be written as

$$\dot{x}_1(t) = x_2(t)$$

$$\dot{x}_2(t) = x_3(t)$$

$$\ldots\ldots\ldots\ldots$$

$$\dot{x}_{n-1}(t) = x_n(t)$$

$$\dot{x}_n(t) = g(t, x_1(t), x_2(t), \ldots, x_n(t))$$

and the initial conditions reduce to

$$x_1(t_0) = x_{01}, \ x_2(t_0) = x_{02}, \cdots, x_n(t_0) = x_{0n}.$$

In vector notation, we get

$$\dot{\mathbf{x}}(t) = \mathbf{f}(t, \mathbf{x}(t)), \ \mathbf{x}(t_0) = \mathbf{x}_0,$$

where

$$\mathbf{x}_0 = [x_{01}, x_{02}, \ldots, x_{0n}]^T$$

$$\mathbf{x}(t) = [x_1(t), x_2(t), \ldots, x_n(t)]^T$$

$$\mathbf{f}(t, \mathbf{x}) = [f_1(t, \mathbf{x}), f_2(t, \mathbf{x}), \ldots, f_n(t, \mathbf{x})]^T$$

with

$$f_1(t, \mathbf{x}) = x_2(t), \ f_2(t, \mathbf{x}) = x_3(t), \ \ldots f_{n-1}(t, \mathbf{x}) = x_n(t), \ f_n(t, \mathbf{x}) = g(t, \mathbf{x}).$$

4.6.1 Existence and uniqueness results for systems

Let D be an open connected set in $\mathbb{R} \times \mathbb{R}^n$, whose points are denoted by $(t, \mathbf{x})$ with $t \in \mathbb{R}$ and $\mathbf{x} \in \mathbb{R}^n$. Assume $\mathbf{f} : D \to \mathbb{R}^n$ is a continuous function. The results on existence and uniqueness of solutions to systems (4.6.1) are very similar to the ones discussed in Section 4.3 for a single equation. Here we restrict ourselves to a description of the method of Picard's iterates. The only difference is that Picard's iterates are now vectors and hence we need to take the vector norm in $\mathbb{R}^n$ instead of absolute values. Choose positive real numbers a and b such that

$$\mathscr{R} = \{(t,\mathbf{x}) \in \mathbb{R} \times \mathbb{R}^n : |t - t_0| \leq a, \ |\mathbf{x} - \mathbf{x}_0| \leq b\}$$

is a subset of D. Let us recall the definition of the Lipschitz condition of the vector $\mathbf{f}$. The vector function $\mathbf{f}$ is said to be Lipschitz continuous in $\mathbf{x}$ on $\mathscr{R}$ if there exists a constant $\alpha > 0$ such that

$$|\mathbf{f}(t,\mathbf{x}_1) - \mathbf{f}(t,\mathbf{x}_2)| \leq \alpha \, |\mathbf{x}_1 - \mathbf{x}_2|, \text{ for all } (t,\mathbf{x}_1), \ (t,\mathbf{x}_2) \in \mathscr{R}.$$

Note that we are using the norm for any vector $\mathbf{x} \in \mathbb{R}^n$ as $|\mathbf{x}| = \sum_{i=1}^{n} |x_i|$. Now suppose that the components f_i of $\mathbf{f}$ are Lipschitz with Lipschitz constants α_i in $\mathbf{x}$ on $\mathscr{R}$, that is,

$$|f_i(t,\mathbf{x}_1) - f_i(t,\mathbf{x}_2)| \leq \alpha_i \, |\mathbf{x}_1 - \mathbf{x}_2|, \text{ for all } (t,\mathbf{x}_1), \ (t,\mathbf{x}_2) \in \mathscr{R}.$$

Then, it is easy to see that $\mathbf{f}$ is Lipschitz in $\mathbf{x}$ on $\mathscr{R}$ with Lipschitz constant $\alpha \leq |\alpha_1| + \cdots + |\alpha_n|$. Conversely, if $\mathbf{f}$ is Lipschitz with Lipschitz constant α, then each component f_i is also Lipschitz whose Lipschitz constant is bounded by α.

Now consider IVP (4.6.1). This system is equivalent to the following system of integral equations

$$\mathbf{x}(t) = \mathbf{x}_0 + \int_{t_0}^{t} \mathbf{f}(\tau,\mathbf{x}(\tau)) \, d\tau,$$

or, component-wise,

$$x_i(t) = x_{0i} + \int_{t_0}^{t} f_i(\tau,\mathbf{x}(\tau)) \, d\tau$$

for $i = 1,2,\ldots,n$. Now define Picard's iterates as

$$\mathbf{x}_{n+1}(t) = \mathbf{x}_0 + \int_{t_0}^{t} \mathbf{f}(\tau,\mathbf{x}_n(\tau)) \, d\tau, \quad n = 0,1,2,\cdots.$$

Following exactly the procedure used for a single equation, see Theorem 4.3.1, we can show that this Picard's iterates converge uniformly to the unique solution of the IVP for any arbitrary $\mathbf{x}_0 \in \mathbb{R}^n$, in some interval $[t_0 - h, t_0 + h]$ for some $h > 0$. Thus, we have the following theorem.

Theorem 4.6.2

Assume that $\mathbf{f}$ is continuous in $\mathscr{R}$ and each f_i is Lipschitz with respect to the space variable $\mathbf{x}$ on $\mathscr{R}$. Then, there exists a unique solution to IVP (4.6.1) in an interval $|t - t_0| \leq h$, for some $h > 0$.

The interval of existence obtained from the theorem need not be the best possible interval. One can also prove the results on continuation of solutions to a larger interval as in the case of a single equation in a similar fashion. We can also introduce the maximal interval of existence and eventually obtain global solutions under the assumption that $\mathbf{f}$ is a continuous bounded function and is globally Lipschitz with respect to the $\mathbf{x}$ variable in the domain of definition.

Example 4.6.3

Consider the nonlinear system

$$\dot{x}_1 = 2\cos(x_2), \ \dot{x}_2 = 3\sin(x_1)$$
$$x_1(0) = a, \ x_2(0) = b.$$

Let $\mathbf{x} = \begin{bmatrix} x_1, x_2 \end{bmatrix}^T$, $\ \tilde{\mathbf{x}} = \begin{bmatrix} \tilde{x}_1, \tilde{x}_2 \end{bmatrix}^T$, $\ \mathbf{f}(t, \mathbf{x}) = \begin{bmatrix} 2\cos x_2 \\ 3\sin x_1 \end{bmatrix}$. Then

$$|\mathbf{f}(t, \ \mathbf{x}) - \mathbf{f}(t, \tilde{\mathbf{x}})| \ = \ \left| \left| \begin{bmatrix} 2\cos x_2 & -2\cos \tilde{x}_2 \\ 3\sin x_1 & -3\sin \tilde{x}_1 \end{bmatrix} \right| \right|$$

$$\leq \ 3|\mathbf{x} - \tilde{\mathbf{x}}|.$$

Thus, $\mathbf{f}(t, \mathbf{x})$ is globally Lipschitz continuous with Lipschitz constant less than or equal to 3. Hence, by the existence and uniqueness theorem, there exists a unique solution for the given differential system around the given initial data.

Let $\mathbf{A} = [a_{ij}]$ be a constant $n \times n$ matrix. Then, $\mathbf{f}(t, \mathbf{x}) = \mathbf{A}\mathbf{x}$ is obviously Lipschitz continuous with Lipschitz constant $\alpha = \|\mathbf{A}\|$. Thus, the linear system with constant coefficients $\dot{\mathbf{x}} = \mathbf{A}\mathbf{x}$, $\mathbf{x}(t_0) = \mathbf{x}_0$ has a unique solution, and the solution is global. A detailed study of the system will be carried out in Chapter 5. More generally, we can consider the non-autonomous system

$$\dot{\mathbf{x}}(t) = \mathbf{A}(t)\,\mathbf{x}$$

$$\mathbf{x}(t_0) = \mathbf{x}_0,$$

where $\mathbf{A}(t)$ is a matrix valued continuous function defined on a compact interval. Then, the linear function $\mathbf{f}(t,x) = \mathbf{A}(t)\mathbf{x}$ will satisfy the required conditions globally, and hence, we get the unique solution to the system. The compactness of the interval may be removed if all the components of $\mathbf{A}(t)$ are bounded functions of t.

4.7 Exercises

1. Discuss the existence and uniqueness of the solution of the following IVPs

   (a) $\dot{y} = \frac{2}{t}y, \quad y(t_0) = y_0, t_0 \neq 0.$
   (b) $\dot{y} = (\cot t)y, \quad y(1) = 0.$

2. Discuss the existence and uniqueness of the following systems

   (a) $\dot{x}_1 = x_2, \dot{x}_2 = \dfrac{-g}{l} \sin x_1 - \dfrac{k}{m} x_2, \ x_1(0) = 0, \ x_2(0) = 0.$
   (b) $\dot{x}_1 = x_2 - x_1(x_1^2 + x_2^2), \ \dot{x}_2 = -x_1 - x_2(x_1^2 + x_2^2) \ x_1(0) = 0 = x_2(0).$

3. Discuss the solvability of the differential equation in Example 4.1.2 with a different initial condition $y(0) = y_0 > 0$. (Hint: $y(t) = y_0 - t$ is a valid solution only for the interval $(-\infty, y_0)$).

4. Apply Picard's iterates to obtain the solution of IVP

   $$\frac{dy}{dt} = 4ty, \quad y(0) = 3.$$

5. Discuss the existence and uniqueness of solutions of the following IVP

   (i) $\dot{y} = \dfrac{1}{2y}, \quad y(1) = 1$

(ii) $\dot{y} = \dfrac{1}{y}, \quad y(0) = 0$

(iii) $\dot{y} = |y|^{1/2}, \quad y(0) = 0$

6. By suitable change of variables, transform the nth order linear non-homogeneous equation

$$y^{(n)}(t) + a_1(t)y^{(n-1)}(t) + \cdots + a_{n-1}(t)y^{(1)} + a_n(t)y(t) = b(t)$$

$$(4.7.1)$$

with initial condition $y^{(i)}(t_0) = x_{0i+1}, \quad i = 0,1,2,\cdots,n-1$ to the linear system (with appropriate $\mathbf{A}$ and $\mathbf{f}$)

$$\dot{\mathbf{x}} = \mathbf{A}(t)\,\mathbf{x}(t) + \mathbf{f}(t), \ \mathbf{x}(t_0) = \mathbf{x}_0.$$

7. Obtain a state-space representation of the gyro system

$$I\,\ddot{\theta} - H\,w\,\cos\theta = b\,\dot{\theta} - k\,\theta$$

$$\theta(0) = a_0, \dot{\theta}(0) = a_1.$$

8. Reduce the following system of second order equations

$$m_1\ddot{x} = k_2(y-x) + b(\dot{y}-\dot{x}) + k_1(u-x)$$

$$m_2\ddot{y} = -k_2(y-x) - b(\dot{y}-\dot{x})$$

to a first order system.

9. State the conditions under which the following differential equation will have unique solutions

(i) The nth order nonlinear equation in Example 4.6.1.

(ii) The nth order linear non-homogeneous equation in Exercise (6) above.

10. In the following exercises, choosing the appropriate domain of definition of the given function and corresponding initial condition, find the maximal interval (α,β) of existence of the solution and find its limits as t approaches α,β

(a) $\dot{y} = \dfrac{1}{1+y^2}$.

(b) $\dot{y} = \dfrac{1}{1-y^2}$.

(c) $\dot{y} = \dfrac{\sin t}{1-y^2}$.

(d) $\dot{y} = \dfrac{1}{y(1-y)}$.

11. Prove the continuity of the solution of the equation given below, in appropriate norm with respect to the initial data $\mathbf{x}_0$ and $\mathbf{f}$

$$\dot{\mathbf{x}} = \mathbf{f}(t,\mathbf{x}), \ \mathbf{x}(t_0) = \mathbf{x}_0,$$

assuming that $\mathbf{f}$ is continuous and Lipschitz continuous with respect to $\mathbf{x}$ variable, in the domain of definition.

12. Consider the n-dimensional control system

$$\dot{\mathbf{x}} = \mathbf{f}(t,\mathbf{x}(t),\mathbf{u}(t)), \ \mathbf{x}(t_0) = \mathbf{x}_0,$$

where the function $\mathbf{f} : \mathbb{R} \times \mathbb{R}^n \times \mathbb{R}^m \to \mathbb{R}^n$ is continuous and Lipschitz continuous with respect to $\mathbf{x}$ and $\mathbf{u}$, where the continuous function $\mathbf{u}(t)$ is an external control input applied to the system. Prove that the system has a unique solution for a given initial condition $\mathbf{x}_0$ and a given control function $\mathbf{u}(t)$. Also, prove the following:

(a). Let $\mathbf{x}$ be the unique solution with initial state $\mathbf{x}_0$ and $\tilde{\mathbf{x}}$ be the unique solution with initial condition $\tilde{\mathbf{x}}_0$, for a fixed control input $\mathbf{u}$. Then there exists $K_1 > 0$ such that

$$||\mathbf{x} - \tilde{\mathbf{x}}|| \le K_1 ||\mathbf{x}_0 - \tilde{\mathbf{x}}_0||.$$

(b). Let $\mathbf{x}_\mathbf{u}$ be the unique solution with a control $\mathbf{u}$ and $\mathbf{x}_{\tilde{\mathbf{u}}}$ be the unique solution with a control $\tilde{\mathbf{u}}$ for a fixed initial state $\mathbf{x}_0$. Then prove that there exists $K_2 > 0$ such that

$$||\mathbf{x}_\mathbf{u} - \mathbf{x}_{\tilde{\mathbf{u}}}|| \le K_2 ||\mathbf{u} - \tilde{\mathbf{u}}||.$$

Thus, we see that the solution varies continuously in the"Lipschitz norm", both with respect to the initial data and control data. If there is higher order smoothness in the data f, we can get the corresponding higher order smoothness in the solution.

4.8   Notes

This chapter deals with some important topics on existence and uniqueness of a solution to an ODE. The significance of these topics are explained through several examples so that a beginner starts appreciating the topics. We have included three results on existence–Cauchy–Peano existence theorem, existence using Picard's iterates and existence using fixed point theorem. The first one requires only the minimal assumption of continuity, but uniqueness is not guaranteed. The other two results require the assumption of Lipschitz continuity and the uniqueness is guaranteed. Gronwall's inequality is stated and proved, which in turn is used to prove the uniqueness of solutions. Gronwall's inequality is also useful for comparison of different solutions with different coefficients and/ or initial data. There are other types of uniqueness results; see for example [AO12]. For general theory, there are many good books, see for example [CL72, Sim91, SK07, Tay11, MU78, HSD04]. Continuous dependence on the data is also discussed in detail. Also discussed is the topic on continuation of solutions to larger intervals; this leads to the concept of maximal interval of existence of a solution. In particular, the conditions for global existence of a solution is dealt with. An application of these results is seen in the proof of Perron's theorem in Chapter 9. A brief discussion on systems is also carried out.

5

Linear Systems and Qualitative Analysis

5.1 General n^{th} Order Equations and Linear Systems

In Chapter 2, we have studied first and second order regular linear ODE. We have seen that the first order case can be completely resolved to get a representation of the solution, whereas for the second order case, we do not have a general method to get two linearly independent solutions. However, we have seen that the solution space of linear second order homogeneous equations is of two dimensions. A general n^{th} order regular linear ODE can be written in the form

$$\frac{d^n y}{dt^n} + p_1(t)\frac{d^{n-1}y}{dt^{n-1}} + \cdots + p_{n-1}(t)\frac{dy}{dt} + p_n(t)y = g(t) \qquad (5.1.1)$$

with n conditions such as initial or boundary conditions. For example, initial conditions can be of the form

$$y(0) = y_0,\ y^{(1)}(0) = y_1, \cdots, y^{(n-1)}(0) = y_{n-1}.$$

If $g(t) \equiv 0$, (5.1.1) is called *homogeneous*, otherwise it is *non-homogeneous*. A class of singular n^{th} order equations, called *Euler equations*, consists of equations of the following type:

$$t^n\frac{d^n y}{dt^n} + p_1 t^{n-1}\frac{d^{n-1}y}{dt^{n-1}} + \cdots + p_{n-1}t\frac{dy}{dt} + p_n y = 0,$$

where $p_1, \cdots, p_n$ are real constants. This equation can be reduced to a constant coefficient equation by changing the independent variable t to τ satisfying $t\frac{d}{dt} = \frac{d}{d\tau}$. However, caution should be exercised while seeking a solution around $t = 0$.

By introducing new variables:

$$x_1 = y, \ x_2 = y^{(1)}, \ldots, x_n = y^{(n-1)} = \frac{d^{n-1}y}{dt^{n-1}},$$

we can convert (5.1.1) into a system of n first order equations which having a matrix representation as $\dot{\mathbf{x}}(t) = \mathbf{A}(t)\mathbf{x}(t) + \mathbf{g}(t)$, where

$$\mathbf{A} = \mathbf{A}(t) = \begin{bmatrix} 0 & 1 & 0 & \cdots & 0 \\ 0 & 0 & 1 & \cdots & 0 \\ \cdots & \cdots & \cdots & \cdots & \cdots \\ 0 & 0 & 0 & \cdots & 1 \\ -p_n(t) & -p_{n-1}(t) & \cdots & \cdots & -p_1(t) \end{bmatrix}$$

and

$$\mathbf{x}(t) = \begin{bmatrix} x_1(t) \\ x_2(t) \\ \cdot \\ \cdot \\ x_{n-1}(t) \\ x_n \end{bmatrix}, \ \mathbf{g}(t) = \begin{bmatrix} 0 \\ 0 \\ \cdot \\ \cdot \\ 0 \\ g(t) \end{bmatrix}.$$

This motivates us to consider the following general system

$$\dot{\mathbf{x}}(t) = \mathbf{A}(t)\mathbf{x}(t) + \mathbf{g}(t), \ \mathbf{x}(0) = \mathbf{x}_0, \tag{5.1.2}$$

where $\mathbf{A}(t)$, is an $n \times n$ matrix whose elements are functions of $t \in \mathbb{R}$ and $\mathbf{g}(t)$ is a vector valued function of $t \in \mathbb{R}$. If the elements of $\mathbf{A}$ and $\mathbf{g}$ are continuous functions of t, then the standard existence theorem (see Chapter 4) will give the uniqueness and existence of the solution $\mathbf{x}(t)$ of the IVP (5.1.2), in any finite interval $\mathbb{R}$. See also Section 4.6.

Definition 5.1.1

If the coefficient matrix $\mathbf{A}$ depends on t, (5.1.2) is called a *non-autonomous system*; otherwise, it is called an *autonomous system*.

Indeed, the operator $L = L(\mathbf{x}, \dot{\mathbf{x}}) = \dot{\mathbf{x}} - \mathbf{A}\mathbf{x}$ is linear in $\mathbf{x}$ and $\dot{\mathbf{x}}$. We study the representation of the solution of (5.1.2) when $\mathbf{A}$ is a constant matrix and

g identically zero. This is more easily done in the case of an autonomous system. In the case of a homogeneous, autonomous system, the solution can be represented as

$$\mathbf{x}(t) = e^{t\mathbf{A}} \mathbf{x}_0, \tag{5.1.3}$$

analogous to the one-dimensional case when $\mathbf{A} = a$, is a scalar. Though (5.1.3) gives the unique solution, it does not give us much information about the trajectories. It is to be remarked that in practical applications, the solution of the system may correspond to the motion of particles. Hence, understanding the trajectories and their behaviour is of paramount importance. So one of our major aims in this chapter is to get the geometry behind (5.1.3) using the spectral analysis of $\mathbf{A}$. Hence, this chapter is also a precursor to Chapter 8 on the qualitative analysis of nonlinear systems. We give a complete description in 2-dimensional systems; it is more complicated in higher dimensional cases. Indeed, we can obtain a good amount of information using the Jordan decomposition of $\mathbf{A}$. Again, the computation of $e^{t\mathbf{A}}$, in general, is not that easy and spectral decomposition helps us to compute $e^{t\mathbf{A}}$. We will also study the non-homogeneous system and represent the solution using *variation of parameters*. We will further introduce the concept of a *transition matrix* for representation of solutions to non-autonomous systems.

5.2   Autonomous Homogeneous Systems

Given a real matrix $\mathbf{A}$ of order n, we may think it as a linear operator from $\mathbb{R}^n \to \mathbb{R}^n$ defined by $\mathbf{x} \mapsto \mathbf{A}\mathbf{x}$ whose operator norm is given by $||\mathbf{A}|| = \sup_{|\mathbf{x}|=1} |\mathbf{A}\mathbf{x}|$, where $|\cdot|$ is the Euclidean norm (or modulus) in $\mathbb{R}^n$. Consider the partial sums of operators $\mathbf{A}_m = \sum_{k=0}^{m} t^k \dfrac{\mathbf{A}^k}{k!}$. It is easy to see that $||\mathbf{A}_m|| \leq e^{||\mathbf{A}|| |t|}$, for all t and for $m > l$, $||\mathbf{A}_m - \mathbf{A}_l|| \leq \sum_{k=l+1}^{m} t^k \dfrac{||\mathbf{A}||^k}{k!} \to 0$ as $l, m \to \infty$. Hence, the sequence of operators $\{\mathbf{A}_m\}$ converges to a linear operator from $\mathbb{R}^n \to \mathbb{R}^n$, whose matrix representation is denoted by $e^{t\mathbf{A}}$. In other words,

$$e^{t\mathbf{A}} = \sum_{k=0}^{\infty} t^k \frac{\mathbf{A}^k}{k!}.$$

Consider the linear homogeneous autonomous system

$$\dot{\mathbf{x}} = \mathbf{A}\mathbf{x}(t), \mathbf{x}(t_0) = \mathbf{x}_0 \tag{5.2.1}$$

With the aid of the matrix exponential, it is now straightforward to verify that a unique solution for the IVP (5.2.1), is given by $\mathbf{x}(t) = e^{(t-t_0)\mathbf{A}}\mathbf{x}_0$. The uniqueness follows from a similar argument for a single equation; see Chapter 3. If $\mathbf{y}(t)$ is any other solution of (5.2.1), then by considering the expression $e^{-(t-t_0)\mathbf{A}}\mathbf{y}(t))$, a differentiation yields uniqueness.

Thus, we have the unique representation of the solution of IVP (5.2.1) as $\mathbf{x}(t) = e^{(t-t_0)\mathbf{A}}\mathbf{x}_0$. Next, we use the spectral decomposition of the matrix $\mathbf{A}$ to elicit more information about the behaviour of the solution, especially of the various orbits or trajectories.

5.2.1   Computation of $e^{t\mathbf{A}}$ in special cases

- If the matrix $\mathbf{A}$ is diagonal denoted by $\mathbf{A} = \mathrm{diag}(\lambda_1,\ldots,\lambda_n)$, then it is an easy exercise to verify that

$$e^{\mathbf{A}} = \mathrm{diag}(e^{\lambda_1},\cdots,e^{\lambda_1}).$$

Note that saying that the matrix is diagonal is equivalent to the fact that the system (5.2.1) is a decoupled system in the sense that we have n separate equations for each component $x_i(t)$ for all $1 \leq i \leq n$. Hence, each equation can be solved to get $x_i(t) = e^{\lambda_i(t-t_0)}x_{0i}$, where x_{0i} is the ith component of the initial value $\mathbf{x}_0$. Thus, $\mathbf{x}(t)$ can be represented as

$$\mathbf{x}(t) = \mathrm{diag}(e^{\lambda_1(t-t_0)},\cdots,e^{\lambda_n(t-t_0)})\mathbf{x}_0 = e^{(t-t_0)\mathbf{A}}\mathbf{x}_0.$$

- We have $e^{\mathbf{A}+\mathbf{B}} = e^{\mathbf{A}}.e^{\mathbf{B}}$, if $\mathbf{A}$ and $\mathbf{B}$ commute. In particular, $e^{\mathbf{A}}$ is always invertible and $(e^{\mathbf{A}})^{-1} = e^{-\mathbf{A}}$. In general, $e^{\mathbf{A}+\mathbf{B}} \neq e^{\mathbf{A}}.e^{\mathbf{B}}$ for non-commutative matrices $\mathbf{A}$ and $\mathbf{B}$.

- (Similarity of matrices) Suppose $\mathbf{B} = \mathbf{P}^{-1}\mathbf{A}\mathbf{P}$ for some invertible $\mathbf{P}$, Then, $e^{\mathbf{B}} = \mathbf{P}^{-1}e^{\mathbf{A}}\mathbf{P}$. This follows easily from the observation $\mathbf{B}^2 = (\mathbf{P}^{-1}\mathbf{A}\mathbf{P})(\mathbf{P}^{-1}\mathbf{A}\mathbf{P}) = \mathbf{P}^{-1}\mathbf{A}^2\mathbf{P}$ and hence, by induction, $\mathbf{B}^k = \mathbf{P}^{-1}\mathbf{A}^k\mathbf{P}$ for any positive integer k.

Using similarity transformation, we can convert the system corresponding to $\mathbf{A}$ to a system corresponding to $\mathbf{B}$ as follows. Put $\mathbf{y} = \mathbf{P}^{-1}\mathbf{x}$. Since $\mathbf{P}$ is a constant matrix, $\mathbf{y}$ satisfies the system

$$\dot{\mathbf{y}}(t) = \mathbf{B}\mathbf{y}(t), \mathbf{y}(t_0) = \mathbf{y}_0 = \mathbf{P}^{-1}\mathbf{x}_0.$$

This leads to the concept of linear equivalence of two linear systems.

Definition 5.2.1

A linear system of ODE

$$\dot{\mathbf{x}} = \mathbf{A}\mathbf{x} \qquad (5.2.2)$$

is said to be linearly equivalent to a system

$$\dot{\mathbf{y}} = \mathbf{B}\mathbf{y} \qquad (5.2.3)$$

if $\mathbf{A}$ and $\mathbf{B}$ are similar, that is, there exists an invertible matrix $\mathbf{P}$ such that $\mathbf{B} = \mathbf{P}^{-1}\mathbf{A}\mathbf{P}$.

□

The idea is to find $\mathbf{B}$ so that (5.2.3) is much simpler to analyze than (5.2.2). Such a $\mathbf{B}$ is provided by the Jordan decomposition of $\mathbf{A}$. We will also see that there is no change in the qualitative behaviour of solutions of (5.2.2) and (5.2.3); in particular, the nature of equilibrium points (which are the same for both (5.2.2) and (5.2.3)) is unaltered. We also remark that the transformation $\mathbf{x} \to \mathbf{y} = \mathbf{P}^{-1}\mathbf{x}$ or $\mathbf{y} \to \mathbf{x} = \mathbf{P}\mathbf{y}$ are linear operators and these transformations are essentially a coordinate change or change of basis. In effect, we view the two systems (5.2.2) and (5.2.3) in two different coordinate systems. We will see examples later.

In a particular case when $\mathbf{A}$ is diagonalizable, that is, $\mathbf{B} = \mathrm{diag}(\lambda_1, \lambda_2, \ldots, \lambda_n)$, then, $e^{\mathbf{A}} = e^{\mathbf{PBP}^{-1}} = \mathbf{P}e^{\mathbf{B}}\mathbf{P}^{-1} = \mathbf{P}\mathrm{diag}(e^{\lambda_1}, e^{\lambda_2}, \cdots, e^{\lambda_n})\mathbf{P}^{-1}$. In this case, the system (5.2.2), which is a coupled system to begin with, is converted to an uncoupled system (5.2.3) which can be solved immediately. Hence, we also get the solution to the uncoupled system (5.2.1) as

$$\mathbf{x}(t) = e^{(t-t_0)\mathbf{A}}\mathbf{x}_0 = \mathbf{P}\mathrm{diag}(e^{\lambda_1(t-t_0)}, e^{\lambda_2(t-t_0)}, \cdots, e^{\lambda_n(t-t_0)})\mathbf{P}^{-1}\mathbf{x}_0.$$
$$(5.2.4)$$

So, we ask a natural question. Is every matrix diagonalizable? In Chapter 2, we have seen that this is not always true. In general, diagonalizability is not possible even in the two-dimensional case. But the two-dimensional case is not that difficult, and we will do a complete analysis separately in this case to bring out the aspects of diagonalizability. The general case is more involved and Jordan decomposition is the best possible reduction.

We already know that diagonalizability is equivalent to the existence of n independent eigenvectors. Note that we do not demand n distinct eigenvalues. However, the existence of n distinct eigenvalues implies the existence of n independent eigenvectors and hence, diagonalizability (it is a sufficient condition) is guaranteed. In general, if the algebraic and geometric multiplicities are equal for all the eigenvalues of a matrix, then that matrix is diagonalizable.

The reader can show that the matrix $\begin{bmatrix} 1 & 0 \\ 1 & 1 \end{bmatrix}$ is not diagonalizable.

5.3 Two-dimensional Systems

In this section, we consider the 2×2 system, that is $\mathbf{A}$ is a 2×2 matrix. To begin with, we also assume that $\mathbf{A}$ is non-singular. Our aim is to do the *phase plane* analysis and classify various types of equilibrium points. We have the following theorem.

Theorem 5.3.1

Every 2×2 system $\dot{\mathbf{x}} = \mathbf{A}\mathbf{x}$ is *linearly equivalent* to only one of the systems $\dot{\mathbf{y}} = \mathbf{B}_i\mathbf{y}$, $i = 1,2,3$, where

$$\mathbf{B}_1 = \begin{bmatrix} \lambda & 0 \\ 0 & \mu \end{bmatrix}, \ \mathbf{B}_2 = \begin{bmatrix} \lambda & 1 \\ 0 & \lambda \end{bmatrix}, \ \mathbf{B}_3 = \begin{bmatrix} a & -b \\ b & a \end{bmatrix}, \tag{5.3.1}$$

where λ, μ, a, b are real numbers and $b \neq 0$.

Proof: If the two eigenvalues λ and μ of $\mathbf{A}$ are real and distinct, then the corresponding eigenvectors $\mathbf{x}$ and $\mathbf{y}$ are linearly independent. If $\lambda = \mu$ is the double real eigenvalue of $\mathbf{A}$ and the corresponding eigenspace is two-dimensional, then also we obtain two linearly independent eigenvectors $\mathbf{x}$ and $\mathbf{y}$. In either of these cases, we obtain, with $\mathbf{P} = [\mathbf{x} \ \ \mathbf{y}] = \begin{bmatrix} x_1 & y_1 \\ x_2 & y_2 \end{bmatrix}$, $\mathbf{AP} = \mathbf{P}\text{diag}(\lambda, \mu) = \mathbf{P}\begin{bmatrix} \lambda & 0 \\ 0 & \mu \end{bmatrix}$. From now onwards, note that we may represent matrices in the form $[\mathbf{x} \ \ \mathbf{y}]$, where $\mathbf{x}$ and $\mathbf{y}$ are the column vectors of the matrix in question. Since $\mathbf{x}$ and $\mathbf{y}$ are independent, the matrix $\mathbf{P}$ is invertible, and we get the first form $\mathbf{B}_1$.

The second case is more delicate, where $\lambda = \mu$ is the double real eigenvalue of $\mathbf{A}$ whose eigenspace is one-dimensional, say spanned by an eigenvector $\mathbf{x}$. In this case, we need one more vector linearly independent of $\mathbf{x}$, to form a basis for $\mathbb{R}^2$. This can come from the notion of *generalized eigenvector*. We have $\ker(\mathbf{A} - \lambda\mathbf{I}) \subset \ker(\mathbf{A} - \lambda\mathbf{I})^2$; the latter is two-dimensional space since $(\mathbf{A} - \lambda\mathbf{I})^2 = 0$ as the characteristic equation of $\mathbf{A}$ is $\det(\mathbf{A} - \lambda\mathbf{I}) = 0$. Hence, $\ker(\mathbf{A} - \lambda\mathbf{I})^2 = \mathbb{R}^2$. Thus, choose a vector $\mathbf{v} \notin \ker(\mathbf{A} - \lambda\mathbf{I})$ and $\mathbf{v} \in \ker(\mathbf{A} - \lambda\mathbf{I})^2$. It is not difficult to verify that the vectors $\mathbf{u} = (\mathbf{A} - \lambda\mathbf{I})\mathbf{v}$ and $\mathbf{v}$ are linearly independent and therefore, the matrix $\mathbf{P} = \begin{bmatrix} \mathbf{u} & \mathbf{v} \end{bmatrix}$ is non-singular. Further, $\mathbf{AP} = \mathbf{PB}_2$.

In the third case, suppose $\mathbf{A}$ has complex eigenvalues. Since $\mathbf{A}$ is real, these eigenvalues are complex conjugates of each other. Let $\lambda = a + ib$ and $\mu = a - ib$ be the complex eigenvalues, with $b \neq 0$. Let $\mathbf{w} = \mathbf{u} + i\mathbf{v}$ be an eigenvector corresponding to λ, that is, $\mathbf{Aw} = \lambda\mathbf{w}$. Here $\mathbf{u}$ and $\mathbf{v}$ are real vectors. Since $\mathbf{A}$ is real, we have $\mathbf{A}\bar{\mathbf{w}} = \bar{\lambda}\bar{\mathbf{w}} = \mu\bar{\mathbf{w}}$, where $\bar{}$ denotes the complex conjugate. It is not difficult to show that the real vectors $\mathbf{u}$ and $\mathbf{v}$ are independent and $\mathbf{AP} = \mathbf{PB}_3$ with $\mathbf{P} = \begin{bmatrix} \mathbf{v} & \mathbf{u} \end{bmatrix}$. The details are left as an exercise to the reader. $\qquad\qquad\square$

5.3.1   Computation of $e^{\mathbf{B}_j}$ and $e^{t\mathbf{B}_j}$

Since $\mathbf{B}_1$ is diagonal, we have already seen that $e^{\mathbf{B}_1} = \mathrm{diag}(e^\lambda, e^\mu)$ and $e^{t\mathbf{B}_1} = \mathrm{diag}(e^{t\lambda}, e^{t\mu})$. In the case of $\mathbf{B}_2$ and $\mathbf{B}_3$, we leave it to the reader to check that

$$e^{\mathbf{B}_2} = e^\lambda \begin{bmatrix} 1 & 1 \\ 0 & 1 \end{bmatrix}, \quad e^{t\mathbf{B}_2} = e^{t\lambda} \begin{bmatrix} 1 & t \\ 0 & 1 \end{bmatrix} \tag{5.3.2}$$

and

$$e^{\mathbf{B}_3} = e^a \begin{bmatrix} \cos b & -\sin b \\ \sin b & \cos b \end{bmatrix}, \quad e^{t\mathbf{B}_2} = e^{ta} \begin{bmatrix} \cos(tb) & -\sin(tb) \\ \sin(tb) & \cos(tb) \end{bmatrix}. \tag{5.3.3}$$

Thus, if we consider the system (5.2.3) with $\mathbf{B} = \mathbf{B}_i$, $i = 1, 2, 3$, the corresponding solutions to the IVP with $\mathbf{y}(0) = \mathbf{y}_0$ can be, respectively, written as

$$\mathbf{y}(t) = \mathrm{diag}(e^{t\lambda}, e^{t\mu})\mathbf{y}_0. \tag{5.3.4}$$

Equivalently, $y_1(t) = e^{t\lambda}y_{01}$, $y_2(t) = e^{t\mu}y_{02}$, for $i = 1$;

$$\mathbf{y}(t) = e^{t\lambda}\begin{bmatrix} 1 & t \\ 0 & 1 \end{bmatrix}\mathbf{y}_0. \tag{5.3.5}$$

Equivalently, $y_1(t) = e^{t\lambda}(y_{01} + ty_{02})$, $y_2(t) = e^{t\lambda}y_{02}$, for $i = 2$; and for $i = 3$,

$$\mathbf{y}(t) = e^{ta}\begin{bmatrix} \cos(tb) & -\sin(tb) \\ \sin(tb) & \cos(tb) \end{bmatrix}\mathbf{y}_0. \tag{5.3.6}$$

Equivalently,

$$y_1(t) = e^{ta}(y_{01}\cos(tb) - y_{02}\sin(tb)), \quad y_2(t) = e^{ta}(y_{01}\sin(tb)$$
$$+ y_{02}\cos(tb)).$$

When we go back to the original system corresponding to $\mathbf{A}$, we may write the solution to the IVP (5.2.1), respectively as

$$\mathbf{x}(t) = e^{t\mathbf{A}}\mathbf{x}_0 = \mathbf{P}e^{t\mathbf{B}_1}\mathbf{P}^{-1}\mathbf{x}_0 = \mathbf{P}\mathrm{diag}(e^{t\lambda}, e^{t\mu})\mathbf{P}^{-1}\mathbf{x}_0 \text{ for } i = 1;$$
$$\tag{5.3.7}$$

$$\mathbf{x}(t) = e^{t\lambda}\mathbf{P}\begin{bmatrix} 1 & t \\ 0 & 1 \end{bmatrix}\mathbf{P}^{-1}\mathbf{x}_0 \text{ for } i = 2; \tag{5.3.8}$$

and for $i = 3$,

$$\mathbf{x}(t) = e^{ta}\mathbf{P}\begin{bmatrix} \cos(tb) & -\sin(tb) \\ \sin(tb) & \cos(tb) \end{bmatrix}\mathbf{P}^{-1}\mathbf{x}_0. \tag{5.3.9}$$

It is interesting to observe that the general solution of any 2×2 system is a linear combination of the product of the elementary functions; namely, polynomials, exponentials and trigonometric functions. We will see later that this statement continues to hold for any general n dimensional system too. We will now have a few examples.

Example 5.3.2

Consider the system

$$\dot{x}_1 = -x_1 - 3x_2, \quad \dot{x}_2 = 2x_2. \tag{5.3.10}$$

The matrix $A = \begin{bmatrix} -1 & -3 \\ 0 & 2 \end{bmatrix}$ has eigenvalues $\lambda_1 = -1$, $\lambda_2 = 2$ with

corresponding eigenvectors $\begin{bmatrix} 1 \\ 0 \end{bmatrix}$ and $\begin{bmatrix} -1 \\ 1 \end{bmatrix}$. Hence, $\mathbf{P} = \begin{bmatrix} 1 & -1 \\ 0 & 1 \end{bmatrix}$ and

$\mathbf{P}^{-1} = \begin{bmatrix} 1 & 1 \\ 0 & 1 \end{bmatrix}$. Further, $\mathbf{B} = \mathbf{P}^{-1}\mathbf{AP} = \begin{bmatrix} -1 & 0 \\ 0 & 2 \end{bmatrix}$. This implies that a

linearly equivalent system is given by $\dot{y}_1 = -y_1$, $\dot{y}_2 = y_2$, which is a diagonal system.

Example 5.3.3

Consider the system

$$\dot{x}_1 = 7x_1 - 4x_2, \quad \dot{x}_2 = x_1 + 3x_2 \tag{5.3.11}$$

The coefficient matrix $\mathbf{A} = \begin{bmatrix} 7 & -4 \\ 1 & 3 \end{bmatrix}$ has a double eigenvalue $\lambda = 5$ with

a corresponding eigenvector $\mathbf{x} = \begin{bmatrix} 2 \\ 1 \end{bmatrix}$; any other eigenvector is a multiple

of this vector. Now choose any vector which is independent of $\mathbf{x}$, say

$\mathbf{v} = \begin{bmatrix} 1 \\ 0 \end{bmatrix}$, then $\mathbf{u} = (\mathbf{A} - 5I)\mathbf{v} = \begin{bmatrix} 2 \\ 1 \end{bmatrix}$ is a generalized eigenvector for $\lambda = 5$.

Thus, the matrix $\mathbf{P}$ is given by $\mathbf{P} = \begin{bmatrix} 2 & 1 \\ 1 & 0 \end{bmatrix}$ and $\mathbf{P}^{-1} = \begin{bmatrix} 0 & 1 \\ 1 & -2 \end{bmatrix}$. The

reader is advised to verify these facts and see that $\mathbf{A}$ is equivalent to $\mathbf{B}_2 = \begin{bmatrix} 5 & 1 \\ 0 & 5 \end{bmatrix}$. The final solution is given by

$$\mathbf{x}(t) = e^{5t}\mathbf{P}\begin{bmatrix} 1 & t \\ 0 & 1 \end{bmatrix}\mathbf{P}^{-1}\mathbf{x}_0.$$

Example 5.3.4

In this example, we consider a second order equation

$$\ddot{z} + 2\dot{z} + 3z = 0. \tag{5.3.12}$$

By introducing the variables $x_1 = z, x_2 = \dot{z}$, the equation reduces to a system of the form $\dot{\mathbf{x}} = \mathbf{Ax}$, where $\mathbf{A}$ is given by

$\mathbf{A} = \begin{bmatrix} 0 & 1 \\ -3 & -2 \end{bmatrix}$. Observe that $\mathbf{A}$ has the complex eigenvalues given by $\lambda = -1 + i\sqrt{2}$ and $\mu = -1 - i\sqrt{2}$. Thus, $a = -1$ and $b = \sqrt{2}$ in (5.3.1). A complex eigenvector $\mathbf{x}$ corresponding to eigenvalue $-1 + i\sqrt{2}$ can be computed as $\mathbf{x} = \begin{bmatrix} 1 \\ \lambda \end{bmatrix} = \begin{bmatrix} 1 \\ -1 \end{bmatrix} + \begin{bmatrix} 0 \\ i\sqrt{2} \end{bmatrix}$. Thus, $\mathbf{P} = \begin{bmatrix} 0 & 1 \\ \sqrt{2} & -1 \end{bmatrix}$ and $\mathbf{P}^{-1} = \frac{1}{\sqrt{2}} \begin{bmatrix} 1 & 1 \\ \sqrt{2} & 0 \end{bmatrix}$. Finally, the solution to the system is given by

$$\mathbf{x}(t) = e^{-t} \mathbf{P} \begin{bmatrix} \cos(\sqrt{2}t) & -\sin(\sqrt{2}t) \\ \cos(\sqrt{2}t) & \sin(\sqrt{2}t) \end{bmatrix} \mathbf{P}^{-1} \mathbf{x}_0.$$

and the solution z to (5.3.12) is the first component $x_1(t)$ of $\mathbf{x}(t)$.

5.4 Stability Analysis

We begin this section with a description of several notions like equilibrium points, phase plane and phase portrait; the concept of a dynamical system, a flow and a vector field. Further classification is done among the equilibrium points, namely stable and unstable nodes; saddles; stable and unstable foci and centres. These notions are also relevant to nonlinear systems. In fact, these notions are at the heart of understanding ODE, especially in the context of applications. We restrict the discussion to the 2×2 systems for describing the aforementioned notions. We remark that 2×2 linear systems are particularly simple to analyse as there are at most two eigenvalues and corresponding eigenvectors to deal with. Nonetheless, the features that we present here, are generally there for higher dimensional systems also.

5.4.1 Phase plane and phase portrait

Consider a general n dimensional autonomous system

$$\dot{\mathbf{x}} = \mathbf{f}(\mathbf{x}), \; \mathbf{x}(t_0) = \mathbf{x}_0, \tag{5.4.1}$$

where $\mathbf{x}_0 \in \mathbb{R}^n$ and in general, $\mathbf{f} : \mathbb{R}^n \to \mathbb{R}^n$ is a smooth map; $f(\mathbf{x}) = \mathbf{A}\mathbf{x}$, where $\mathbf{A}$ is an $n \times n$ matrix, will give an n dimensional linear system.

In mechanics, the motion of a particle is described by Newton's second law and may be formulated as a second order ODE: $m\ddot{x} + F(x, \dot{x}) = 0$, where m is the mass of the particle. The position $x(t)$ and its velocity

$\dot{x}(t)$ are called the *phases* of the system under consideration. We adopt same terminology for a first order system (5.4.1) and call the components $x_1, x_2, \cdots, x_n$ of the solution $\mathbf{x}$ as phases of (5.4.1). Indeed, if Newton's law transformed to a first order system, we get, $x_1 = x$ and $x_2 = \dot{x}$. If $\mathbf{x}$ is a solution of (5.4.1) in some interval $I \subset \mathbb{R}$, containing t_0, the set $\{\mathbf{x}(t) \in \mathbb{R}^n : t \in I\}$ is called a *trajectory* or an *orbit* passing through $\mathbf{x}_0$. In this scenario, $\mathbb{R}^n$ is referred to as the *phase space* (*phase plane* if $n = 2$ and *phase line* if $n = 1$) of (5.4.1). Thus, the phase space contains all the trajectories of (5.4.1) passing through different points of the phase space. Description of all the trajectories of (5.4.1) in the phase space (or phase plane or phase line) is referred to as the *phase portrait* of (5.4.1) and the analysis involved in this process may be called as the phase space (or phase plane or phase line) analysis of (5.4.1).

We now give an example.

Example 5.4.1

Consider the decoupled system given by

$$\dot{x}_1 = -x_1, \ \dot{x}_2 = 2x_2, \ x_1(0) = x_{01}, \ x_2(0) = x_{02}. \tag{5.4.2}$$

The solution is given by $\mathbf{x}(t) = \begin{bmatrix} e^{-t} & 0 \\ 0 & e^{2t} \end{bmatrix} \mathbf{x}_0$, where $\mathbf{x}_0 = \begin{bmatrix} x_{01} \\ x_{02} \end{bmatrix}$. Even though it is a system of two separate equations (that is, uncoupled), we would like to consider it as a system. From a dynamical system point of view, $\mathbf{x}(t)$ can be thought of as the motion of a particle in the $x_1 x_2$-plane which is the phase plane for (5.4.2), as t varies. At $t = 0$, the particle is at $\mathbf{x}_0$ and the particle moves to $\mathbf{x}(t)$ at time t. Figure 5.1 gives the various trajectories of the particle under motion, where the arrow indicates the direction of motion for positive (increasing) time and different initial positions.

We will see later that what is observed in Fig. 5.1 will be the feature of any system having eigenvalues with different signs, except that the direction of arrows and coordinates might change. The reader can see the phase portrait of Example 5.3.1.

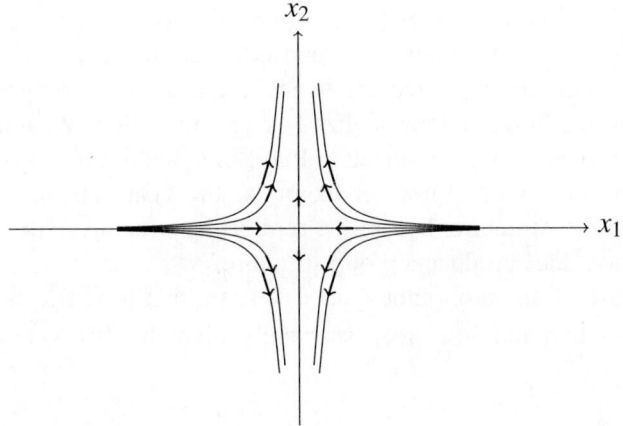

Fig. 5.1   Saddle point equilibrium

5.4.2   Dynamical system, flow, vector fields

Dynamical System: The *dynamical system* of (5.4.1) is a mapping Φ : $\mathbb{R} \times \mathbb{R}^n \to \mathbb{R}^n$ given by the solution $\mathbf{x}(t) = \mathbf{x}(t, \mathbf{x}_0)$, that is, $\Phi(t, \mathbf{x}_0) = \mathbf{x}(t, \mathbf{x}_0)$. Geometrically, the dynamical system describes the motion of the points in the phase space along the solution curves. For fixed $\mathbf{x}_0$, $\Phi(\cdot, \mathbf{x}_0)$ represents the solution passing through the point $\mathbf{x}_0$.

Flow: Given the dynamical system Φ as described earlier for any fixed t, introduce the map $\phi_t : \mathbb{R}^n \to \mathbb{R}^n$ by $\phi_t(\mathbf{x}_0) = \Phi(t + t_0, \mathbf{x}_0) = \mathbf{x}(t + t_0, \mathbf{x}_0)$. Then, the collection $G = \{\phi_t : t \in \mathbb{R}\}$ is called the *flow* of the system (5.4.1).

   The notion of a flow gives an entirely different perspective which is quite useful in applications. For example, when we watch fluid flowing, we normally do not see the trajectory lines (stream lines); rather, we see a body of fluid moving. This is the view incorporated in the concept of flow. More precisely, we would like to see a neighbourhood, say U of $\mathbf{x}_0$ moving with time. Thus, $\phi_t(U)$ is the position of all particles at time t, whose initial position is in U. The collection G satisfies

$$\phi_0(\mathbf{x}) = \mathbf{x}, \ \phi_s(\phi_t(\mathbf{x})) = \phi_{s+t}(\mathbf{x}), \ \phi_{-t}(\phi_t(\mathbf{x})) = \phi_t(\phi_{-t}(\mathbf{x})) = \mathbf{x},$$

$$(5.4.3)$$

for all $\mathbf{x} \in \mathbb{R}^n$. The first property comes from the initial condition, whereas the second property follows from the uniqueness of a solution to the IVP. This is known as *semigroup property*. In other words, the flow is a semigroup. The last property is a consequence of the second one, and asserts the existence of inverse and hence, the flow G has properties of a group. Thus, the flow can be visualized as a group action. We remark that this notion can even be generalized to PDE. In general, every system may not have the group structure; for example, the heat equation does not produce a group structure, but only a semigroup structure. However, the wave equation does produce a group structure.

In the case of an autonomous linear system, that is, if $\mathbf{f}(x) = \mathbf{A}\mathbf{x}$, the dynamical system and flow are, respectively, given by $\Phi(t, \mathbf{x}_0) = e^{(t-t_0)\mathbf{A}}\mathbf{x}_0$ and $\phi_t = e^{t\mathbf{A}}$.

Vector Field: A point $\mathbf{x} \in \mathbb{R}^n$ can also be viewed as a vector, namely, the *position vector*. When we view it as a vector, it has got magnitude as well as direction and it is irrelevant where the initial position was. In other words, any vector having the same magnitude and direction as that of $\mathbf{x}$, is considered as the same vector. Both these views are helpful in understanding the system. For example, consider the motion of a particle in three-dimensional space. It is natural to view the position as a point, whereas it is better to view the velocity, which is also a point in $\mathbb{R}^3$, as a vector placed at the position.

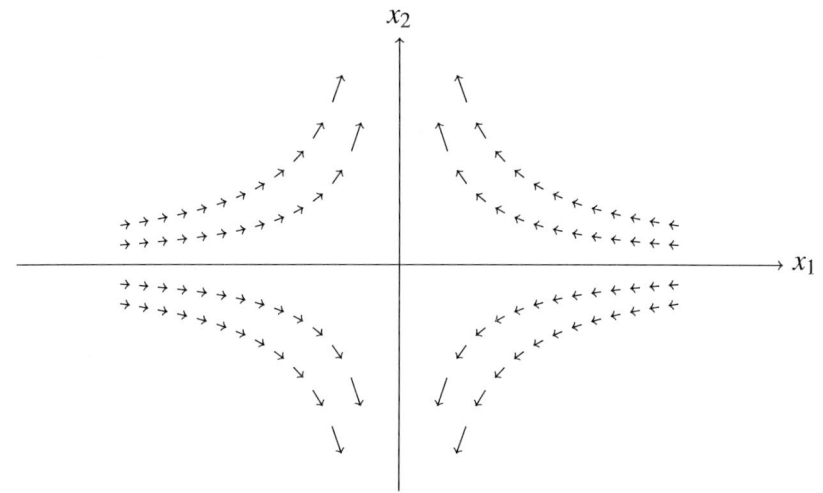

Fig. 5.2   Vector field

Now consider the system (5.4.1). Here $\mathbf{f}(\mathbf{x})$ is the given information or data which we view as a vector located at the point $\mathbf{x}$ and thus producing a vector at each point in a domain $\Omega \subset \mathbb{R}^n$, where the ODE is described. This we name as a *vector field*. Thus, a vector field X in a domain $\Omega \subset \mathbb{R}^n$ is a mapping such that a vector $X(\mathbf{x}) \in \mathbb{R}^n$ is associated with every $\mathbf{x} \in \Omega$. We say that the vector field is smooth if this mapping is smooth. The vector field associated with the system in Example 5.4.1 is represented in Fig. 5.2. We may ask a question: what is the connection between the vector field and the solution of a system? It is easy to see that the tangent to the solution curves gives the vector field and conversely, any curve whose tangents are from the vector field will be the solution to the system.

5.4.3   Equilibrium points and stability

The analysis of a general 2×2 linear system is done by changing it to a linearly equivalent system via Theorem 5.3.1. Thus, there are essentially three different types of matrices to be considered for the analysis. We prefer to do the analysis by taking an example of each of such matrices, as there is no change in the qualitative behaviour of solutions in the general case. We begin with a definition.

Definition 5.4.2

For a general autonomous system $\dot{\mathbf{x}} = \mathbf{f}(\mathbf{x})$, a point $\bar{\mathbf{x}}$ is called an *equilibrium point*, if $\mathbf{f}(\bar{\mathbf{x}}) = 0$. □

We would like to state an important point at this stage. Note that $\mathbf{x}(t) = \bar{\mathbf{x}}$ for all t is a solution to the system $\dot{\mathbf{x}} = \mathbf{f}(\mathbf{x})$ if $\bar{\mathbf{x}}$ is an equilibrium point. This means that if the motion starts from the equilibrium point, the trajectory will remain there forever. In physical problems, especially in mechanics, it represents the steady state solution, the one that does not change with time. Hence, we not only view an equilibrium point as a *point*, but also as a solution to the system.

Since equilibrium point is a steady state solution, we would be interested in the behaviour of solutions which start close to an equilibrium point. This is very important since when we make small errors, we would like to know whether the trajectory also remains in a neighbourhood of the equilibrium point. This is the motivation behind stability analysis of equilibrium points.

For a linear system, that is, $\mathbf{f}(x) = \mathbf{A}x$, observe that $\bar{\mathbf{x}} = 0$ is always an equilibrium point and in addition, if $\mathbf{A}$ is invertible, this is the only equilibrium point. In general, the set of all equilibrium points is given by $\ker(\mathbf{A})$. In this section, we will characterize various types of equilibrium points for a 2×2 system. We will do this via various examples. Stability of nonlinear systems will be studied in Chapter 8.

Saddle Point, Node, Focus and Center: Note that in Example 5.4.1, the first component $x_1(t) \to 0$, whereas the second component $x_2(t) \to \pm\infty$ as $t \to \infty$ depending on the initial condition. This equilibrium point is called a *saddle point* and it is classified as *unstable*. In fact, this will be the feature for any system having two real non-zero eigenvalues with opposite sign except that the trajectories will remain in four parts separated by a different set of coordinate axes given by the eigenvectors, and which need not be the standard coordinate axes (see Example 5.3.2).

Example 5.4.3

Let $\mathbf{A} = \begin{bmatrix} \lambda & 0 \\ 0 & \lambda \end{bmatrix}$, $\lambda > 0$.

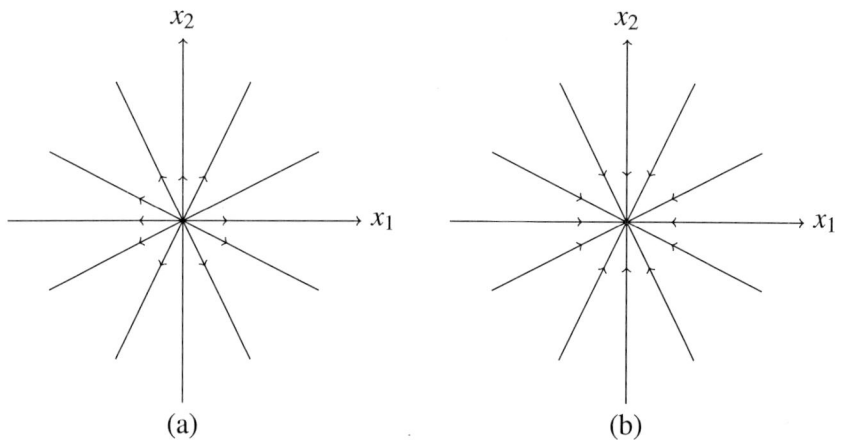

(a)   (b)

Fig. 5.3   (a) Unstable node, (b) Stable node

In this case, both trajectories $\mathbf{x}_j(t) \to \infty$ as $t \to \infty$. The phase portrait is depicted in Fig. 5.3(a).   The direction of the arrows will change if we consider $A = \begin{bmatrix} -\lambda & 0 \\ 0 & -\lambda \end{bmatrix}$ with $\lambda > 0$ and both the trajectories will now approach 0 as $t \to \infty$. See Fig. 5.3(b). This equilibrium in both the cases is referred as a *node*. In the first case, we have an *unstable node* and the second case corresponds to a *stable node*.

Example 5.4.4

Take $\mathbf{A} = \begin{bmatrix} 2 & 0 \\ 0 & 1 \end{bmatrix}$.

This has two distinct eigenvalues, 2 and 1, having the same sign and the solution is given by $x_1(t) = x_{01}e^{2t}$, $x_2(t) = x_{02}e^{t}$. Eliminating t, we will get $x_1 = cx_2^2$ (see Fig. 5.4(b)) and an unstable node. Again the arrows will get reversed if we take negative numbers in $\mathbf{A}$ and we get a *stable node* (see Fig. 5.4(a)). The situation is exactly the same if we replace 2 and 1 by any two real numbers with the same sign. More generally, the behaviour of the trajectories will remain the same for any system having two distinct real eigenvalues of the same sign except that the trajectories will remain in four parts separated by a different coordinate system.

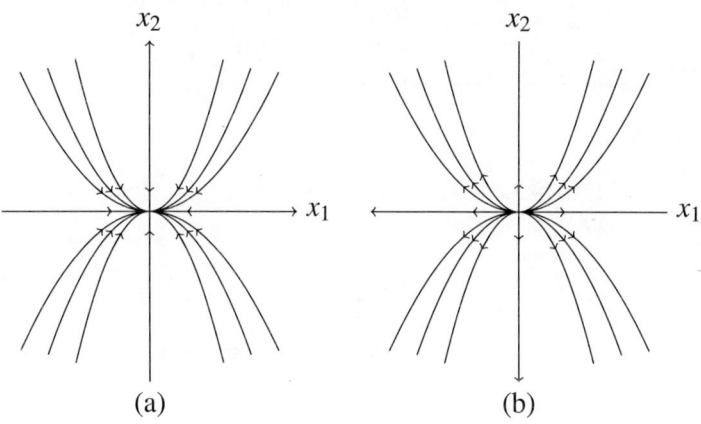

(a)   (b)

Fig. 5.4   (a) Stable node, (b) Unstable node

These two examples cover matrices of the form $\mathbf{B}_1$ in Theorem 5.3.1. Now, we consider an example with a double real eigenvalue, but which is not diagonalizable.

Example 5.4.5

Now, consider $\mathbf{A} = \begin{bmatrix} \lambda & 1 \\ 0 & \lambda \end{bmatrix}$.

The solution is given by

$$x_1(t) = (x_{01} + x_{02}t)e^{\lambda t}, \; x_2(t) = x_{02}e^{\lambda t}.$$

Note that both the solutions tend to $\pm\infty$ or 0 according as $\lambda > 0$ or $\lambda < 0$, respectively. For very large t, the coefficient factor t does not matter in analysing the behaviour of the solution (why?). But near origin, the shape may vary depending on the initial conditions; however, this will not affect the stability. This equilibrium is again known as *node* and it is stable/unstable according as $\lambda < 0/\lambda > 0$. The comments regarding the change in direction as well as for any system linearly equivalent to this system remain the same. A phase portrait is depicted as in Fig. 5.5.

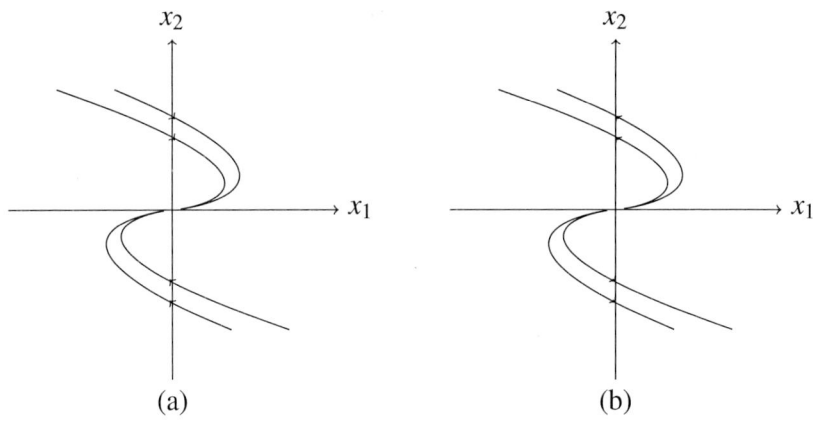

Fig. 5.5   (a) Stable node, (b) Unstable node

Now consider the case with complex eigenvalues.

Example 5.4.6

Let $\mathbf{A} = \begin{bmatrix} a & -b \\ b & a \end{bmatrix}$

The solution is $\mathbf{x}(t) = e^{at} \begin{bmatrix} \cos(bt) & -\sin(bt) \\ \sin(bt) & \cos(bt) \end{bmatrix} \mathbf{x}_0$. Indeed, the sign of a will determine the stability; the components of the matrix appearing in the solution are periodic with the sign of b determining the orientation of the rotation. Of course, we take $b \neq 0$ to get the complex (non-real) eigenvalues.

Case (i), $a = 0$:   Note that the matrix $\mathbf{C} = \begin{bmatrix} \cos(bt) & -\sin(bt) \\ \sin(bt) & \cos(bt) \end{bmatrix}$ shows the periodic nature of the trajectories, rotating around the origin. In fact, $\mathbf{C}$ is a rotation matrix with determinant 1. Thus, we have $|\mathbf{x}(t)| = |\mathbf{C}\mathbf{x}_0| = |\mathbf{x}_0|$ for all t. In other words, $\mathbf{x}(t)$ rotates around the origin along the circle of radius $|\mathbf{x}_0|$ as t increases or decreases. The rotation will be clockwise if $b < 0$ and it is counter-clockwise if $b > 0$. In this case, the equilibrium point $\mathbf{0}$ is referred to as a *center*. See Fig. 5.6.

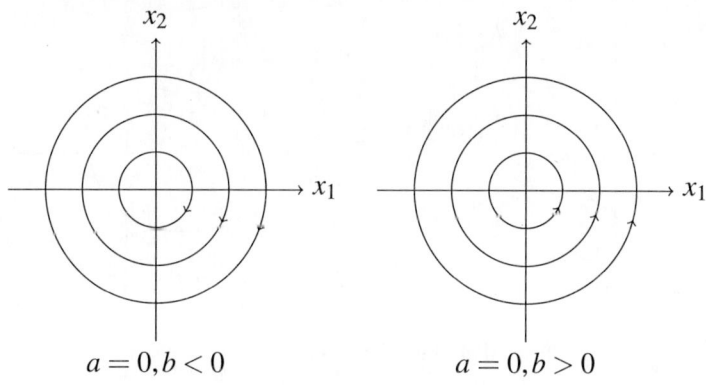

$$a = 0, b < 0 \qquad\qquad\qquad a = 0, b > 0$$

Fig. 5.6   Centre

Case (ii), $a \neq 0$:   Here also the rotation matrix $\mathbf{C}$ acts the same way, but the presence of e^{at} changes amplitude making a spiral around the origin. The spiral moves towards infinity as time increases if $a > 0$ and it tends to the origin, if $a < 0$. The situation leads to four different cases and these are depicted in Fig. 5.7 and Fig. 5.8. The equilibrium point in this case is referred to as a *focus*. It is a stable/unstable focus depending on whether $a < 0 / a > 0$, respectively.

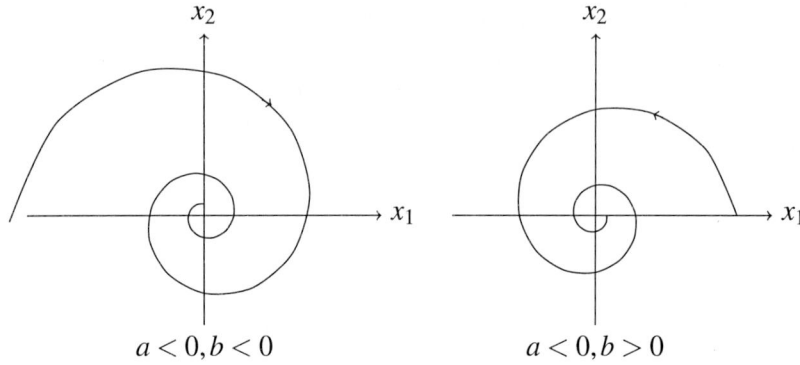

Fig. 5.7   Stable focus

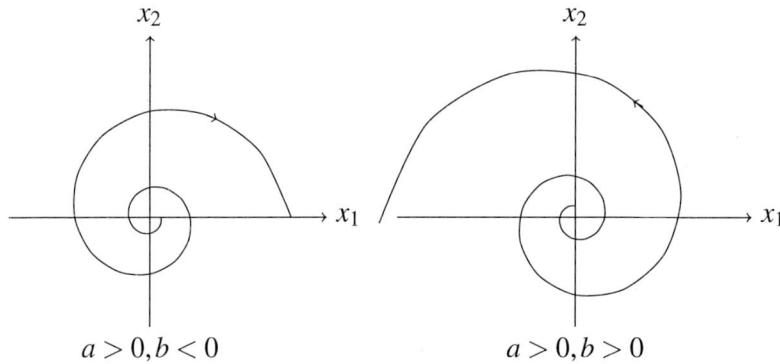

Fig. 5.8   Unstable focus

So far, we have considered only cases of non-zero eigenvalues, that is, when the determinant of $\mathbf{A}$ is non-zero. We now consider the degenerate case, namely, $\det(\mathbf{A}) = 0$. The set of equilibrium points is $\ker(\mathbf{A})$, whose dimension is one or two. If $\dim(\ker(\mathbf{A})) = 2$, then $\mathbf{A}$ is the zero matrix. In this case, every point in the plane is an equilibrium point. We will now describe two examples when $\dim(\ker(\mathbf{A})) = 1$ exhibiting different behaviours.

Example 5.4.7

Take the matrix $\mathbf{A} = \begin{bmatrix} 0 & 0 \\ 0 & -2 \end{bmatrix}$. Then, $x_1(t) = x_{01}$, $x_2(t) = x_{02}e^{-2t}$.

In this degenerate case, where one eigenvalue is zero, all the points on the x_1-axis are equilibrium points. Figure 5.9 is self explanatory. However, note that since the eigenvalues are distinct, **A** has two linearly independent eigenvectors. Now, consider $\mathbf{A} = \begin{bmatrix} 0 & 1 \\ 0 & 0 \end{bmatrix}$. The double eigenvalue 0 has geometric multiplicity one. The reader should work out the further details and observe the different behaviour in this case compared to the previous example.

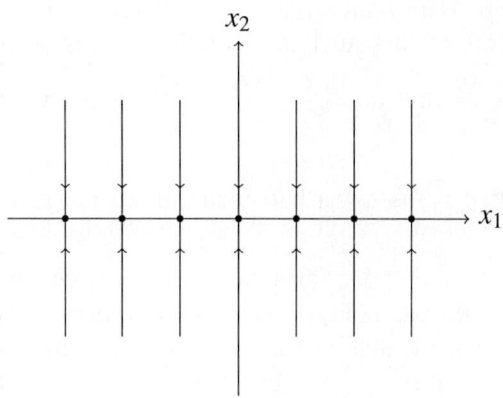

Fig. 5.9   Degenerate case

We now summarize the various types of equilibrium points discussed so far in the following definition.

Definition 5.4.8

Let the 2×2 matrix **A** be linearly equivalent to **B**. Then, the equilibrium point 0 of the linear system (5.2.2) is said to be a

1. saddle point if $\mathbf{B} = \begin{bmatrix} \lambda & 0 \\ 0 & \mu \end{bmatrix}$ with $\lambda\mu < 0$;

2. node if $\mathbf{B} = \begin{bmatrix} \lambda & 0 \\ 0 & \mu \end{bmatrix}$ with $\lambda\mu > 0$ or $\mathbf{B} = \begin{bmatrix} \lambda & 1 \\ 0 & \lambda \end{bmatrix}$, $\lambda \neq 0$;

3. focus if $\mathbf{B} = \begin{bmatrix} a & -b \\ b & a \end{bmatrix}$ with $a \neq 0$, $b \neq 0$;

4. center if $\mathbf{B} = \begin{bmatrix} 0 & -b \\ b & 0 \end{bmatrix}$, $b \neq 0$.

A stable node or a focus is also called a **sink** and an unstable node or focus is called a **source**. If $\det(\mathbf{A}) = 0$, then the origin is called a **degenerate equilibrium point.**

Bifurcation Diagram: We now describe an interesting bifurcation diagram in the $\alpha\delta$ plane, where $\alpha = \text{tr}(\mathbf{A}) = \lambda + \mu$ and $\delta = \det(\mathbf{A}) = \lambda\mu$. Here λ and μ are the eigenvalues of $\mathbf{A}$. We know that λ, μ are the roots of the quadratic equation $z^2 - \alpha z + \delta = 0$, that is, $\lambda = \dfrac{\alpha + \sqrt{\alpha^2 - 4\delta}}{2}$ and $\mu = \dfrac{\alpha - \sqrt{\alpha^2 - 4\delta}}{2}$. It is easy to analyse the various cases.

1. If $\delta < 0$, that is the lower half plane in $\alpha\delta$ plane, we have saddle points.

2. Consider now the upper half plane, which divides into two regions separated by the parabola $\Delta \equiv \alpha^2 - 4\delta = 0$. In the region $\alpha^2 - 4\delta > 0$, we get a node which, in fact, is separated by the δ-axis to obtain two regions. For $\alpha < 0$, we get a stable node and for $\alpha > 0$, it is an unstable node.

3. Similarly, the region $\alpha^2 - 4\delta < 0$ with $\delta > 0$ has two components, where the equilibrium point is a focus, giving us a stable focus for $\alpha < 0$ and an unstable focus for $\alpha > 0$.

4. In the positive δ-axis, we have the center equilibrium, whereas the α-axis is the case where at least one of the eigenvalues is zero, thus falling into a degenerate case.

A schematic representation (bifurcation diagram) is shown in Fig. 5.10.

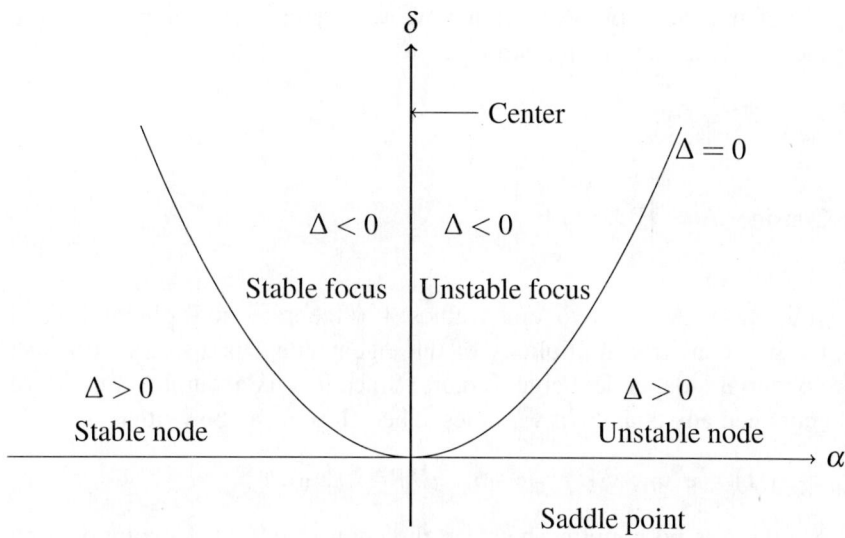

Fig. 5.10   Bifurcation diagram

5.5   Higher Dimensional Systems

In a higher dimensional system, the number of possibilities, depending on the nature of eigenvalues of the matrix, are naturally many more compared to a two-dimensional system. This naturally makes the analysis more involved. Some such possibilities are: the real and complex eigenvalues may occur *together*; the deficiency of eigenvalues may be zero or positive. Accordingly, **A** will be linearly equivalent to a matrix **B** consisting of different blocks. This is given by the Jordan decomposition. For example, when $n = 3$, the following possibilities occur:

$$\begin{bmatrix} \lambda_1 & 0 & 0 \\ 0 & \lambda_2 & 0 \\ 0 & 0 & \lambda_3 \end{bmatrix}, \begin{bmatrix} \lambda_1 & 0 & 0 \\ 0 & \lambda_2 & 1 \\ 0 & 0 & \lambda_2 \end{bmatrix}, \begin{bmatrix} \lambda_1 & 1 & 0 \\ 0 & \lambda_1 & 1 \\ 0 & 0 & \lambda_1 \end{bmatrix}, \begin{bmatrix} \lambda_1 & 0 & 0 \\ 0 & a & -b \\ 0 & b & a \end{bmatrix}$$

according to the case when **A** has eigenvalues; 3 real (need not be distinct) with 3 independent eigenvectors, 3 real with only two independent eigenvectors, 3 real with only a single independent eigenvector, a real and two complex eigenvalues, respectively.

It is not very difficult to consider all possibilities with different signs and sketch various phase portraits. We will see a few examples and the reader can work out the remaining cases.

Example 5.5.1

Consider $\mathbf{A} = \begin{bmatrix} 1 & 0 & 0 \\ 0 & 1 & 0 \\ 0 & 0 & -1 \end{bmatrix}$.

In this case, $\mathbf{A}$ has two eigenvalues 1 (algebraic multiplicity 2) and -1. The geometric multiplicity of the eigenvalue 1 is also 2 and hence, we obtain 3 independent eigenvectors. Since it is a decoupled system, we do not need any change of variables. The solution can be written as

$$x_1(t) = e^t x_{01}, \; x_2(t) = e^t x_{02}, \; x_3(t) = e^{-t} x_{03}.$$

Since there is no coupling between the components of the solution, any trajectory whose initial condition is in any of the axes or any of the planes determined by the axes, will remain in the same axis or plane. So, if we consider the $x_1 x_2$-plane, the trajectories appear like the planar trajectories corresponding to an unstable node (unstable equilibrium), whereas in $x_1 x_3$- or $x_2 x_3$-planes, the origin appears like a saddle type equilibrium. The phase portrait is shown in Fig. 5.11. Further, any trajectory starting from the x_1- or x_2-axis, will remain there and tends to $\pm\infty$, as $t \to \infty$, whereas the trajectories originating from x_3-axes will tend to the origin.

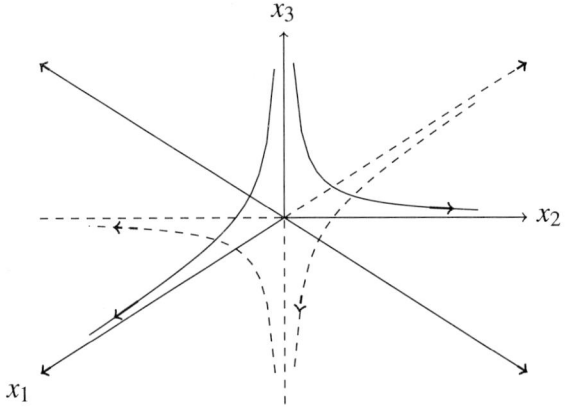

Fig. 5.11   Phase portrait of a 3×3 system

Example 5.5.2

Consider an example, again, a three-dimensional system, having one real and one complex eigenvalue; of course, the complex eigenvalues occur in conjugate pairs. For simplicity, we consider the system in the transformed form, namely $\mathbf{A} = \begin{bmatrix} a & -b & 0 \\ b & a & 0 \\ 0 & 0 & \lambda \end{bmatrix}$.

The matrix $\mathbf{A}$ has a real eigenvalue λ and two complex eigenvalues $a \pm ib$, with $b \neq 0$. As the x_3 component is separated from the other coupled components, the system is reduced to a single equation and a 2×2 system. The solution can be written as

$$\begin{bmatrix} x_1(t) \\ x_2(t) \end{bmatrix} = e^{ta} \begin{bmatrix} \cos(tb) & -\sin(tb) \\ \sin(tb) & \cos(tb) \end{bmatrix} \begin{bmatrix} x_{01} \\ x_{02} \end{bmatrix}, \quad x_3(t) = e^{\lambda t} x_{03},$$

that is,

$$x_1(t) = e^{ta}(x_{01}\cos(tb) - x_{02}\sin(tb)), \ x_2(t) = e^{ta}(x_{01}\sin(tb) + x_{02}\cos(tb)).$$

Now consider the same with special eigenvalues. The reader may work out the same problem with different signs for a, b and λ. Let $a > 0, b < 0$ and $\lambda > 0$. If we take the initial point in the $x_1 x_2$-plane, that is, $x_{03} = 0$, then the entire trajectory will remain in the $x_1 x_2$-plane and it is similar to a planar trajectory corresponding to an unstable focus, rotating clockwise as $b < 0$, with increasing amplitude of the spiral as $a > 0$. On the other hand, if the initial point is on the x_3-axis, that is, $x_{01} = 0 = x_{02}$, then the trajectory will remain on the x_3-axis and, since $\lambda > 0$, tend to $\pm\infty$ along the x_3-axis as $t \to \infty$, according to whether x_{03} is positive or negative. So, if we put both these arguments together, for a general initial point, the trajectory will spiral around the x_3-axis, increasing the distance of the spiral from the x_3-axis, but moving towards $\pm\infty$. See Fig. 5.12(b). As another case, if we take $a = 0$ and $b > 0$, we get spirals around the x_3-axis, maintaining the same distance from the x_3-axis, moving towards $\pm\infty$, since $\lambda > 0$, but in the counter-clockwise direction since $b > 0$. See Fig. 5.12(a).

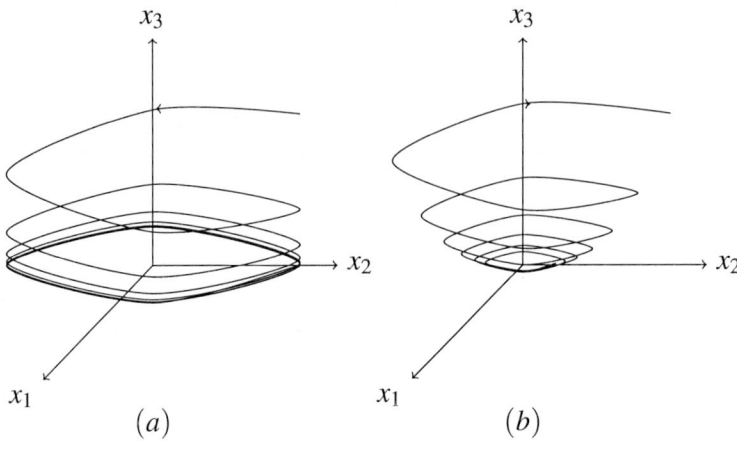

Fig. 5.12   Phase portrait of a 3×3 system

In general, in the case of distinct eigenvalues, real or complex, we can transform a matrix to a diagonal matrix or block diagonal matrix with 2×2 blocks on the diagonal. Note that if the complex eigenvalue is $a + ib$, we obtain the 2×2 block $\begin{bmatrix} a & -b \\ b & a \end{bmatrix}$.

Theorem 5.5.3

Suppose **A** has n distinct real eigenvalues $\lambda_1, \lambda_2, \cdots, \lambda_n$ with eigenvectors $\mathbf{v}_1, \mathbf{v}_2, \cdots, \mathbf{v}_n$, then, the matrix $\mathbf{P} = \begin{bmatrix} \mathbf{v}_1 & \mathbf{v}_2 & \cdots & \mathbf{v}_n \end{bmatrix}$ is invertible, $\mathbf{P}^{-1}\mathbf{A}\mathbf{P} = \text{diag}[\lambda_1, \lambda_2, \cdots \lambda_n]$ and the solution $\mathbf{x}(t)$ of the system (5.2.1) is given by

$$\mathbf{x}(t) = \mathbf{P} \, \text{diag} \left[e^{\lambda_1 t}, e^{\lambda_2 t}, \cdots e^{\lambda_n t} \right] \mathbf{P}^{-1}\mathbf{x}_0.$$

Theorem 5.5.4

Suppose **A** is of even order $n = 2k$ and all the eigenvalues are complex and distinct given by $\lambda_j = a_j + ib_j$, $\bar{\lambda}_j = a_j - ib_j$, $j = 1, \cdots, k$, with complex eigenvectors $\mathbf{w}_j = \mathbf{u}_j + i\mathbf{v}_j$, $\overline{\mathbf{w}}_j = \mathbf{u}_j - i\mathbf{v}_j$. Then, $\mathbf{P} = \begin{bmatrix} \mathbf{v}_1 & \mathbf{u}_1 & \mathbf{v}_2 & \mathbf{u}_2 & \cdots & \mathbf{v}_n & \mathbf{u}_n \end{bmatrix}$ is invertible, $\mathbf{P}^{-1}\mathbf{A}\mathbf{P} = \text{diag}\begin{bmatrix} \mathbf{B}_1, \cdots, \mathbf{B}_n \end{bmatrix}$, where each $\mathbf{B}_j$ is a 2×2 matrix given by $\mathbf{B}_j = \begin{bmatrix} a_j & -b_j \\ b_j & a_j \end{bmatrix}$.

In this case, the solution of (5.2.1) is given by

$$\mathbf{x}(t) = \mathbf{P} \operatorname{diag} \left[e^{t\mathbf{B}_1}, \cdots, e^{t\mathbf{B}_n} \right] \mathbf{P}^{-1}\mathbf{x}_0.$$

Note that $e^{t\mathbf{B}_j} = e^{a_j t} \begin{bmatrix} \cos(b_j t) & -\sin(b_j t) \\ \sin(b_j t) & \cos(b_j t) \end{bmatrix}.$

Further, observe that $\begin{bmatrix} \cos(b_j t) & -\sin(b_j t) \\ \sin(b_j t) & \cos(b_j t) \end{bmatrix}$ represents a pure rotation.

Sylvester's Formula: When the eigenvalues are distinct, we have an easier formula to compute $e^{t\mathbf{A}}$. Let $\lambda_1, \lambda_2, \cdots, \lambda_n$ be the distinct eigenvalues of $\mathbf{A}$. Let $\mathbf{P}_i$ be the projection operator from $\mathbb{R}^n$ onto the kernel, $\ker(\mathbf{A} - \lambda_i \mathbf{I})$ for $i = 1, \cdots, n$. Then, it is easy to see that

$$\mathbf{P}_1 + \cdots + \mathbf{P}_n = \mathbf{I}$$
$$\lambda_1 \mathbf{P}_1 + \cdots + \lambda_n \mathbf{P}_n = \mathbf{A}$$
$$\cdots \quad \cdots \quad \cdots$$
$$\lambda_1^{n-1} \mathbf{P}_1 + \cdots + \lambda_n^{n-1} \mathbf{P}_n = \mathbf{A}^{n-1}.$$

Since the coefficient matrix of this system of equations is a Vandermonde matrix, $\mathbf{P}_1, \cdots, \mathbf{P}_n$ can be obtained as

$$\mathbf{P}_i = \prod_{j=1, j \neq i}^{n} \frac{\mathbf{A} - \lambda_j \mathbf{I}}{\lambda_i - \lambda_j}$$

for $i = 1, \cdots, n$. Finally,

$$e^{t\mathbf{A}} = \exp\left(t \sum_{j=1}^{n} \lambda_j \mathbf{P}_j \right).$$

We leave it as an exercise to the reader to show (using the properties of projection like $\mathbf{P}_j = \mathbf{P}_j^k$ for any $k = 2, \cdots$) that

$$e^{t\mathbf{A}} = \sum_{j=1}^{n} e^{t\lambda_j} \mathbf{P}_j.$$

This is known as Sylvester's formula.

Multiple Eigenvalues: In the case of multiple eigenvalues, a diagonalization may not be possible. One of the major difficulties in dealing with the eigenvalues of multiplicity > 1, is that we may not get enough (as many as the algebraic multiplicity) number of eigenvectors to form a basis of $\mathbb{R}^n$. To see this, let $\mathbf{A} = \begin{bmatrix} 1 & 0 \\ 0 & 1 \end{bmatrix}$ have the eigenvalue 1 with multiplicity (algebraic) 2 with two independent eigenvectors (correct number). But for the matrix $\begin{bmatrix} 1 & 1 \\ 0 & 1 \end{bmatrix}$, the eigenvalue 1 has the same algebraic multiplicity 2; however, it has only one independent eigenvector. In other words, the dimension of $\ker(\mathbf{A} - \lambda I)$ may not be the same as the algebraic multiplicity. The $\dim(\ker(\mathbf{A} - \lambda I))$ is known as the *geometric multiplicity* of the eigenvalue λ. The difference between *algebraic multiplicity* and *geometric multiplicity* is called the *deficiency index* of the eigenvalue.

However, every matrix $\mathbf{A}$ can be transformed into a matrix $\mathbf{B}$ which can be decomposed into a *diagonalizable part* and a *nilpotent part*.

Definition 5.5.5

A matrix $\mathbf{N}$ is said to be *nilpotent* of order k if there exists an integer $k \geq 1$ such that $\mathbf{N}^{k-1} \neq \mathbf{0}$ and $\mathbf{N}^k = \mathbf{0}$.

For a nilpotent matrix $\mathbf{N}$ of order k, we have $e^{\mathbf{N}} = \sum_{i=0}^{k-1} \dfrac{\mathbf{N}^i}{i!}$. For example, the matrix $\mathbf{N} = \begin{bmatrix} 0 & 1 \\ 0 & 0 \end{bmatrix}$ is nilpotent of order 2 and $e^{\mathbf{N}} = I + \mathbf{N} = \begin{bmatrix} 1 & 1 \\ 0 & 1 \end{bmatrix}$. The $n \times n$ matrix whose only non-zero entries are 1 along the first off-diagonal, that is

$$\mathbf{N} = \begin{bmatrix} 0 & 1 & 0 & \cdots & 0 \\ 0 & 0 & 1 & \cdots & 0 \\ \cdots & \cdots & \cdots & \cdots & 1 \\ 0 & \cdots & \cdots & \cdots & 0 \end{bmatrix}$$

is nilpotent of order n. More generally, any strictly upper (lower) triangular matrix is also nilpotent.

If one or more eigenvalues of a matrix $\mathbf{A}$ have positive deficiency indices, then, we will not get sufficient number of eigenvectors of $\mathbf{A}$ to form a basis for $\mathbb{R}^n$. This makes it necessary to find some more linearly independent vectors, of course related to $\mathbf{A}$ so that the eigenvectors and the additional vectors put together form a basis for $\mathbb{R}^n$. This possibly may transform the matrix $\mathbf{A}$ to a simpler form though not in a diagonal form. More precisely, we may not be able to decouple the system completely, but it is still possible to convert the system into decoupled sub-systems. Further, we would like to have these smaller sub-systems in a form as simple as possible. This is essentially carried out in the Jordan decomposition. The main concept is the introduction of *generalized eigenvectors* and the choice of these vectors in a clever order.

Definition 5.5.6

[Generalized eigenvector] Let λ be an eigenvalue, then any vector $\mathbf{v}$ satisfying $(\mathbf{A} - \lambda \mathbf{I})^k \mathbf{v} = 0$ for some $k \geq 1$ is called a generalized eigenvector; if $k = 1$, it is the usual eigenvector.

It is a known fact that the smallest such k is less than or equal to the algebraic multiplicity of λ. For example, in $\begin{bmatrix} 1 & 1 \\ 0 & 1 \end{bmatrix}$, the vector $\begin{bmatrix} 1 \\ 0 \end{bmatrix}$ is an eigenvector. Since, we are in the second dimension and the algebraic multiplicity of the eigenvalue 1 is 2, the matrix $\mathbf{A}$ satisfies $(\mathbf{A} - \mathbf{I})^2 \mathbf{v} = 0$ for any vector $\mathbf{v}$. Hence, any vector can be taken as a generalized eigenvector. We now state the following theorem without proof.

Theorem 5.5.7

Let $\lambda_1, \lambda_2, \cdots, \lambda_n$ be real eigenvalues of an $n \times n$ matrix $\mathbf{A}$ counted according to their (algebraic) multiplicity. Then, there exists an invertible matrix $\mathbf{P} = \begin{bmatrix} \mathbf{v}_1 & \mathbf{v}_2 & \cdots & \mathbf{v}_n \end{bmatrix}$ consisting of generalized eigenvectors of $\mathbf{A}$ such that $\mathbf{A} = \mathbf{S} + \mathbf{N}$, where $\mathbf{S}$ is diagonalizable using $\mathbf{P}$, that is $\mathbf{P}^{-1}\mathbf{S}\mathbf{P} = \text{diag}[\lambda_1, \cdots, \lambda_n]$ and $\mathbf{N} = \mathbf{A} - \mathbf{S}$ is nilpotent of order k less than or equal to n. Further, $\mathbf{SN} = \mathbf{NS}$.

Since $\mathbf{S}$ and $\mathbf{N}$ commute, the solution to the linear system (5.2.1) is given by

$$\mathbf{x}(t) = e^{t(\mathbf{S}+\mathbf{N})}\mathbf{x}_0 = e^{t\mathbf{S}}e^{t\mathbf{N}}\mathbf{x}_0$$

$$= \mathbf{P}\,\text{diag}\left[e^{\lambda_1 t}, \cdots, e^{\lambda_n t}\right]\mathbf{P}^{-1}\left[I + t\mathbf{N} + \cdots + \frac{t^{k-1}\mathbf{N}^{k-1}}{(k-1)!}\right]\mathbf{x}_0.$$

$$(5.5.1)$$

Example 5.5.8

Solve the linear system with $A = \begin{bmatrix} 3 & 1 \\ -1 & 1 \end{bmatrix}$.

We leave the details to the reader. The eigenvalues are $\lambda_1 = \lambda_2 = 2$. It is easy to see that the geometric multiplicity of the eigenvalue 2 is one and we may choose $\begin{bmatrix} 1 \\ -1 \end{bmatrix}$ as a corresponding eigenvector. Any other vector is a generalized eigenvector, but we choose it so that it is linearly independent of $\begin{bmatrix} 1 \\ -1 \end{bmatrix}$, say, $\begin{bmatrix} 1 \\ 0 \end{bmatrix}$. Hence, $\mathbf{P} = \begin{bmatrix} 1 & 1 \\ -1 & 0 \end{bmatrix}$ and $\mathbf{P}^{-1} = \begin{bmatrix} 0 & -1 \\ 1 & 1 \end{bmatrix}$.

Now $\mathbf{S}$ is given by $\mathbf{S} = \mathbf{P} \begin{bmatrix} 2 & 0 \\ 0 & 2 \end{bmatrix} \mathbf{P}^{-1} = \begin{bmatrix} 2 & 0 \\ 0 & 2 \end{bmatrix}$ and hence, $\mathbf{N} = \mathbf{A} -$

$\mathbf{S} = \begin{bmatrix} 1 & 1 \\ -1 & -1 \end{bmatrix}$ and $\mathbf{N}^2 = 0$, that is, $\mathbf{N}$ is nilpotent of order 2. Hence,

$$\mathbf{x}(t) = \mathbf{P} e^{2t} \mathbf{P}^{-1} [I + t\mathbf{N}] \mathbf{x}_0 = e^{2t} \begin{bmatrix} 1+t & t \\ -t & 1-t \end{bmatrix} \mathbf{x}_0.$$

Component-wise, we have

$$x_1(t) = e^{2t}((1+t)x_{01} + x_{02}), \quad x_2(t) = e^{2t}(-tx_{01} + (1-t)x_{02}).$$

Example 5.5.9

Let $A = \begin{bmatrix} 1 & 0 & 0 \\ -1 & 2 & 0 \\ 1 & 1 & 2 \end{bmatrix}$.

The eigenvalues are $\lambda_1 = 1, \lambda_2 = \lambda_3 = 2$ and the algebraic multiplicities of the eigenvalues 1 and 2 are, respectively one and two. Choose $\mathbf{v}_1 = \begin{bmatrix} 1 \\ 1 \\ -2 \end{bmatrix}$ and $\mathbf{v}_2 = \begin{bmatrix} 0 \\ 0 \\ 1 \end{bmatrix}$ as the corresponding eigenvectors 1 and 2. It is not hard to verify that the geometric multiplicity of the eigenvalue 2 is one. It is straightforward to verify that $\dim(\ker(\mathbf{A} - 2\mathbf{I})^2) = 2$, and thus, another vector from $\ker(\mathbf{A} - 2\mathbf{I})^2$, say $\mathbf{v}_3 = \begin{bmatrix} 0 \\ 1 \\ 0 \end{bmatrix}$ has to be chosen linearly independent of $\mathbf{v}_2$; of course, it is also linearly independent of $\mathbf{v}_1$.

Thus, $\mathbf{P} = \begin{bmatrix} 1 & 0 & 0 \\ 1 & 0 & 1 \\ -2 & 1 & 0 \end{bmatrix}$, $\mathbf{P}^{-1} = \begin{bmatrix} 1 & 0 & 0 \\ 2 & 0 & 1 \\ -1 & 1 & 0 \end{bmatrix}$. Now compute

$\mathbf{S} = \mathbf{P} \begin{bmatrix} 1 & 0 & 0 \\ 0 & 2 & 0 \\ 0 & 0 & 2 \end{bmatrix} \mathbf{P}^{-1} = \begin{bmatrix} 1 & 0 & 0 \\ -1 & 2 & 0 \\ 2 & 0 & 2 \end{bmatrix}$ and $\mathbf{N} = \mathbf{A} - \mathbf{S} = \begin{bmatrix} 0 & 0 & 0 \\ 0 & 0 & 0 \\ -1 & 1 & 0 \end{bmatrix}$,

$\mathbf{N}^2 = 0$. The solution is given by

$$\mathbf{x}(t) = \begin{bmatrix} e^t & 0 & 0 \\ e^t - e^{2t} & e^{2t} & 0 \\ -2e^t + (2-t)e^{2t} & te^{2t} & e^{2t} \end{bmatrix} \mathbf{x}_0.$$

If all the generalized eigenvectors are complex, we have the following theorem.

Theorem 5.5.10

Let $\mathbf{A}$ be a $2k \times 2k$ matrix with complex eigenvalues $\lambda_j = a_j + ib_j$, $\bar{\lambda}_j = a_j - ib_j$, $j = 1, \cdots, k$. Then, there exist generalized complex eigenvectors $\mathbf{w}_j = \mathbf{u}_j + i\mathbf{v}_j$, $\overline{\mathbf{w}}_j = \mathbf{u}_j - i\mathbf{v}_j$, $j = 1, 2, ..., k$ such that $\{\mathbf{v}_1, \mathbf{u}_1, \mathbf{v}_2, \mathbf{u}_2, \cdots, \mathbf{v}_k, \mathbf{u}_k\}$ form a basis of $\mathbb{R}^{2k}$ and

$$\mathbf{P} = \begin{bmatrix} \mathbf{v}_1 & \mathbf{u}_1 & \mathbf{v}_2 & \mathbf{u}_2 & \cdots & \mathbf{v}_k & \mathbf{u}_k \end{bmatrix}$$

is invertible. Further, $\mathbf{A} = \mathbf{S} + \mathbf{N}$, where $\mathbf{P}^{-1}\mathbf{SP} = \text{diag}[\mathbf{B}_1, \cdots, \mathbf{B}_k]$, where each $\mathbf{B}_j$ is a 2×2 matrix given by $\mathbf{B}_j = \begin{bmatrix} a_j & -b_j \\ b_j & a_j \end{bmatrix}$ and $\mathbf{N} = \mathbf{A} - \mathbf{S}$ is nilpotent of order m less than or equal to $2k$ and $\mathbf{SN} = \mathbf{NS}$.

Thus, in the case of all eigenvalues complex, the solution to (5.2.1) is given by

$$\mathbf{x}(t) = \mathbf{P}\, \text{diag} \begin{bmatrix} e^{t\mathbf{B}_1}, \cdots, e^{t\mathbf{B}_k} \end{bmatrix} \mathbf{P}^{-1} \left[\mathbf{I} + t\mathbf{N} + \cdots + \frac{t^{m-1}\mathbf{N}^{m-1}}{(m-1)!} \right] \mathbf{x}_0.$$

Here $e^{t\mathbf{B}_j} = e^{a_j t} \begin{bmatrix} \cos(b_j t) & -\sin(b_j t) \\ \sin(b_j t) & \cos(b_j t) \end{bmatrix}$.

The final result is the Jordan form for a general matrix $\mathbf{A}$.

[The Jordan Canonical Form] Let $\mathbf{A}$ be a real matrix of order $n = k + 2m$ with real eigenvalues $\lambda_1, \lambda_2, \cdots, \lambda_k$ and complex eigenvalues $\lambda_j = a_j + ib_j$, $\bar{\lambda}_j = a_j - ib_j$, $j = k + 1, \cdots, m$. Then, there exists a basis $\{\mathbf{v}_1, \mathbf{v}_2, \cdots, \mathbf{v}_k, \mathbf{v}_{k+1}, \mathbf{u}_{k+1}, \cdots, \mathbf{v}_{k+m}, \mathbf{u}_{k+m}\}$ of $\mathbb{R}^n$, where $\mathbf{v}_j$, $j = 1, 2, \cdots, k$, $\mathbf{w}_j = \mathbf{u}_j + i\mathbf{v}_j$, $j = k + 1, \cdots, k + m$ are generalized eigenvectors corresponding to the eigenvalues λ_j, such that

$$\mathbf{P} = \begin{bmatrix} \mathbf{v}_1 & \mathbf{v}_2 & \cdots & \mathbf{v}_k & \mathbf{v}_{k+1} & \mathbf{u}_{k+1} & \cdots & \mathbf{v}_{k+m} & \mathbf{u}_{k+m} \end{bmatrix}$$

is invertible and $\mathbf{P}^{-1}\mathbf{A}\mathbf{P} = \mathrm{diag}[\mathbf{B}_1, \mathbf{B}_2, \cdots, \mathbf{B}_r]$ is block diagonal with Jordan blocks $\mathbf{B}_j$, $j = 1, \cdots, r$ for some r. Further, $\mathbf{B}_j$ takes one of the following two forms

$$\begin{bmatrix} \lambda & 1 & 0 & \cdots, & 0 & 0 \\ 0 & \lambda & 1 & \cdots & 0 & 0 \\ \cdots & \cdots, & \cdots & \cdots & \cdots & \cdots \\ 0 & \cdots & \cdots & \cdots & \lambda & 1 \\ 0 & \cdots & \cdots & \cdots & 0 & \lambda \end{bmatrix} \quad \text{or} \quad \begin{bmatrix} \mathbf{D} & \mathbf{I}_2 & \mathbf{0}_2 & \cdots & \mathbf{0}_2 & \mathbf{0}_2 \\ \mathbf{0}_2 & \mathbf{D} & \mathbf{I}_2 & \cdots & \mathbf{0}_2 & \mathbf{0}_2 \\ \cdots & \cdots & \cdots & \cdots & \cdots & \cdots \\ \mathbf{0}_2 & \cdots & \cdots & \cdots & \mathbf{D} & \mathbf{I}_2 \\ \mathbf{0}_2 & \cdots & \cdots & \cdots & \mathbf{0}_2 & \mathbf{D} \end{bmatrix}$$

depending on whether $\lambda = \lambda_j$ is real or $\lambda = a_j + ib_j$ is complex, respectively; in the latter case, we have $\mathbf{D} = \begin{bmatrix} a_j & -b_j \\ b_j & a_j \end{bmatrix}$, $\mathbf{I}_2 = \begin{bmatrix} 1 & 0 \\ 0 & 1 \end{bmatrix}$ and $\mathbf{0}_2$ is the 2×2 zero matrix.

Hence, the solution to the IVP (5.2.1) can be written as

$$\mathbf{x}(t) = \exp((t - t_o)\mathbf{A})\mathbf{x}_0 = \mathbf{P} \, \mathrm{diag}\left[\exp((t - t_0)\mathbf{B}_1), \cdots, \right.$$

$$\left. \exp((t - t_0)\mathbf{B}_r)\right] \mathbf{P}^{-1}\mathbf{x}_0.$$

The number r is the sum of the geometric multiplicities of the eigenvalues. Using the Jordan form, we can write the solution component-wise. If $\mathbf{B}$ has the first form, then $\mathbf{B} = \lambda\mathbf{I} + \mathbf{N}$ and for those components

$$
e^{t\mathbf{B}} = e^{t\lambda} e^{t\mathbf{N}} = e^{t\lambda}
\begin{bmatrix}
1 & t & \dfrac{t^2}{2!} & \cdots & \dfrac{t^{m-1}}{(m-1)!} \\
0 & 1 & t & \cdots & \dfrac{t^{m-2}}{(m-2)!} \\
\cdots & \cdots & \cdots & \cdots & \cdots \\
\cdots & \cdots & \cdots & \cdots & \cdots \\
0 & \cdots & \cdots & \cdots & 1
\end{bmatrix} .
$$

On the other hand, if $\mathbf{B}$ has the second form, then

$$
e^{t\mathbf{B}} = e^{ta_j}
\begin{bmatrix}
\mathbf{R} & t\mathbf{R} & \cdots & \cdots & \cdots & \dfrac{t^{m-1}\mathbf{R}}{(m-1)!} \\
\cdots & \cdots & \cdots & \cdots & \cdots & \cdots \\
\cdots & \cdots & \cdots & \cdots & \cdots & \cdots \\
\mathbf{0}_2 & \mathbf{0}_2 & \cdots & \cdots & \mathbf{R} & t\mathbf{R} \\
\mathbf{0}_2 & \mathbf{0}_2 & \cdots & \cdots & \mathbf{0}_2 & \mathbf{R}
\end{bmatrix} ,
$$

where $\mathbf{R} = \begin{bmatrix} \cos(b_j t) & -\sin(b_j t) \\ \sin(b_j t) & \cos(b_j t) \end{bmatrix}$.

Note that each component of the solution $\mathbf{x}(t)$ of the initial value problem is a linear combination of the form $t^k e^{ta} \cos(bt)$ or $t^k e^{at} \sin(bt)$, where $\lambda = a + ib$ is an eigenvalue of $\mathbf{A}$, $0 \le k \le n-1$ (b may be zero or non-zero). This is an important feature of linear systems with constant coefficients.

5.6   Invariant Subspaces under the Flow $e^{t\mathbf{A}}$

We have seen in the examples earlier that when a trajectory starts with initial conditions in certain planes or axes, the solution remained in the same plane or axes without going out of that region. This essentially means that the action of the group given by the flow $e^{t\mathbf{A}}$ on these subspaces is invariant. In other words, these are *invariant subspaces* under group action. This allows us to decompose the entire phase space into smaller subspaces which are invariant under the flow and we can restrict the problem to these smaller subspaces, if necessary, for further analysis.

Definition 5.6.1

Let $\mathbf{A}$ be a given $n \times n$ matrix. A subspace $E \subset \mathbb{R}^n$ is said to be invariant with respect to the flow $e^{t\mathbf{A}}$ if $e^{t\mathbf{A}}(E) \subset E$ for all t. □

Recall Example 5.4.1, where x_1 and x_2 axes are two invariant subspaces, namely $E_1 = \{(x_1, 0), x_1 \in \mathbb{R}\}$ and $E_2 = \{(0, x_2), x_2 \in \mathbb{R}\}$. For any initial condition $(x_{01}, 0) \in E_1$, we have the solution $\mathbf{x}(t) = (x_1(t), 0) \in E_1$ for all t. Further, $x_1(t) \to 0$ as $t \to \infty$. This subspace is referred to as a *stable subspace*. On the other hand, for E_2, any solution which starts in E_2, remains there for all t, but now it goes to $\pm\infty$ as $t \to \infty$. In this case, the subspace is called an *unstable subspace*.

In Example 5.4.4, both axes are unstable invariant subspaces and hence, the entire $\mathbb{R}^2$ space is unstable. In Example 5.5.1, the $x_1 x_2$-plane is the unstable subspace, whereas the x_3-axis is the stable subspace.

The subspace generated by the generalized eigenvectors of an eigenvalue λ of a matrix $\mathbf{A}$ is called the generalized eigenspace corresponding to λ.

Proposition 5.6.2

Let E be a generalized eigenspace corresponding to an eigenvalue λ of a matrix $\mathbf{A}$. Then, E is invariant under the matrix $\mathbf{A}$, that is, $\mathbf{A}(E) \subset E$.

Proof: Let $E = \text{span}\{\mathbf{v}_1, \cdots, \mathbf{v}_k\}$, $\mathbf{v}_i$s are generalized eigenvectors of λ. Hence, there exist positive integers k_j such that $(\mathbf{A} - \lambda \mathbf{I})^{k_j}\mathbf{v}_j = 0$. Now, for any $\mathbf{v} \in E$, we have $\mathbf{v} = \sum_{j=1}^{k} c_j \mathbf{v}_j$ and $\mathbf{A}\mathbf{v} = \sum_{j=1}^{k} c_j \mathbf{A}\mathbf{v}_j$. We will show that $\mathbf{A}\mathbf{v}_j \in E$ for all j and hence, $\mathbf{A}\mathbf{v} \in E$. If $k_j = 1$, then $\mathbf{A}\mathbf{v}_j = \lambda \mathbf{v}_j \in E$. To apply induction assume the result is true for $k_j \geq 2$. If we define $\bar{\mathbf{v}}_j = (\mathbf{A} - \lambda \mathbf{I})\mathbf{v}_j$, then

$$\bar{\mathbf{v}}_j \in \ker(\mathbf{A} - \lambda \mathbf{I})^{k_j - 1}$$

and thus, $\bar{\mathbf{v}}_j$ is a generalized eigenvector and $\bar{\mathbf{v}}_j \in E$. Finally, it follows that $\mathbf{A}\mathbf{v}_j = \lambda \mathbf{v}_j + \bar{\mathbf{v}}_j \in E$. Hence the proposition. □

Now, the entire space $\mathbb{R}^n$ can be decomposed into stable, unstable and center spaces. This is the content of the following theorem. Given an $n \times n$ real matrix $\mathbf{A}$, denote by E^s, E^u and E^c the subspaces spanned by the generalized eigenvectors corresponding to the eigenvalues with negative real parts, positive real parts and zero real parts, respectively. Then, by the generalized spectral theorem, it follows that

$$\mathbb{R}^n = E^s \oplus E^u \oplus E^c.$$

Theorem 5.6.3

The subspaces E^s, E^u and E^c defined earlier, are invariant under the flow e^{tA}, that is, $e^{tA}E^s \subset E^s$, $e^{tA}E^u \subset E^u$ and $e^{tA}E^c \subset E^c$. Further, for any $\mathbf{x}_0 \in E^s$, the solution $e^{tA}\mathbf{x}_0 \to 0$ as $t \to \infty$ and for any $\mathbf{x}_0 \in E^u$, the solution $e^{tA}\mathbf{x}_0 \to 0$ as $t \to -\infty$.

Proof: To see the invariance under the flow, it suffices to consider one of the subspaces, say for E^s. Let $\mathbf{z}$ be a generalized eigenvector, then, by Proposition 5.6.2, it follows that $\mathbf{Az} \in E^s$ and hence, $\mathbf{A}^k\mathbf{z} \in E^s$ for any positive integer k. Therefore, it follows that

$$e^{tA}\mathbf{z} = \lim_{k \mapsto \infty} \sum_{j=0}^{k} \frac{t^j \mathbf{A}^j \mathbf{z}}{k!} \in E^s.$$

Now, any $\mathbf{x} \in E^s$ is a finite linear combination of generalized eigenvectors and since e^{tA} is a linear operator, it follows that $e^{tA}\mathbf{x} \in E^s$. The other statements in the theorem, now follow from the explicit representation of the solution. □

Remark 5.6.4

The subspaces E^s, E^u and E^c are respectively called *stable*, *unstable* and *center* subspaces of the flow. In general, we cannot make any conclusion on the limit of the solution $e^{tA}\mathbf{x}_0$, $\mathbf{x}_0 \in E^c$ as $t \to \pm\infty$. There is an analogous result for nonlinear systems as well and is known as the *stable manifold theorem*. □

5.7   Non-homogeneous, Autonomous Systems

We now consider the non-homogeneous system

$$\dot{\mathbf{x}}(t) = \mathbf{Ax}(t) + \mathbf{g}(t), \quad \mathbf{x}(t_0) = \mathbf{x}_0, \tag{5.7.1}$$

where $\mathbf{g} : \mathbb{R}^n \mapsto \mathbb{R}^n$. We assume $\mathbf{g}$ is continuous. By taking $\mathbf{x}_0 = \mathbf{e}_i$ (canonical basis elements), it is easy to see that the matrix $\mathbf{\Phi}(t) \equiv e^{t\mathbf{A}}$ satisfies the matrix differential equation

$$\dot{\mathbf{\Phi}}(t) = \mathbf{A}\mathbf{\Phi}(t), \ \mathbf{\Phi}(0) = \mathbf{I}, \tag{5.7.2}$$

where $\mathbf{I}$ is the identity matrix. Clearly, $\tilde{\mathbf{\Phi}}(t, t_0) = \mathbf{\Phi}(t - t_0) = e^{(t-t_0)\mathbf{A}}$ will satisfy the same differential system with the initial time at t_0, that is

$$\dot{\tilde{\mathbf{\Phi}}}(t, t_0) = \mathbf{A}\tilde{\mathbf{\Phi}}(t, t_0), \ \tilde{\mathbf{\Phi}}(t_0, t_0) = \mathbf{I}, \tag{5.7.3}$$

and the ith column of $\tilde{\mathbf{\Phi}}$ will satisfy (5.7.1) with $\mathbf{g} = 0$ and $\mathbf{x}(t_0) = \mathbf{e}_i$.

Definition 5.7.1

Any non-singular matrix $\mathbf{\Psi}$ which satisfies the matrix system $\dot{\mathbf{\Psi}} = \mathbf{A}\mathbf{\Psi}$ is called a fundamental matrix. The special fundamental matrix $\tilde{\mathbf{\Phi}}(t, t_0)$ is called the transition matrix corresponding to the linear system (5.2.2). $\square$

Let $\mathbf{\Phi}(t)$ be the fundamental matrix given in (5.7.2) and $\mathbf{C}$ be a non-singular scalar matrix. Then, it is readily seen that $\mathbf{\Psi}(t) = \mathbf{\Phi}(t)\mathbf{C}$ is also a fundamental matrix. Conversely, if $\mathbf{\Psi}$ is a fundamental matrix, it is easy to check that $\mathbf{\Psi}(t) = \mathbf{\Phi}(t)\mathbf{C}$, where $\mathbf{C} = \mathbf{\Psi}(t_0)$, using the uniqueness of the solution of the linear matrix system. Thus, we have the following proposition.

Proposition 5.7.2

If $\mathbf{\Psi}(t) = \mathbf{\Phi}(t)\mathbf{C}$, where $\mathbf{C}$ is a non-singular scalar matrix and $\mathbf{\Phi}$ is the transition matrix, then $\mathbf{\Psi}$ is a fundamental matrix. Conversely, every fundamental matrix $\mathbf{\Psi}$ is of the form $\mathbf{\Psi}(t) = \mathbf{\Phi}(t)\mathbf{C}$, where $\mathbf{C}$ is a non-singular matrix. $\square$

5.7.1 Solution to Non-homogeneous system (Variation of parameters)

We know that $e^{t\mathbf{A}}$ applied to any constant vector, say $\mathbf{v}$ satisfies the homogeneous system (5.2.2). So we cannot expect a solution to the non-homogeneous system (5.7.1) of the form $e^{t\mathbf{A}}\mathbf{v}$. Hence, we can only expect to get a solution to the non-homogeneous system by applying the flow $e^{t\mathbf{A}}$ to a varying vector. Thus, look for a solution of the form $\mathbf{x}(t) = e^{t\mathbf{A}}\mathbf{y}(t)$, where $\mathbf{y}(t)$ is to be determined so that $\mathbf{x}(t)$ satisfies (5.7.1).

A simple computation yields

$$\dot{\mathbf{x}}(t) = \mathbf{A}e^{t\mathbf{A}}\mathbf{y}(t) + e^{t\mathbf{A}}\dot{\mathbf{y}}(t) = \mathbf{A}\mathbf{x}(t) + e^{t\mathbf{A}}\dot{\mathbf{y}}(t).$$

Thus, we need to choose $\mathbf{y}$ which satisfies $e^{t\mathbf{A}}\dot{\mathbf{y}}(t) = \mathbf{g}(t)$. In other words,

$$\mathbf{y}(t) = \mathbf{y}(t_0) + \int_{t_0}^{t} e^{-s\mathbf{A}}\mathbf{g}(s)ds = e^{-t_0\mathbf{A}}\mathbf{x}_0 + \int_{t_0}^{t} e^{-s\mathbf{A}}\mathbf{g}(s)ds.$$

Thus, the solution to (5.7.1) is given by

$$\mathbf{x}(t) = e^{(t-t_0)\mathbf{A}}\mathbf{x}_0 + \int_{t_0}^{t} e^{(t-s)\mathbf{A}}\mathbf{g}(s)\,ds = \mathbf{\Phi}(t-t_0)\mathbf{x}_0 + \int_{t_0}^{t} \mathbf{\Phi}(t-s)\mathbf{g}(s)\,ds.$$

$$(5.7.4)$$

In the case of the finite dimensional linear control theory, we have $\mathbf{g}(t) = \mathbf{B}\mathbf{u}(t)$, where $\mathbf{B}$ is an $n \times r$ matrix and $\mathbf{u}$ is an $r \times 1$ control vector. In this case,

$$\mathbf{x}(t) = \mathbf{\Phi}(t-t_0)\mathbf{x}_0 + \int_{t_0}^{t} \mathbf{\Phi}(t-s)\mathbf{B}\mathbf{u}(s)\,ds. \qquad (5.7.5)$$

Indeed, $\mathbf{u}(t)$ needs to satisfy certain integrability conditions. The formula (5.7.4) also gives a definition of the solution of (5.7.1) in a weak or mild form. This form does not require the differentiability assumption of the solution. This is important in control theory as the strong form of (5.7.1) demands the continuity of the control $\mathbf{u}$, whereas (5.7.5) does not demand smoothness. It is a great advantage to consider non-smooth control in applications like the *bang-bang control*.

5.7.2   Non-autonomous systems

The equation we consider here is

$$\dot{\mathbf{x}}(t) = \mathbf{A}(t)\mathbf{x}(t), \ \mathbf{x}(t_0) = \mathbf{x}_0 \qquad (5.7.6)$$

and its non-homogeneous counterpart is

$$\dot{\mathbf{x}}(t) = \mathbf{A}(t)\mathbf{x}(t) + \mathbf{g}(t), \quad \mathbf{x}(t_0) = \mathbf{x}_0. \tag{5.7.7}$$

Here, $\mathbf{A}(t) = [a_{ij}(t)]$ is a continuous, real $n \times n$ matrix valued function defined on a compact interval $[a,b]$; $\mathbf{g}$ is also a vector valued continuous function defined on $[a,b]$. With $\mathbf{f}(t,\mathbf{x}) = \mathbf{A}(t)\mathbf{x}$, we see that $\mathbf{f}$ is Lipschitz continuous in $\mathbf{x}$ on the $n+1$ dimensional set
$$D = \{(t,\mathbf{x}) : t \in [a,b], \mathbf{x} \in \mathbb{R}^n\}.$$

For, if $(t,\mathbf{x}_1)$ and $(t,\mathbf{x}_2)$ in D, then

$$|\mathbf{f}(t,\mathbf{x}_1) - \mathbf{f}(t,\mathbf{x}_2)| \leq k|\mathbf{x}_1 - \mathbf{x}_2|,$$

where $k = \max\{|\mathbf{A}(t)| : t \in [a,b]\}$ with $|\mathbf{A}(t)| = \max\limits_{1 \leq j \leq n} \sum\limits_{i=1}^{n} |a_{ij}(t)|$.

The existence and uniqueness of a solution to (5.7.7) now follows from the results in Chapter 4. However, the representation like in the autonomous system is not possible, in general, in a simple way like $e^{t\mathbf{A}}$; but the methodology discussed in Section 5.7 can be adapted.

More precisely, the local existence and uniqueness theorem gives a unique solution $\mathbf{x}$ defined in some interval $I = [t_0 - h, t_0 + h] \subset [a,b]$. We claim that $I = [a,b]$. For, if $t \in I$, we have $\mathbf{x}(t) = \mathbf{x}_0 + \int_{t_0}^{t} \mathbf{A}(s)\mathbf{x}(s)\,ds$.
Therefore,

$$|\mathbf{x}(t)| \leq |\mathbf{x}_0| + \int_{t_0}^{t} |\mathbf{A}(s)||\mathbf{x}(s)|\,ds, \quad t \geq t_0.$$

Therefore, by Gronwall's inequality, it follows that

$$|\mathbf{x}(t)| \leq |\mathbf{x}_0| \exp\left(\int_{t_0}^{t} |\mathbf{A}(s)|\,ds\right), \quad t \geq t_0,$$

$$\leq |\mathbf{x}_0| \exp(k(b-a)), \tag{5.7.8}$$

for all $t \in I$. Therefore, the points $(t,\mathbf{x}(t)), t \in I$ remain in a bounded subset of D and the solution can be continued to $[a,b]$, using similar arguments as in Chapter 4.

Remark 5.7.3

Suppose $A(t)$ is defined for all $t \in \mathbb{R}$ and is continuous. Then, the earlier arguments show that the solution to IVP in any *compact* interval is bounded. Therefore, the solution is defined for all $t \in \mathbb{R}$. Uniqueness of the solution follows from the local Lipschitz condition. In particular, if $A(t)$ is bounded, say $|A(t)| \leq k$, for all $t \in \mathbb{R}$, then the solution satisfies the following estimate (see (5.7.8)):

$$|\mathbf{x}(t)| \leq |\mathbf{x}_0| \exp(k|t - t_0|),$$

for all $t \in \mathbb{R}$. However, if $A(t)$ is not bounded, the solution need not be bounded by an exponential, with a linear factor in t in the exponent. For example, the solution of the equation $\dot{x} = 2tx$ is given by $x(t) = ce^{t^2}$, where c is an arbitrary constant. See also Example 5.7.5. □

The same definition of a fundamental matrix can be introduced here as well. Let $\boldsymbol{\Phi}_i(t,t_0)$ be the unique solution of (5.7.6) with $\mathbf{x}(t_0) = \mathbf{e}_i$, $i = 1, \cdots, n$. We remark that $\boldsymbol{\Phi}_i(t,t_0)$ do not have an exponential representation as in the autonomous system. Indeed, $\{\boldsymbol{\Phi}_1(t,t_0), \boldsymbol{\Phi}_2(t,t_0), \cdots, \boldsymbol{\Phi}_n(t,t_0)\}$ is a set of linearly independent vectors in $\mathbb{R}^n$ for fixed t,t_0 and hence, the matrix $\boldsymbol{\Phi}(t,t_0) = [\boldsymbol{\Phi}_1(t,t_0) \quad \boldsymbol{\Phi}_2(t,t_0) \quad \cdots \boldsymbol{\Phi}_n(t,t_0)]$ is invertible. Further, it satisfies $\dot{\boldsymbol{\Phi}}(t,t_0) = A(t)\boldsymbol{\Phi}(t,t_0)$, $\boldsymbol{\Phi}(t_0,t_0) = I$. An important point to note here is that, unlike in the case of an autonomous system, one cannot write the solution at any initial time t_0 by translating the solution at the initial time 0.

A matrix function $\boldsymbol{\Psi}(t)$ is called a *fundamental matrix* if it satisfies the matrix system $\dot{\boldsymbol{\Psi}}(t) = A(t)\boldsymbol{\Psi}(t)$. We immediately see that the matrix $\boldsymbol{\Phi}(t,t_0)$ is a fundamental matrix satisfying the initial condition $\boldsymbol{\Phi}(t_0,t_0) = I$. Further, any fundamental matrix $\boldsymbol{\Psi}$ is given by $\boldsymbol{\Psi}(t) = \boldsymbol{\Phi}(t,t_0)C$, where C is a non-singular constant matrix. In fact, $C = \boldsymbol{\Psi}(t_0)$, that is, $\boldsymbol{\Psi}(t) = \boldsymbol{\Phi}(t,t_0)\boldsymbol{\Psi}(t_0)$ or $\boldsymbol{\Phi}(t,t_0) = \boldsymbol{\Psi}(t)\boldsymbol{\Psi}^{-1}(t_0)$. Indeed, when $A(t) = A$, we have $\boldsymbol{\Phi}(t,t_0) = e^{(t-t_0)A}$ and $\boldsymbol{\Phi}(t,s) = e^{(t-s)A}$. In general, we do not have this advantage. But, we have the following properties

$$\boldsymbol{\Phi}^{-1}(t,t_0) = \boldsymbol{\Phi}(t_0,t)$$

and

$$\boldsymbol{\Phi}(t,s)\boldsymbol{\Phi}(s,t_0) = \boldsymbol{\Phi}(t,t_0), \quad \boldsymbol{\Phi}(t_0,t_0) = I.$$

The last two properties together are known as *semi-group properties* and in this particular ODE system, we have group structure due to the first property explained here. As remarked earlier, when we consider a PDE, the heat equation, for example, we may not get a group structure, but a semi-group structure. We have also noted earlier that it is possible to study ODEs in infinite dimensional spaces such as a Hilbert space, a Banach space, etc.

Now the solution to (5.7.6) is given by

$$\mathbf{x}(t) = \mathbf{\Phi}(t,t_0)\mathbf{x}_0 = \mathbf{\Psi}(t)\mathbf{\Psi}^{-1}(t_0)\mathbf{x}_0.$$

The matrix $\mathbf{\Phi}(t,t_0)$ is known as the *transition matrix*. The solution of the *non-homogeneous system* (5.7.7) is given by

$$
\begin{aligned}
\mathbf{x}(t) &= \mathbf{\Phi}(t,t_0)\mathbf{x}_0 + \int_{t_0}^{t} \mathbf{\Phi}(t,t_0)\mathbf{\Phi}^{-1}(s,t_0)\mathbf{g}(s)\,ds \\
&= \mathbf{\Phi}(t,t_0)\mathbf{x}_0 + \int_{t_0}^{t} \mathbf{\Phi}(t,s)\mathbf{g}(s)\,ds.
\end{aligned}
\tag{5.7.9}
$$

As we have remarked earlier, in general, we do not have an exponential representation of the solution for the non-autonomous system. However, in a special case, we do have such a representation which is given in the following proposition.

Proposition 5.7.4

Let the matrices $\mathbf{A}(t)$, $t \in [a,b]$ satisfy the commutative property $\mathbf{A}(t)\mathbf{A}(s) = \mathbf{A}(s)\mathbf{A}(t)$ for all $s,t,\in [a,b]$. Then, the transition matrix has the representation

$$\mathbf{\Phi}(t,t_0) = \exp\left(\int_{t_0}^{t} \mathbf{A}(s)\,ds \right),$$

for all $t \in [a,b]$ and $t_0 \in [a,b]$.   □

The result is not true in general, in the absence of the aforementioned commutative property.

Example 5.7.5

Let $A(t) = \begin{bmatrix} 1 & 1+t \\ 0 & t \end{bmatrix}$.

It is straightforward to verify that $A(t)$ and $A(s), t \neq s$ do not commute in general. The matrix $B(t) = \int_0^t A(s) \, ds = \begin{bmatrix} t & t+t^2/2 \\ 0 & t^2/2 \end{bmatrix}$.

It can be verified that

$$\exp(B(t)) = \begin{bmatrix} e^t & \mu(t)(e^t - e^{t^2/2}) \\ 0 & e^{t^2/2} \end{bmatrix}$$

with $\mu(t) = \dfrac{2+t}{2-t}$, when $t \neq 2$. When $t = 2$, $\exp(B(t)) = \begin{bmatrix} e^2 & 4e^2 \\ 0 & e^2 \end{bmatrix}$.

The solution of the system $\dot{x}(t) = A(t)x(t)$ is given by $x_2(t) = e^{t^2/2}x_{02}$ and

$$x_1(t) = e^t x_{01} + e^t x_{02} \int_0^t (1+s)e^{(-s+s^2/2)} \, ds,$$

which is different from $\exp(B(t)) \begin{bmatrix} x_{01} \\ x_{02} \end{bmatrix}$.

Wronskian Type Result: In the case of second order linear equations, we have defined the Wronskian of two functions and we have seen that either the Wronskian vanishes identically (dependent functions) or never vanishes (independent functions). In this section, we give a formula connecting the determinant of the fundamental matrix and the trace of $A(t)$ known as *Abel's formula*. Consider $\det \Phi(t, t_0)$, the determinant of $\Phi(t, t_0)$. This is called the *Wronskian* of the solution. Let $\text{tr}(A(t))$ denote the trace of $A(t)$. From the DE satisfied by $\Phi(t, t_0)$, we can derive, for small Δt,

$$\begin{aligned} \Phi(t+\Delta t, t_0) &= \Phi(t, t_0) + \Delta t \dot{\Phi}(t, t_0) + O((\Delta t)^2) \\ &= \Phi(t, t_0) + \Delta t A(t) \Phi(t, t_0) + O((\Delta t)^2) \\ &= (I + \Delta t A(t))\Phi(t, t_0) + O((\Delta t)^2), \end{aligned}$$

which gives

$$\det(\boldsymbol{\Phi}(t+\Delta t, t_0)) = \det(\mathbf{I} + \mathbf{A}(t)\Delta t)\det\boldsymbol{\Phi}(t, t_0) + O((\Delta t)^2).$$

For the first term,[1] we have

$$\det(\mathbf{I} + \mathbf{A}(t)\Delta t) = 1 + \Delta t\,\mathrm{tr}(\mathbf{A}(t)) + O((\Delta t)^2).$$

Thus, as $\Delta t \to 0$, we see that $\dfrac{d}{dt}\det(\boldsymbol{\Phi}(t, t_0)) = \mathrm{tr}(\mathbf{A}(t))\det(\boldsymbol{\Phi}(t, t_0))$. This upon integration, produces Abel's formula

$$\det\boldsymbol{\Phi}(t, t_0) = \exp\left(\int_{t_0}^{t} \mathrm{tr}(\mathbf{A}(s))ds\right).$$

Periodic Coefficients and Floquet Theorem: In many practical applications, the coefficient matrix will have an additional property of periodicity. For example, in the *Mathieu equation* $\ddot{z} + (\lambda - 16d\cos 2t)z = 0$, the coefficient matrix $\mathbf{A}(t)$ (convert to a first order system) is periodic. More generally, equations of *Hill's type*: $\ddot{z} + p(t)z = 0$, $p(t + \pi) = p(t)$ are also periodic equations.

We would like to know whether the solution to the system (5.7.6) is also periodic or not. We do not, in general, get periodicity, but we may get quasi-periodicity. For example, any non-trivial solution of the first order equation $\dot{y} + (a + b\cos t)y = 0$ with a, b real constants, is not periodic unless $a = 0$. On the other hand, any solution of the second order equation $\ddot{y} + a^2 y = 0\,(a \neq 0)$, is periodic.

Assume $\mathbf{A}(t)$ is periodic with period T, that is, $\mathbf{A}(t + T) = \mathbf{A}(t)$ for all t.

Let λ be an eigenvalue of the transition matrix $\boldsymbol{\Phi}(t_0 + T, t_0)$ and $\mathbf{v}$ be a corresponding eigenvector. Then, $\mathbf{x}(t) = \boldsymbol{\Phi}(t, t_0)\mathbf{v}$ is the unique solution to the ODE system in (5.7.6) with the initial condition $\mathbf{x}(t_0) = \mathbf{v}$. Now $\mathbf{x}(t_0 + T) = \boldsymbol{\Phi}(t_0 + T, t_0)\mathbf{v} = \lambda\mathbf{v}$. If $\lambda = 0$, then $\mathbf{x}$ will be a solution to the linear system in (5.7.6) with a zero initial condition at $t_0 + T$ and by uniqueness $\mathbf{x} \equiv 0$. This would imply that $\mathbf{v} = \mathbf{0}$. This contradiction shows that $\lambda \neq 0$.

Now consider $\mathbf{y}(t) = \mathbf{x}(t + T)$. By periodicity of $\mathbf{A}(t)$, $\mathbf{y}$ will satisfy the same linear system with the initial condition $\mathbf{y}(t_0) = \mathbf{x}(t_0 + T) = \lambda\mathbf{v}$,

---

[1]If $\mathbf{A}$ is any $n \times n$ matrix with eigenvalues $\lambda_1, \cdots, \lambda_n$, not necessarily distinct, then $\det(\mathbf{I} + a\mathbf{A}) = (1 + a\lambda_1)\cdots(1 + a\lambda_n)$, for any scalar a.

but $\lambda \mathbf{x}$ will also satisfy the same system with the same initial condition. Hence, by uniqueness, we get $\mathbf{x}(t + T) = \lambda \mathbf{x}(t)$ for all t. This is *quasi-periodicity*. Moreover, since $\lambda \neq 0$, one can choose an α so that $\lambda = e^{\alpha T}$. Now, it is easy to see that $\mathbf{x}(t) = e^{\alpha t} \mathbf{z}(t)$, where $\mathbf{z}$ is a periodic function of period T. Thus, we have the following theorem.

Theorem 5.7.6

[Floquet Theorem] Let the coefficient matrix $\mathbf{A}(t)$ be periodic with period T and the transition matrix of the system (5.7.6) has r independent eigenvectors, then the ODE system in (5.7.6) (without initial conditions) has r linearly independent solutions $\mathbf{x}_i$ which can be represented as $\mathbf{x}_i(t) = e^{\alpha_i t} \mathbf{z}_i(t)$ for some scalars α_i and $\mathbf{z}_i$s are periodic functions of period T. □

Note that the eigenvalues, may, in general, be complex and hence solutions appear to be complex valued; but this is not the case. This requires little more work. The interested reader is referred to [Inc26, Lef77] for further reading.

5.8 Exercises

1. Let $\mathbf{A}$ be an $n \times n$ matrix with an eigenvalue λ_0 of multiplicity n. Show that the standard basis can be chosen as the basis of generalized eigenvectors so that $\mathbf{B} = \mathbf{I}$ which allows us to write $\mathbf{A} = \mathbf{S} + \mathbf{N}$ in the appropriate theorem and then represent the solution.

2. Use the decomposition in Exercise 1 and solve the system $\dot{\mathbf{x}} = \mathbf{A}\mathbf{x}$, where $\mathbf{A} = \begin{bmatrix} 2 & 0 & 0 \\ -1 & 2 & 0 \\ 1 & 1 & 2 \end{bmatrix}$.

3. Find the general solution and phase portraits of the following systems

   (a) $\dot{x}_1 = -x_1 + x_2,\ \dot{x}_2 = -x_2$.
   (b) $\dot{x}_1 = x_1,\ \dot{x}_2 = 5x_2$.
   (c) $\dot{x}_1 = -x_1 - 3x_2,\ \dot{x}_2 = -2x_2$.
   (d) $\dot{x}_1 = -x_2,\ \dot{x}_2 = -x_1,\ \dot{x}_3 = x_3$.

(e) $\dot{x}_1 = x_1 - x_2, \dot{x}_2 = x_1 + x_2, \dot{x}_3 = x_3$.

(f) $\dot{x}_1 = x_1 + x_2, \dot{x}_2 = x_1 + x_2, \dot{x}_3 = -x_3$.

4. Find a basis of generalized eigenvectors and solve the systems

   (a) $\dot{x}_1 = x_1, \dot{x}_2 = -x_1 + x_2, \dot{x}_3 = x_1 + x_2 + 2x_3$.

   (b) $\dot{x}_1 = -3x_1, \dot{x}_2 = 3x_2 - 2x_3, \dot{x}_3 = x_2 + 2x_3$.

   Draw the phase portraits of the transformed as well as the original system.

5. In the previous problems, analyse the stability and write down the stable, unstable and center subspaces

6. The purpose of this exercise is to establish the Lipschitz continuity of the linear mapping $\mathbf{x} \mapsto \mathbf{A}(t)\mathbf{x}$, used in Section 5.7.2.

   (a) Let $\mathbf{A} = [a_{ij}]$ be a real $n \times n$ matrix. For $\mathbf{x} = (x_1, \ldots, x_n) \in \mathbb{R}^n$, define the norm

   $$|\mathbf{x}| = |x_1| + \cdots + |x_n|.$$

   Show that $|\mathbf{Ax} - \mathbf{Ay}| \leq |\mathbf{A}||\mathbf{x} - \mathbf{y}|$ for any $\mathbf{x}, \mathbf{y} \in \mathbb{R}^n$, where $|\mathbf{A}| = \max\limits_{1 \leq j \leq n} |a_{ij}|$.

   (b) Suppose $\mathbf{A}(t)$ is a bounded, continuous, real $n \times n$ matrix valued function defined for $t \in \mathbb{R}$. Show that the function $\mathbf{f}$ defined by $\mathbf{f}(t, \mathbf{x}) = \mathbf{A}(t)\mathbf{x}$, for $t \in \mathbb{R}$ and $\mathbf{x} \in \mathbb{R}^n$ is Lipschitz continuous (in the norm defined earlier) with Lipschitz constant less than or equal to k, where $k = \sup\limits_{t \in \mathbb{R}} |\mathbf{A}(t)|$.

7. Derive Abel's formula (Section 5.7.2) directly, by differentiating the determinant of the transition matrix: $\det \mathbf{\Phi}(t, t_0)$ and obtaining the first order linear equation satisfied by it.

8. Solve the linear system

   $$\dot{x}_1 = x_1 + x_2$$
   $$\dot{x}_2 = x_2 + x_3$$
   $$\ldots\ldots$$
   $$\dot{x}_{n-1} = x_{n-1} + x_n$$
   $$\dot{x}_n = x_n$$

by computing the appropriate exponential matrix.

9. Consider the system

$$\dot{x}_1 = x_1 + x_2, \ \dot{x}_2 = x_1 - x_2$$

Transform the system to a diagonal system by finding the matrix **B** whose columns are eigenvectors. Solve the transformed and original systems and sketch the solutions in the phase plane. Write down the stable, unstable and center subspaces.

10. Consider the 6×6 system

$$\dot{x}_1 = x_1, \quad \dot{x}_2 = x_5$$
$$\dot{x}_3 = -x_3 - x_6, \quad \dot{x}_4 = x_4$$
$$\dot{x}_5 = x_5, \quad \dot{x}_6 = x_3 - x_6.$$

Show that this system has trajectories (solutions) which behave like saddle point trajectories, center (periodic) and stable focus type trajectories depending on the initial condition.

11. Work out all the details in Examples 5.3.2–5.3.4

12. Sketch the phase-portrait for different signs and values of λ in Example 5.4.5 and in Example 5.4.6 for different values of a and b with different signs.

13. Solve the systems in Example 5.5.8 and Example 5.5.9 by constructing appropriate eigenvectors and generalized eigenvectors

14. Solve and draw the phase portraits in Example 5.5.2 in the following cases with a, b, λ respectively as;

$$2, 1, 2; \quad 2, 1, -2; \quad -2, 1, 2; \quad -2, -1, -2; \quad 0, -1, -2.$$

15. Find the Jordan canonical form of the following 2×2 and 3×3 matrices by finding generalized eigenvalues and generalized eigenvectors. Convert the system $\dot{x} = \mathbf{A}x$ to $\dot{y} = \mathbf{B}y$ and solve both the systems and draw the phase portraits, wherever possible, in $\mathbb{R}^2$ or $\mathbb{R}^3$:

(a)

$$\begin{bmatrix} 0 & -1 \\ 1 & 1 \end{bmatrix}, \quad \begin{bmatrix} 0 & 1 \\ 1 & 0 \end{bmatrix}, \quad \begin{bmatrix} 1 & 1 \\ 0 & -1 \end{bmatrix}, \quad \begin{bmatrix} 1 & 1 \\ 1 & 1 \end{bmatrix}, \quad \begin{bmatrix} 1 & 1 \\ -1 & 1 \end{bmatrix}, \quad \begin{bmatrix} 1 & -1 \\ 0 & 1 \end{bmatrix}.$$

(b)

$$\begin{bmatrix} 1 & 0 & 0 \\ 0 & 0 & 1 \\ 0 & 1 & 0 \end{bmatrix}, \quad \begin{bmatrix} 1 & 0 & 0 \\ 0 & 1 & 1 \\ 0 & 0 & -1 \end{bmatrix}, \quad \begin{bmatrix} 1 & 0 & 0 \\ 0 & 0 & -1 \\ 0 & 1 & 0 \end{bmatrix},$$

$$\begin{bmatrix} 1 & 1 & 0 \\ 0 & 1 & 1 \\ 0 & 0 & -1 \end{bmatrix}, \quad \begin{bmatrix} 1 & 0 & 0 \\ 1 & 2 & 0 \\ 1 & 2 & 3 \end{bmatrix}, \quad \begin{bmatrix} 1 & 0 & 0 \\ -1 & 2 & 0 \\ 1 & 0 & 2 \end{bmatrix}.$$

(c)

$$\begin{bmatrix} -1 & 0 & 0 & 0 \\ 1 & -2 & 0 & 0 \\ 1 & -2 & 3 & 0 \\ 1 & 2 & 3 & -4 \end{bmatrix}, \quad \begin{bmatrix} -1 & 0 & 0 & 0 \\ 1 & 2 & 0 & 0 \\ 1 & 0 & 2 & 0 \\ 1 & 1 & 0 & 2 \end{bmatrix},$$

$$\begin{bmatrix} -3 & 1 & 4 & 0 \\ 0 & -3 & 1 & 0 \\ 0 & 0 & -3 & 0 \\ 0 & 0 & 0 & -3 \end{bmatrix}, \quad \begin{bmatrix} 2 & 1 & 4 & 0 \\ 0 & 2 & 1 & -1 \\ 0 & 0 & 2 & 1 \\ 0 & 0 & 0 & 2 \end{bmatrix}.$$

16. List all possible upper Jordan canonical forms of a 4×4 matrix with a real eigenvalue λ of multiplicity 4 and find the corresponding deficiency index in each case. Do the same with repeated complex eigenvalues.

17. If the matrices $\mathbf{A}(t)$ and $\int_{t_0}^{t} \mathbf{A}(s)\, ds$ commute for all t in an interval, show that the transition matrix has the representation

$$\mathbf{\Phi}(t, t_0) = \exp \left(\int_{t_0}^{t} \mathbf{A}(s)\, ds \right).$$

18. The result in Exercise 18, in general, is not true if $\mathbf{A}(t)$ and $\int_{t_0}^{t} \mathbf{A}(s)\, ds$ do not commute. To see this (see Example 5.7.5), work out the details with the following matrix: $\mathbf{A}(t) = \begin{bmatrix} 1 & 1+t \\ 0 & t \end{bmatrix}$. Also find the solution to the corresponding IVP.

5.9 Notes

Qualitative analysis of linear systems is the main concept in this chapter. This chapter is also a precursor to the study of stability analysis of nonlinear systems carried out in Chapter 8. A good reference for this chapter among others is [Per01]; see also [Tay11, Sim91, SK07, CL72, HSD04]. A detailed study of 2×2 systems is done here by directly developing the required linear algebra. However, for higher order systems, the analysis is done by borrowing the Jordan decomposition theorem from linear algebra. The other notions that have been introduced are dynamical systems, flow, invariant subspaces, which will also be useful in the study of nonlinear systems. Non-homogeneous and non-autonomous systems are studied by introducing the concepts of fundamental matrix and transition matrix. A brief mentioning of Floquet theory is also done; this concerns non-autonomous systems with periodic coefficients.

6

Series Solutions: Frobenius Theory

6.1 Introduction

In Chapter 3, we have seen that the solutions of linear first order equations can be obtained in explicit form by converting the problem essentially to an integral calculus problem. We have also seen that there is no general procedure to obtain the solutions of linear second order equations with variable coefficients, in explicit form. Nevertheless, we could obtain valuable information about the solutions by exploiting the linearity, superposition principle, etc. In this chapter, we consider a class of linear second order equations whose solutions may be written down in explicit form. Since the solutions will be in the form of an infinite (power) series, eliciting the qualitative behavior of solutions will be difficult. The results of this chapter are collectively called *Frobenius theory*. Some important equations such as Bessel's equation, Hermite equation, Chebyshev equation, Laguerre equation, etc., are included in the class of equations considered here. Owing to the importance of these equations, which appear in applications frequently, the major properties of their solutions have been tabulated in mathematical handbooks. The interested reader may refer to [AS72]. We restrict our discussion to the real domain. There are also very interesting and important results for equations in the complex domain and the reader is referred to [Inc26].

6.2 Real Analytic Functions

The class of equations we consider will have *analytic* coefficients. Roughly speaking, analyticity means convergent power series. We are familiar with power series in the context of Taylor's series and Maclaurin's series in calculus.

Definition 6.2.1

[Analyticity] A function $f : (a,b) \to \mathbb{R}$, where (a,b) is an open interval in $\mathbb{R}$, is said to be **(real) analytic** at $t_0 \in (a,b)$ if there exists $\delta > 0$ such that $(t_0 - \delta, t_0 + \delta) \subset (a,b)$ and

$$f(t) = \sum_{n=0}^{\infty} a_n (t - t_0)^n,$$

for all $t \in (t_0 - \delta, t_0 + \delta)$, where a_ns are real numbers, that is, $f(t)$ is represented as a convergent power series in $t - t_0$ in a neighborhood of t_0. If f is analytic at every point in the interval (a,b), we say that f is analytic in (a,b). $\square$

We now recall certain facts about convergent power series which will be needed in what follows. For details, see [Apo11, Rud76].

Consider a real power series $\sum_{n=0}^{\infty} a_n t^n$ and put $R^{-1} = \limsup_{n \to \infty} \sqrt[n]{|a_n|}$.

Then, the given power series converges for all t satisfying $|t| < R$ and diverges for $|t| > R$; the case of $|t| = R$ is, in general, inconclusive. The number R is called the *radius of convergence* of the power series. Note that R can also take the value 0 or ∞. Put $f(t) = \sum_{n=0}^{\infty} a_n t^n$ for $t \in (-R, R)$.

The following statements hold:

1. The series converges uniformly in any compact subset of $(-R, R)$.

2. The function f is infinitely differentiable in $(-R, R)$ and

$$f^{(k)}(t) = \sum_{n=0}^{\infty} (n+1)(n+2) \cdots (n+k) a_{n+k} t^n,$$

   for $k = 1, 2, \cdots$. The series converges for all $t \in (-R, R)$.

3. In particular, $f^{(k)}(0) = k! a_k$, for $k = 0, 1, 2, \ldots$.

Remark 6.2.2

If $a_n \neq 0$ after a certain stage and $\lim_{n \to \infty} \dfrac{|a_{n+1}|}{|a_n|} = l$, then it is well known that $\lim_{n \to \infty} \sqrt[n]{|a_n|}$ also equals l. Thus, we have an alternative way of calculating the radius of convergence, when applicable. $\square$

We may replace t by $t - t_0$ in the earlier discussion. We now give several examples of analytic functions, which are familiar to us from calculus. The functions $\sin t, \cos t, e^t$ are analytic in $\mathbb{R}$. The function $\log t$ is analytic in $(0, \infty)$ and $t^{1/3}$ is analytic at any $t_0 > 0$ as follows from the binomial series. The function $(t - a)^{-1}$ is analytic everywhere except at a. A point where a function is not analytic is termed as a *singular point* of the function.

Denote by $\mathscr{A}(a,b)$, the set of all analytic functions in (a,b). This is a real vector space. It is also closed under multiplication and thus becomes an *algebra*. The composition of two analytic functions, when defined, is also analytic. If f is analytic at t_0 and $f^{(k)}(t_0) = 0$, for $k = 0, 1, 2, \cdots$, it follows that $f \equiv 0$ in a neighborhood of t_0. This property distinguishes an analytic function from a mere infinitely differentiable function. For example, if f is analytic in (a,b), then f cannot be compactly supported in (a,b), that is, f cannot vanish outside any $[c,d] \subset (a,b)$.

Example 6.2.3

Consider the function $f : \mathbb{R} \to \mathbb{R}$ defined by

$$f(t) = \begin{cases} \exp(-1/t^2) & \text{if } t > 0, \\ 0 & \text{if } t \le 0. \end{cases}$$

It is not difficult to check that f is in $C^\infty(\mathbb{R})$ (verification needed only at $t = 0$), but f is not analytic at $t = 0$.

The natural question that arises is: which C^∞ functions are analytic? We state the following result without proof.

Theorem 6.2.4

A function f defined in a neighborhood of t_0 is analytic at t_0 if and only if

1. f is C^∞ in a neighborhood of t_0; and

2. there exist positive δ and M such that for any $t \in (t_0 - \delta, t_0 + \delta)$, the inequality

$$|f^{(k)}(t)| \le M \frac{k!}{\delta^k}, \tag{6.2.1}$$

holds for $k = 0, 1, 2, \cdots$.

Remark 6.2.5

If we replace $k!$ in (6.2.1) by a weaker condition, $(k!)^s$ where $s \geq 1$, we obtain a class of C^∞ functions called the *Gevrey class* of *index s*. For $s = 1$, we recover the functions which are analytic. When $s > 1$, it is interesting to note that the Gevrey class contains functions with compact support. The definition of the Gevrey class of functions may easily be extended to open subsets of $\mathbb{R}^n$. These functions play an important role in obtaining the regularity of weak solutions to linear parabolic equations and weakly hyperbolic systems in the theory of partial differential equations. $\qquad\qquad\qquad\qquad\qquad\qquad\square$

6.3 Equations with Analytic Coefficients

We begin with a familiar example. Consider the following equation

$$\ddot{y} + y = 0. \qquad (6.3.1)$$

We seek an analytic solution y of (6.3.1) around $t = 0$ in the form

$$y(t) = \sum_{n=0}^{\infty} a_n t^n. \qquad (6.3.2)$$

Assuming the convergence of this series in an interval $(-R, R)$, we obtain by term-by-term differentiation that

$$\dot{y}(t) = \sum_{n=0}^{\infty} (n+1)a_{n+1} t^n, \qquad (6.3.3)$$

and

$$\ddot{y}(t) = \sum_{n=0}^{\infty} (n+1)(n+2)a_{n+2} t^n, \qquad (6.3.4)$$

Substituting the expressions in (6.3.4) and (6.3.2) into (6.3.1), we obtain

$$\sum_{n=0}^{\infty} [(n+1)(n+2)a_{n+2} + a_n] t^n = 0. \qquad (6.3.5)$$

Therefore, we have

$$(n+1)(n+2)a_{n+2} + a_n = 0, n = 0, 1, 2, \cdots.$$

We, thus recursively obtain the following:

$$a_2 = -\frac{a_0}{2}, a_3 = -\frac{a_1}{3!}, a_4 = -\frac{a_2}{3 \cdot 4} = \frac{a_0}{4!}, a_5 = -\frac{a_3}{4 \cdot 5} = \frac{a_1}{5!}, \cdots$$

Substituting these expressions in (6.3.2), the expression for y becomes

$$y(t) = a_0 \sum_{n=0}^{\infty} (-1)^n \frac{t^{2n}}{(2n)!} + a_1 \sum_{n=0}^{\infty} (-1)^n \frac{t^{2n+1}}{(2n+1)!}. \tag{6.3.6}$$

The two power series in (6.3.6) are very familiar to us; they represents $\cos t$ and $\sin t$ respectively. Thus,

$$y(t) = a_0 \cos t + a_1 \sin t,$$

where a_0 and a_1 are arbitrary real constants. Thus, the series (6.3.2) for y converges *for all* $t \in \mathbb{R}$. This may be expected as the coefficients in (6.3.1) are analytic in $\mathbb{R}$.

Of course, we would have obtained the aforementioned solution without going through the exercise of power series, as (6.3.1) is an equation with constant coefficients. Nevertheless, this exercise contains all the ingredients of a general procedure to obtain series solutions to linear equations with analytic coefficients. In general, we will not be as lucky as in this example to recognize the power series in terms of familiar functions.

We consider one more example before stating the general result. The second order equation

$$\ddot{y} - 2t\dot{y} + 2py = 0, \tag{6.3.7}$$

where p is a real constant, is termed as *Hermite's equation*. If we again assume the solution in the form (6.3.2), then we obtain from (6.3.7), after the substitution of expressions in (6.3.2), (6.3.3) and (6.3.4),

$$\sum_{n=0}^{\infty} [(n+1)(n+2)a_{n+2} - 2na_n + 2pa_n]t^n = 0. \tag{6.3.8}$$

Equating each coefficient to zero in this series, we get

$$(n+1)(n+2)a_{n+2} = -2(p-n)a_n, n = 0, 1, 2, \cdots,$$

similar to the expression obtained earlier. Therefore, the solution y may be written as

$$y(t) = a_0 y_1(t) + a_1 y_2(t), \qquad\qquad (6.3.9)$$

where y_1 and y_2 are given by the following series

$$y_1(t) = 1 - \frac{2p}{2!}t^2 + \frac{2^2 p(p-2)}{4!}t^4 - \frac{2^3 p(p-2)(p-4)}{6!}t^6 + \cdots$$

$$\qquad\qquad (6.3.10)$$

and

$$y_2(t) = t - \frac{2(p-1)}{3!}t^3 + \frac{2^2(p-1)(p-3)}{5!}t^5$$

$$\qquad\qquad - \frac{2^3(p-1)(p-3)(p-5)}{7!}t^7 + \cdots \qquad (6.3.11)$$

By the simple ratio test, it is straightforward to verify that both these series converge for all $t \in \mathbb{R}$. It is also not difficult to see that they are linearly independent and hence they span the solution space of Hermite's equation. We now make the following observations.

First, note that unless p is a non-negative integer, the infinite series for y_1 and y_2 do not terminate. If p is a non-negative even integer, the series for y_1 terminates and y_1 becomes a polynomial of degree p. Similarly, if p is a non-negative odd integer, y_2 becomes a polynomial of degree p. Any other polynomial solution of Hermite's equation is a multiple of one of these polynomials. It is not difficult to compute these polynomials for small p. For example, when $p = 0, 1, 2, 3$, the respective polynomials are given by $1, t, 1 - 2t^2, t - \frac{2}{3}t^3$.

Since any constant multiple of these polynomials is also a solution of Hermite's equation with p a non-negative integer, it is customary to take the coefficient of t^n, the leading term, as 2^n. The resulting polynomials are then termed as *Hermite polynomials* and are denoted by $H_n(t)$. Thus, $H_0(t) = 1$, $H_1(t) = 2t$ and $H_3(t) = 8t^3 - 12t$. Hermite polynomials appear frequently in several applications, especially in quantum mechanics.

The following interesting formula for H_n may be deduced from the expressions (6.3.10) and (6.3.11):

$$H_n(t) = (-1)^n e^{t^2} \frac{d^n}{dt^n} e^{-t^2}.$$

Theorem 6.3.1

Consider the second order linear equation

$$\ddot{y} + P(t)\dot{y} + Q(t) = 0, \tag{6.3.12}$$

where P, Q are analytic functions at $t_0 \in \mathbb{R}$ and are expressed in convergent power series in $t - t_0$, $t \in (t_0 - R, t_0 + R)$ for some $R > 0$. Then, given any arbitrary real numbers a_0 and a_1, there exists a unique analytic solution y of (6.4.7) satisfying $y(t_0) = a_0$ and $\dot{y}(t_0) = a_1$. Further, the solution y may also be expressed as a convergent power series in $t - t_0$ for $t \in (t_0 - R, t_0 + R)$. The point t_0 is referred to as an *ordinary point*.

Proof: The uniqueness question has already been dealt with in detail in Chapter 3. We may take $t_0 = 0$, by changing the variable, if necessary. Suppose

$$P(t) = \sum_{n=0}^{\infty} p_n t^n \text{ and } Q(t) = \sum_{n=0}^{\infty} q_n t^n, \tag{6.3.13}$$

for $t \in (-R, R)$. We seek a solution y of (6.4.7) in the form

$$y(t) = \sum_{n=0}^{\infty} a_n t^n. \tag{6.3.14}$$

Assuming for the moment that the series for y converges in a small interval around 0 and the operations of term-by-term differentiation are legitimate, we obtain

$$\dot{y}(t) = \sum_{n=0}^{\infty} (n+1) a_{n+1} t^n, \tag{6.3.15}$$

and

$$\ddot{y}(t) = \sum_{n=0}^{\infty} (n+1)(n+2) a_{n+2} t^n, \tag{6.3.16}$$

Using the first series in (6.3.13) and (6.3.15), we obtain[1]

$$P(t)\dot{y} = \left(\sum_{n=0}^{\infty} p_n t^n \right) \left(\sum_{n=0}^{\infty} (n+1)a_{n+1}t^n \right)$$

$$= \sum_{n=0}^{\infty} \left[\sum_{k=0}^{n} (k+1)p_{n-k}a_{k+1} \right] t^n \qquad (6.3.17)$$

and, using the second series in (6.3.13) and (6.3.14),

$$Q(t)y = \left(\sum_{n=0}^{\infty} q_n t^n \right) \left(\sum_{n=0}^{\infty} a_n t^n \right)$$

$$= \sum_{n=0}^{\infty} \left[\sum_{k=0}^{n} q_{n-k}a_k \right] t^n. \qquad (6.3.18)$$

Plugging the expressions in (6.3.16),(6.3.17) and (6.3.18) into the equation (6.4.7), we obtain

$$\sum_{n=0}^{\infty} \left[(n+1)(n+2)a_{n+2} + \sum_{k=0}^{n} (k+1)p_{n-k}a_{k+1} + \sum_{k=0}^{n} q_{n-k}a_k \right] t^n = 0.$$

Equating each coefficient in this series to zero, we obtain the following recursion relations for the coefficients a_ns:

$$(n+1)(n+2)a_{n+2} = - \sum_{k=0}^{n} [(k+1)p_{n-k}a_{k+1} + q_{n-k}a_k]. \quad (6.3.19)$$

Through these recursion relations, the a_ns are determined in terms of a_0, a_1 and p_n, q_ns. We now proceed to prove the convergence of the series (6.3.14) for $t \in (-R,R)$. Once this is accomplished, the operation of term-by-term differentiation is justified and completes the proof.

---

[1] Recall that for two convergent series, their product is given by

$$\left(\sum_{n=0}^{\infty} \alpha_n t^n \right) \left(\sum_{n=0}^{\infty} \beta_n t^n \right) = \sum_{n=0}^{\infty} \gamma_n t^n,$$

where $\gamma_n = \sum_{k=0}^{n} \alpha_k \beta_{n-k}$.

However, the task of proving convergence is not easy and it is a good exercise in infinite series. We adapt the ratio test for this purpose.

Fix $r < R$. Since both the series in (6.3.13) for P, Q converge for $t = r$, we can find an $M > 0$ such that

$$|p_n|r^n \leq M \text{ and } |q_n|r^n \leq M,$$

for $n = 0, 1, 2, \ldots$. Using these estimates in the recursion relation (6.3.19), we obtain

$$(n+1)(n+2)|a_{n+2}| \leq \frac{M}{r^n} \sum_{k=0}^{n} [(k+1)|a_{k+1}| + |a_k|]r^k$$

$$\leq \frac{M}{r^n} \sum_{k=0}^{n} [(k+1)|a_{k+1}| + |a_k|]r^k + M|a_{n+1}|r,$$

where the term $M|a_{n+1}|r$ is added for the purpose of what follows. Now define $b_0 = |a_0|$, $b_1 = |a_1|$ and recursively

$$(n+1)(n+2)b_{n+2} = \frac{M}{r^n} \sum_{k=0}^{n} [(k+1)b_{k+1} + b_k]r^k + Mb_{n+1}r, \quad (6.3.20)$$

It follows that $|a_n| \leq b_n$ for all n. We now consider the ratios $\dfrac{b_{n+1}}{b_n}$ for large n for the application of the ratio test. Replace first n by $n-1$ and then, by $n-2$ in (6.3.20) to obtain

$$n(n+1)b_{n+1} = \frac{M}{r^{n-1}} \sum_{k=0}^{n-1} [(k+1)b_{k+1} + b_k]r^k + Mb_n r$$

and

$$(n-1)nb_n = \frac{M}{r^{n-2}} \sum_{k=0}^{n-2} [(k+1)b_{k+1} + b_k]r^k + Mb_{n-1}r.$$

Multiplying the first expression here by r and using the second, we obtain

$$rn(n+1)b_{n+1} = \frac{M}{r^{n-2}} \sum_{k=0}^{n-2} [(k+1)b_{k+1} + b_k]r^k + rM(nb_n + b_{n-1})$$

$$+ Mb_n r^2$$

$$= (n-1)nb_n - Mb_{n-1}r + rM(nb_n + b_{n-1}) + Mb_n r^2$$

$$= [(n-1)n + rMn + Mr^2]b_n.$$

Thus,

$$\frac{b_{n+1}}{b_n} = \frac{(n-1)n + rMn + Mr^2}{rn(n+1)},$$

which tends to $\dfrac{1}{r}$ as $n \to \infty$. Therefore, by the ratio test, the series $\displaystyle\sum_{n=0}^{\infty} b_n t^n$ converges for all t satisfying $|t| < r$. Since $|a_n| \le b_n$, by comparison test, it follows that the series (6.4.7) for the solution y also converges for $|t| < r$. Since $r < R$ is arbitrary, this completes the proof. $\square$

6.4   Regular Singular Points

Many second order linear equations that make frequent appearance in applications do not have analytic coefficients. However, the singularities present are isolated and *regular*, which will be defined later in this section. We have already had an idea in Chapter 4 of how a singularity in the coefficients may influence the solution in question. We consider the following simple example to see what we can expect when singularities are present in the coefficients.

Example 6.4.1

Consider the second order equation

$$\ddot{y} + \frac{k}{t^2}y = 0, t > 0.$$

Here, k is a real constant. Notice that $t = 0$ is a singular point of the coefficient of y. It is not hard to write down the general solution of this equation. We have

$$y(t) = t^{1/2}(c_1 \sin(\mu \log t) + c_2 \cos(\mu \log t)), t > 0, \text{ if } k > 1/4,$$

$$y(t) = t^{1/2}(c_1 \log t + c_2), t > 0, \text{ if } k = 1/4,$$

$$y(t) = t^{1/2}(c_1 t^{\mu} + c_2 t^{-\mu}), t > 0, \text{ if } k < 1/4.$$

Here, $\mu = \sqrt{k - 1/4}$ if $k > 1/4$ and $\mu = \sqrt{1/4 - k}$ if $k < 1/4$; c_1, c_2 are arbitrary constants. Observe that any solution, except one, is defined only for $t > 0$ and possesses a singularity at $t = 0$. One solution, which is a multiple of $t^{1/2}$ when $k = 1/4$ is defined for $t \geq 0$. Even in this case, we cannot arbitrarily prescribe the initial data at $t = 0$.

The general situation is going to be somewhat similar, as we will see later. First, we list some of the important equations which fall in this category.

1. *Bessel's Equation of Order Zero*: $t\ddot{y} + \dot{y} + ty = 0$.
   Observe that $t = 0$ is a singularity.

2. *Bessel's Equation of Order p*: $t^2\ddot{y} + t\dot{y} + (t^2 - p^2)y = 0$,
   where p is a non-negative real number. Again $t = 0$ is a singularity.

3. *Legendre's Equation*: $(1 - t^2)\ddot{y} - 2t\dot{y} + p(p+1)y = 0$,
   where p is a real constant. Here $t = \pm 1$ are the singular points.

4. *Chebyshev's Equation*: $(1 - t^2)\ddot{y} - t\dot{y} + p^2 y = 0$,
   where p is a real constant. Again, $t = \pm 1$ are the singular points.

5. *Gauss's Hypergeometric Equation*: $t(1 - t)\ddot{y} + [c - (a + b + 1)t]\dot{y} - aby = 0$,
   where a, b and c are real constants. Here $t = 0, 1$ are the singular points.

6.4.1   Equations with Regular Singular Points

We now consider the general second order equation

$$\ddot{y} + P(t)\dot{y} + Q(t)y = 0. \tag{6.4.1}$$

We assume that the functions P and Q in (6.4.1) have a singular point t_0 in $\mathbb{R}$. The singular point t_0 is called a *regular singular point* of (6.4.1) if the functions $(t - t_0)P(t)$ and $(t - t_0)^2 Q(t)$ are analytic at $t = t_0$; otherwise, it is called an *irregular singular point*. We are going to obtain a convergent series solution y of (6.4.1) near a regular singular point. However, we will not have freedom to arbitrarily fix initial conditions at a regular singular point. For the ease of writing, we assume that $t = 0$ is a regular singular point of (6.4.1); the general case follows from a change of variable.

From the hypothesis, it follows that P and Q are of the form

$$P(t) = \frac{p_0}{t} + p_1 + p_2 t + \cdots \tag{6.4.2}$$

and

$$Q(t) = \frac{q_0}{t^2} + \frac{q_1}{t} + q_2 + q_3 t + \cdots \tag{6.4.3}$$

We assume a solution y of (6.4.1) in the form of a 'quasi power series'

$$y(t) = t^m (a_0 + a_1 t + a_2 t^2 + \cdots) \tag{6.4.4}$$

where $a_0 \neq 0$. The determination of the exponent m is part of the problem. The equation satisfied by the index m, the *indicial equation*, will be a quadratic equation reflecting the order of (6.4.1). The nature of the indicial equation will produce solutions with different behaviors at $t = 0$.

Assuming that the series for y is convergent in an interval $(0, T)$, we obtain the following (see the previous section):

$$\dot{y} = \sum_{n=0}^{\infty} a_n (m+n) t^{m+n-1},$$

$$\ddot{y} = \sum_{n=0}^{\infty} a_n (m+n)(m+n-1) t^{m+n-2} = t^{m-2} \sum_{n=0}^{\infty} a_n (m+n)(m+n-1) t^n.$$

For the terms $P(t)\dot{y}$ and $Q(t)y$, using (6.4.2) and (6.4.3), we get

$$P(t)\dot{y} = \frac{1}{t} \left(\sum_{n=0}^{\infty} p_n t^n \right) \left[\sum_{n=0}^{\infty} a_n (m+n) t^{m+n-1} \right]$$

$$= t^{m-2} \sum_{n=0}^{\infty} \left[\sum_{k=0}^{n} p_{n-k} a_k (m+k) \right] t^n$$

$$= t^{m-2} \sum_{n=0}^{\infty} \left[\sum_{k=0}^{n-1} p_{n-k} a_k (m+k) + p_0 a_n (m+n) \right] t^n$$

and

$$Q(t)y \;=\; \frac{1}{t^2}\left(\sum_{n=0}^{\infty} q_n t^n\right)\left[\sum_{n=0}^{\infty} a_n t^{m+n}\right]$$

$$= \; t^{m-2}\sum_{n=0}^{\infty}\left[\sum_{k=0}^{n} q_{n-k}a_k\right] t^n$$

$$= \; t^{m-2}\sum_{n=0}^{\infty}\left[\sum_{k=0}^{n-1} q_{n-k}a_k + q_0 a_n\right] t^n.$$

After the substitution of these expressions for $\ddot{y}, P(t)\dot{y}$ and $Q(t)y$ in (6.4.1) and canceling the common factor t^{m-2} throughout, we obtain

$$\sum_{n=0}^{\infty}\left[a_n\{(m+n)(m+n-1)+(m+n)p_0+q_0\}\right.$$

$$\left. + \sum_{k=0}^{n-1} a_k\{(m+k)p_{n-k}+q_{n-k}\}\right] t^n = 0. \qquad (6.4.5)$$

By equating the coefficients of t^n in (6.4.5) to zero, we successively obtain

$$a_0[m(m-1)+mp_0+q_0] = 0,$$

$$a_1[m(m+1)+(m+1)p_0+q_0] + a_0(mp_1+q_1) = 0,$$

$$a_2[(m+1)(m+2)+(m+2)p_0+q_0]+$$

$$+a_0(mp_2+q_2)+a_1[(m+1)p_1+q_1] = 0,$$

$$\cdots \quad \cdots \quad \cdots \quad \cdots \quad \cdots \quad \cdots$$

$$a_n[(m+n-1)(m+n)+(m+n)p_0+q_0]$$

$$+a_0(mp_n+q_n)+\cdots+a_{n-1}[(m+n-1)p_1+q_1] = 0,$$

$$\cdots \quad \cdots \quad \cdots \quad \cdots \quad \cdots \quad \cdots$$

$$(6.4.6)$$

Put $f(m) = m(m-1) + mp_0 + q_0$. Then, (6.4.6) may concisely be written as

$$a_0 f(m) = 0,$$

$$a_1 f(m+1) + a_0(mp_1 + q_1) = 0,$$

$$a_2 f(m+2) + a_0(mp_2 + q_2) + a_1[(m+1)p_1 + q_1] = 0,$$

$$\cdots \quad \cdots \quad \cdots \quad \cdots \quad \cdots \quad \cdots$$

$$a_n f(m+n) + a_0(mp_n + q_n) + \cdots + a_{n-1}[(m+n-1)p_1 + q_1] = 0,$$

$$\cdots \quad \cdots \quad \cdots \quad \cdots \quad \cdots \quad \cdots$$

$$(6.4.7)$$

Since $a_0 \neq 0$, it follows that $f(m) = 0$, that is,

$$m(m-1) + mp_0 + q_0 = 0. \qquad (6.4.8)$$

This is the *indicial equation*, which determines the possible values of the exponent m in the assumed expression for the solution y. Let m_1 and m_2 be the roots of (6.4.8). If we choose $m = m_1$, then, from the aforementioned expressions, we see that a_n is determined in terms of $a_0, a_1, \cdots, a_{n-1}$, successively for $n = 1, 2, \cdots$, provided that $f(m+n) \neq 0$. The process breaks off if $f(m+n) = 0$. Thus, if $m_1 = m_2 + n$, for some positive integer n, the choice $m = m_1$ gives a formal solution, but in general, the choice $m = m_2$ does not, since $f(m_2 + n) = f(m_1) = 0$. If $m_1 = m_2$, then also we obtain only one formal solution. In all the other cases, when the roots of the indicial equation are real, we obtain two linearly independent formal solutions.

The roots of the indicial equation may also be complex, and therefore, this procedure leads to a formal series with complex coefficients. Since we are only interested in real solutions, we need to consider real and imaginary parts of these formal solutions, which in general is quite complicated and requires tools from complex analysis. We will not pursue these topics here and the interested reader may refer to [Inc26] for a discussion on differential equations in the complex domain.

We now state the forgoing discussion in the following theorem.

Theorem 6.4.2

Assume that $t = 0$ is a regular singular point of (6.4.1) and that the power series for $tP(t)$ and $t^2 Q(t)$ given, respectively, by (6.4.2) and (6.4.3) converge for $t \in (-R, R)$ for some $R > 0$. Let the roots m_1 and m_2 of the indicial equation (6.4.8) be real with $m_2 \leq m_1$. Then, (6.4.1) has at least one solution

$$y_1 = t^{m_1} \sum_{n=0}^{\infty} a_n t^n \quad (a_0 \neq 0) \tag{6.4.9}$$

on the interval $0 < t < R$, where a_ns are determined in terms of a_0 by the recursion formula (6.4.7) with m replaced by m_1. Also, the series $\sum_{n=0}^{\infty} a_n t^n$ converges on the interval $(-R, R)$. Furthermore, if $m_1 - m_2$ is not a non-negative integer, then (6.4.1) has a second independent solution

$$y_2 = t^{m_2} \sum_{n=0}^{\infty} a_n t^n \quad (a_0 \neq 0) \tag{6.4.10}$$

on the same interval, where now a_ns are determined using the recursion relation (6.4.7) in terms of a_0 and m replaced by m_2. Again, the series $\sum_{n=0}^{\infty} a_n t^n$ converges on the interval $(-R, R)$.   □

The series in (6.4.9) and (6.4.10) are called *Frobenius series*. In a specific problem, it is much preferable to start with a series of the form (6.4.4) and derive the indicial equation and recursion relations. However, the recursion formula (6.4.7) finds its main application in the proof of Theorem 6.4.2, which is similar to the one in the previous section, but is more delicate because of the presence of the terms $f(m+n)$. We will not present a proof here and the reader is referred to [Sim91] for details.

The theorem leaves unanswered the cases of $m_1 = m_2$ and when $m_1 - m_2$ is a positive integer.

Suppose $m_1 = m_2$ and y_1 is a solution given by the Frobenius series. We may now proceed to find a second independent solution by the procedure described in Chapter 3. Let $y_2 = y_1 v$ be another solution, where v is a non-constant function. Then,

$$\dot{v} = \frac{1}{y_1^2} \exp\left(-\int P(t)\, dt\right)$$

$$= \frac{1}{t^{2m_1}(a_0 + a_1t + \cdots)^2} \exp\left(-\int \left[\frac{p_0}{t} + p_1 + \cdots\right] dt\right)$$

$$= \frac{1}{t^{2m_1}(a_0 + a_1t + \cdots)^2} \exp\left(-p_0 \log t - p_1 t - \cdots\right)$$

$$= \frac{1}{t(a_0 + a_1t + \cdots)^2} \exp\left(-p_1 t - \cdots\right)$$

$$= \frac{1}{t} g(t), \text{ say},$$

where we have used the fact that $2m_1 + p_0 = 1$ when $m_1 = m_2$ and g is an analytic function at $t = 0$ with $g(0) = \frac{1}{a_0^2}$. Therefore, we have

$$v(t) = b_0 \log t + b_1 t + \cdots,$$

where $g(t) = b_0 + b_1 t + \cdots$. Of course, it may not be easy to determine the coefficients in the power series expansion of g. When $m_1 - m_2$ is a positive integer k, then the expressions for $\dot{v}$ is $\frac{1}{t^k} g(t)$. In this case, there may or may not be a logarithmic term in v.

We now illustrate the aforementioned procedure, in a somewhat different way, by considering Bessel's equation of order zero:

$$t\ddot{y} + \dot{y} + ty = 0. \tag{6.4.11}$$

It is easy to see that $t = 0$ is a regular singular point of (6.4.11). Let us consider the Frobenius series

$$y = t^m \sum_{n=0}^{\infty} a_n t^n \tag{6.4.12}$$

for a solution of (6.4.11), with $a_0 \neq 0$. We may take $a_0 = 1$. We obtain, after collecting the like terms,

$$t\ddot{y} + \dot{y} + ty = m^2 t^{m-1} + (m+1)^2 a_1 t^m + \{(m+2)^2 a_2 + 1\} t^{m+1}$$

$$+ \{(m+3)^2 a_3 + a_1\} t^{m+2} + \cdots \tag{6.4.13}$$

Now, let $a_1, a_2, \cdots$ be chosen to satisfy the following relations

$$(m+1)^2 a_1 = 0,$$

$$(m+2)^2 a_2 + 1 = 0,$$

$$(m+3)^2 a_3 + a_1 = 0,$$

$$\cdots \qquad \cdots$$

Then, unless m is a negative integer, we have $a_k = 0$ for *odd* integers k and

$$a_2 = -\frac{1}{(m+2)^2},$$

$$a_4 = -\frac{a_2}{(m+4)^2} = \frac{1}{(m+2)^2(m+4)^2},$$

$$\cdots \cdots \cdots$$

Substituting these values in (6.4.12) and (6.4.13), we infer that if

$$y = t^m \left\{ 1 - \frac{t^2}{(m+2)^2} + \frac{t^4}{(m+2)^2(m+4)^2} - \cdots \right\} \qquad (6.4.14)$$

and if m is not a negative integer, then

$$t\ddot{y} + \dot{y} + ty = m^2 t^{m-1}. \qquad (6.4.15)$$

Choosing $m = 0$ in (6.4.14) and (6.4.15), we see that

$$y = 1 - \frac{t^2}{2^2} + \frac{t^4}{2^2 \cdot 4^2} - \cdots \qquad (6.4.16)$$

is a solution of Bessel's equation

$$t\ddot{y} + \dot{y} + ty = 0. \qquad (6.4.17)$$

The series in (6.4.16) is denoted by $J_0(t)$ and is called *Bessel's function of zero order of the first kind*. It is easy to see that $J_0(t)$ is an even function of t and converges for all $t \in \mathbb{R}$ with $J_0(0) = 1$. We can also see that the indicial equation for Bessel's equation is given by $m^2 = 0$; thus, its roots are equal and equal to zero. We now proceed to find another independent solution of Bessel's equation. The general procedure tells us that the second solution involves a logarithm term. We are going to derive an expression for the same using (6.4.13). Differentiating both the sides of (6.4.13) with respect to m and then choosing $m = 0$, we obtain

$$t\ddot{Y}_0 + \dot{Y}_0 + tY_0 = 0,$$

where $Y_0 = \dfrac{\partial y}{\partial m}$ evaluated at $m = 0$. Now, from (6.4.14),

$$\frac{\partial y}{\partial m} = t^m \log t \left\{ 1 - \frac{t^2}{(m+2)^2} + \frac{t^4}{(m+2)^2(m+4)^2} - \cdots \right\}$$

$$+ t^m \left\{ \frac{2t^2}{(m+2)^2} \frac{1}{m+2} - \frac{2t^4}{(m+2)^2(m+4)^2} \left(\frac{1}{m+2} + \frac{1}{m+4} \right) \right.$$

$$\left. + \frac{2t^6}{(m+2)^2(m+4)^2(m+6)^2} \left(\frac{1}{m+2} + \frac{1}{m+4} + \frac{1}{m+6} \right) - \cdots \right\}.$$

Hence, putting $m = 0$, we obtain

$$Y_0(t) = J_0(t) \log t - \frac{t^2}{2^2} + \frac{t^2}{2^2 \cdot 4^2} \left(1 + \frac{1}{2} \right) + \frac{t^6}{2^2 \cdot 4^2 \cdot 6^2} \left(1 + \frac{1}{2} + \frac{1}{6} \right) - \cdots,$$

$$(6.4.18)$$

which is called Bessel's function of the second kind of order zero. Using

$$1 + \frac{1}{2} + \cdots + \frac{1}{n} = \log n + \gamma + \varepsilon_n,$$

where γ is Euler's constant and $\varepsilon_n \to 0$ as $n \to \infty$, it is straightforward to check that the power series in (6.4.18) (excluding the term $J_0(t) \log t$) converges for all values of t. It follows that the general solution of Bessel's equation is given by

$$y = AJ_0 + BY_0,$$

for arbitrary constants A, B.

Remark 6.4.3

In some situations, the point at infinity plays an important role. This case may be easily handled as follows. Consider a second order linear equation and change the independent variable t to $\tau = \frac{1}{t}$. If the resulting equation after this change of variable has the point $\tau = 0$ as an ordinary point or a singular point, then we say that the point at infinity is an ordinary point or a singular point, respectively, for the original equation. We may use the method of Frobenius for the transformed equation with τ as the independent variable and then go back to the original problem.   □

6.5 Exercises

1. Show that the function f in Example 6.2.3 is in $C^\infty(\mathbb{R})$ and $f^{(n)}(0) = 0$ for $n = 1, 2, \cdots$.

2. Let

$$y_1(t) = t^{r_1} \sum_{n=0}^{\infty} a_n t^n, \; y_2(t) = t^{r_2} \sum_{n=0}^{\infty} b_n t^n,$$

for $t > 0$. Here r_1 and r_2 are real and unequal; a_0, b_0 are non-zero. State and prove a general theorem concerning the linear independence of y_1 and y_2.

3. Consider the second order Euler equation

$$t^2 \ddot{y} + at\dot{y} + by = 0,$$

where a, b are real. Find two linearly independent solutions of the equation in each of the following cases by applying the method of Frobenius.

   (a) $a = \dfrac{1}{2}, b = -\dfrac{1}{2}$.

   (b) $a = -5, b = 9$.

4. Discuss the solution of Legendre's equation

$$(1 - t^2)\ddot{y} - 2t\dot{y} + a(a+1)y = 0$$

in the neighborhoods of $t = 1$ and $t = -1$.

5. For each of the following equations, write the indicial equation and find its roots. Write the form of two linearly independent solutions

without computing the coefficients, and discuss the limiting behavior
of the solutions as $t \to 0$.

(a) $\ddot{y} + \dfrac{5}{2t}\dot{y} + \dfrac{1}{2t}y = 0.$

(b) $t^2\ddot{y} + 4t\dot{y} + (2 - t)y = 0.$

(c) $t^2\ddot{y} + (1 - 6t)y = 0.$

6. In each of the following equations, locate all the singular points by
   describing whether they are regular or irregular.

   (a) $(t - 2)(t + 3)^2\ddot{y} + 3t^2\dot{y} - 2(t + 3)y = 0.$

   (b) $t^2\ddot{y} + (\sin t)\dot{y} + (\cos t)y = 0.$

   (c) $(e^t - 1)^2\ddot{y} + 2(\sin t)\dot{y} + 3y = 0.$

   (d) $\ddot{y} + 3\dot{y} + t^{\frac{1}{2}}y = 0, t \geq 0.$

7. Consider Bessel's equation of order $\dfrac{1}{2}$

$$t^2\ddot{y} + t\dot{y} + \left(t^2 - \dfrac{1}{4}\right)y = 0.$$

   (a) Verify that the functions

$$y_1(t) = \sqrt{\dfrac{2}{\pi t}}\sin t, \quad y_2(t) = \sqrt{\dfrac{2}{\pi t}}\cos t,$$

   are linearly independent solutions for $t > 0$.

   (b) Show that the indicial equation has roots $m_1 = \dfrac{1}{2}$ and $m_2 = -\dfrac{1}{2}$,
   whose difference is the positive integer 1. Show, nevertheless,
   that both the solutions y_1, y_2 can be derived from the method of
   Frobenius without introducing a logarithm term.

8. Determine whether the point at infinity is an ordinary, regular
   singular, or irregular singular point for each of the given equations.

   (a) $t\ddot{y} + 3\dot{y} + 2ty = 0.$

   (b) $t^5\ddot{y} + 2t^4\dot{y} + y = 0.$

   (c) (Bessel's equation of order p) $t^2\ddot{y} + t\dot{y} + (t^2 - p^2)y = 0.$

6.6 Notes

In this chapter, we have considered a couple of classes of linear second order equations with variable coefficients whose solutions can be obtained explicitly, in the form of a power series; see [Sim91]. The analysis mainly involves proving the convergence of the power series, obtained heuristically, of a solution. For power series solutions of a system of linear equations, the reader is referred to [Tay11].

7

Regular Sturm–Liouville Theory

7.1 Introduction

In this chapter, we are going to study certain boundary value problems (BVP) associated with regular second order linear equations containing a *parameter*. More specifically, we will be looking for non-trivial solutions of the following equation

$$\mathscr{L}u(t) \equiv -\frac{d}{dt}\left(p(t)\frac{du}{dt}\right) + q(t)u(t) = \lambda\rho(t)u(t).$$

Here $p \in C^1$, $p > 0$ and $q, \rho \in C$, $\rho > 0$ are defined on a bounded interval $[a, b]$ in $\mathbb{R}$ and λ is a real parameter. Additionally, the solution is required to satisfy boundary conditions at the end points a, b of the given interval. We refer to these BVP as *regular Sturm–Liouville systems* or simply S–L systems. When p vanishes somewhere in the interval or the interval under consideration is unbounded, the problem is termed as *singular*. Singular BVP are more difficult to deal with, and we will not discuss them in this book. The interested reader may refer to [CL72], for example.

S–L systems arise in many physical problems, for example, in the following situation. Consider the longitudinal vibrations of an elastic bar of local stiffness $p(\xi)$ and density $\rho(\xi)$. The mean longitudinal displacement $v(\xi, \tau)$, at position ξ and time τ, of the section of such a bar from its equilibrium satisfies the one-dimensional wave equation

$$\rho(\xi)\frac{\partial^2 v}{\partial \tau^2} = \frac{\partial}{\partial \xi}\left[p(\xi)\frac{\partial v}{\partial \xi}\right].$$

If we now seek the solution in the following form of *simple harmonic* vibrations (or the *normal modes* of vibrations) as

$$v(\xi, \tau) = u(\xi)\cos(k(\tau - \tau_0)),$$

we obtain an S–L system for u, changing ξ variable to t variable, with $q \equiv 0$ and $\lambda = k^2$. For a finite bar, say, $\xi \in [a, b]$, following are some of the natural boundary conditions:

$u(a) = u(b) = 0$ (rigidly fixed ends)

$\dot{u}(a) = \dot{u}(b) = 0$ (free ends)

$\dot{u}(a) + \alpha u(a) = \dot{u}(b) + \beta u(b) = 0$ (elastically held ends)

$u(a) = u(b),\ \dot{u}(a) = \dot{u}(b)$ (periodic conditions).

Similarly, if we consider the *vibrating circular membrane*, we obtain a wave equation in two dimensions. After the introduction of polar coordinates and assuming the radial symmetry of the solution, the equation can be reduced to a one-dimensional wave equation. If we again seek a solution by the separation of variables as in the case of the elastic bar earlier, this will lead to a singular S–L system involving Bessel's equation. For more details, see [BR03].

It will be seen that non-trivial solutions to the BVP exist only for a discrete set of values of the parameter λ, tending to infinity. The situation may thus be compared with eigenvalues of a matrix, considered as a linear operator on a finite dimensional space. The main difference is that we are now working in an infinite dimensional space. In analogy with matrices, the discrete set of the values of the parameter for which the non-trivial solutions exist, are called the *eigenvalues* of the BVP and the corresponding non-trivial solutions are called the *eigenfunctions*.

Again, continuing the similarity with matrices, we know that for a *good* matrix, any vector in the finite dimensional space in question, can be written as a unique linear combination of eigenvectors of the matrix. For example, this happens when the matrix is real, symmetric and the space is $\mathbb{R}^n$ for any positive integer n. In this case, the eigenvectors are also orthogonal; we discuss more regarding orthogonality in the next section. In the case of a BVP too, one may express an *arbitrary* function as a *linear combination* of the corresponding eigenfunctions. Since these are (infinite) power series, we need to discuss the convergence, etc. and the tools required come from functional analysis (Hilbert space), the discussion of these topics is outside the purview of the present book.

Before proceeding further, let us consider a simple example

$$\mathscr{L}u = -\ddot{u}.$$

It is easy to see that non-trivial solutions to the BVP

$$\mathscr{L}u = \lambda u, u(0) = u(\pi) = 0,$$

exist if and only if $\lambda = n^2$, $n \in \mathbb{Z} \setminus \{0\}$. Thus, the eigenvalues are n^2, n a non-zero integer and the corresponding eigenfunctions are $\sin(nt)$. If, instead, we consider the boundary conditions as $u(-\pi) = u(\pi) = 0$, the eigenvalues are now $n^2/4$, $n \in \mathbb{Z} \setminus \{0\}$, and the eigenfunctions are $\sin(nt/2)$ and $\cos(nt/2)$ for n even and n odd respectively. A reader familiar with Fourier (sine and cosine) series recognizes that any suitable function satisfying the given boundary conditions, can be written as an infinite series involving the corresponding eigenfunctions.

If we now consider the *periodic boundary condition* $u(-\pi) = u(\pi)$, we do obtain the situation of the Fourier series. However, the question of convergence, especially that of point-wise convergence, of a Fourier series is a delicate issue.

From this example, we learn that the form of the boundary conditions plays an important role in the determination of the eigenvalues; it is also important in making the operator $\mathscr{L}$ *self-adjoint* and the orthogonality condition of the eigenfunctions, as we will see in the next section.

7.2   Basic Result and Orthogonality

We begin with following basic result.

Theorem 7.2.1

Let

$$\mathscr{L} = -\frac{d}{dt}\left(p(t)\frac{d}{dt}\right) + q(t),$$

where $p \in C^1$, $p > 0$ and $q \in C$ on a given interval $[a,b]$ in $\mathbb{R}$. Consider the following BVPs:

$$\mathscr{L}u = r, \quad u(a) = \alpha, u(b) = \beta, \tag{7.2.1}$$

where α, β are real, $r \in C[a,b]$ and its homogeneous counterpart

$$\mathscr{L}u = 0, \ u(a) = 0, u(b) = 0. \tag{7.2.2}$$

Then, the following alternatives hold:

(1) If (7.2.2) has only trivial solutions, then, there exists a unique solution of (7.2.1).

(2) If (7.2.2) has a non-trivial solution, then, (7.2.1) has infinitely many solutions, provided that it has a solution.   □

We may replace the boundary conditions in (7.2.1) by other linear combinations of $u, \dot{u}$ at the end points a, b, separately; see (7.2.3) later in this section. See also Theorem 9.2.1 for a more general situation and for higher order equations, see [Inc26]. We have the following example concerning the hypothesis of (2) in the theorem.

The BVP, $\ddot{u} + \pi^2 u = 0, u(0) = u(1) = 0$ has a non-trivial solution. The BVP $\ddot{u} + \pi^2 u = 0, u(0) = \alpha, u(1) = \beta$ has a solution if and only if $\alpha + \beta = 0$.

Proof: (of Theorem 7.2.1) Suppose u_1 and u_2 are two linearly independent solutions of the homogeneous equation $\mathscr{L}u = 0$. If we put

$$u = c_1 u_1 + c_2 u_2,$$

for some constants c_1, c_2, then, u is a solution of (7.2.2) if and only if the following equations are satisfied:

$$c_1 u_1(a) + c_2 u_2(a) = 0,$$

$$c_1 u_1(b) + c_2 u_2(b) = 0.$$

Therefore, (7.2.2) has only trivial solutions if and only if the matrix

$$\mathbf{A} = \begin{bmatrix} u_1(a) & u_2(a) \\ u_1(b) & u_2(b) \end{bmatrix}$$

is non-singular. Next, let u_0 be any particular solution of $\mathscr{L}u = r$. Then, the function u defined by $u = u_0 + c_1 u_1 + c_2 u_2$ is a solution of (7.2.1) if and only if the following equations are satisfied:

$$u_0(a) + c_1 u_1(a) + c_2 u_2(a) = \alpha,$$

$$u_0(b) + c_1 u_1(b) + c_2 u_2(b) = \beta.$$

The possibility of obtaining such c_1, c_2 depends on the rank of the augmented matrix

$$\mathbf{B} = \begin{bmatrix} u_1(a) & u_2(a) & \alpha - u_0(a) \\ u_1(b) & u_2(b) & \beta - u_0(b) \end{bmatrix}.$$

We infer that a solution for (7.2.1) exists if and only if the matrices $\mathbf{A}$ and $\mathbf{B}$ have the same rank.

If (7.2.2) has only trivial solutions, then the matrix $\mathbf{A}$ is non-singular and therefore, $\mathbf{A}$ and $\mathbf{B}$ have the same rank. This completes the proof of (1). If u_2 is a non-trivial solution of (7.2.2), then we choose u_1 satisfying $\mathscr{L}u_1 = 0, u_1(a) \neq 0$ such that u_1, u_2 are linearly independent. Then, rank of $\mathbf{A}$ is 1 and therefore, a solution of (7.2.1) exists if and only if rank of $\mathbf{B}$ is also 1. If u_0 is a solution of (7.2.1), so is $u_0 + cu_2$ for any real c. This proves (2) and the proof of the theorem is thus complete.   □

It turns out that the more interesting situation is when (7.2.2) has non-trivial solutions. This will lead to the existence of eigenvalues and eigenfunctions for $\mathscr{L}$.

We are now going to obtain the self-adjointness of $\mathscr{L}$. To this end, we introduce the following *inner product*. For any continuous functions u, v defined on $[a, b]$, define the inner product by

$$\langle u, v \rangle = \int_a^b u(t) v(t) \, dt.$$

(If u, v are complex valued, $v(t)$ should be replaced by $\overline{v(t)}$, the complex conjugate).

We impose the following boundary conditions on functions in $C^2[a, b]$:

$$\left. \begin{array}{l} \alpha_1 u(a) + \alpha_2 \dot{u}(a) = 0, |\alpha_1| + |\alpha_2| > 0, \\ \beta_1 u(b) + \beta_2 \dot{u}(b) = 0, |\beta_1| + |\beta_2| > 0. \end{array} \right\} \tag{7.2.3}$$

Here α_1, α_2, β_1, β_2 are real constants.

Theorem 7.2.2

(1) Suppose $u, v \in C^2[a,b]$ satisfy the boundary conditions (7.2.3). Then, $\mathscr{L}$ is *self-adjoint* in the sense that

$$\langle \mathscr{L}u, v \rangle = \langle u, \mathscr{L}v \rangle.$$

(See Remark 7.2.3).

(2) Suppose $u, v \in C^2[a,b]$ satisfy the following equations

$$\mathscr{L}u = \lambda \rho u, \quad \mathscr{L}v = \mu \rho v$$

and the boundary conditions (7.2.3). If $\lambda \neq \mu$, then

$$\int_a^b \rho(t) u(t) v(t)\, dt = 0.$$

The functions u, v are said to satisfy an *orthogonal condition with weight* ρ.

Proof. We begin by proving *Lagrange's identity*. For *any* two C^2 functions u, v defined on $[a,b]$ and $t \in [a,b]$, we have

$$
\begin{aligned}
v(t)\mathscr{L}u(t) - u(t)\mathscr{L}v(t) &= -v(t)\frac{d}{dt}\left(p(t)\dot{u}(t)\right) + q(t)u(t)v(t) \\
&\quad + u(t)\frac{d}{dt}\left(p(t)\dot{v}(t)\right) - q(t)u(t)v(t) \\
&= -\frac{d}{dt}\left(p(t)v(t)\dot{u}(t)\right) + p(t)\dot{u}(t)\dot{v}(t) \\
&\quad + \frac{d}{dt}\left(p(t)u(t)\dot{u}(t)\right) - p(t)\dot{u}(t)\dot{v}(t) \\
&= \frac{d}{dt}\left[p(t)\left(u(t)\dot{v}(t) - v(t)\dot{u}(t)\right)\right] \\
&= \frac{d}{dt}\left(p(t)W(t)\right), \quad\quad\quad (7.2.4)
\end{aligned}
$$

where W denotes the Wronskian of u and v. Integrating Lagrange's identity (7.2.4), over $[a,b]$, we obtain

$$\langle \mathscr{L}u, v \rangle - \langle u, \mathscr{L}v \rangle = p(b)W(b) - p(a)W(a). \tag{7.2.5}$$

Using the boundary conditions (7.2.3), we see that $W(a) = 0$ and $W(b) = 0$. Thus, the right side of (7.2.5) vanishes. This completes the proof of (1). Now, suppose u and v satisfy the equations in (2). Then, we have, using the self-adjointness of $\mathscr{L}$,

$$0 = \langle \mathscr{L}u, v \rangle - \langle u, \mathscr{L}v \rangle$$

$$= \lambda \int_a^b \rho(t)u(t)v(t)\,dt - \mu \int_a^b \rho(t)u(t)v(t)\,dt$$

$$= (\lambda - \mu) \int_a^b \rho(t)u(t)v(t)\,dt. \tag{7.2.6}$$

Since $\lambda \neq \mu$, we get the weighted orthogonality as in (2) and the proof of the theorem is complete. □

Remark 7.2.3

What actually we have shown is that $\mathscr{L}$ is *symmetric* with respect to inner product $\langle , \rangle$. In *functional analysis*, self-adjointness is more than being symmetric. This will require a detailed description of the domain of $\mathscr{L}$, whose discussion is beyond the scope of this book. With a slight modification of Lagrange's identity, the same proof works for complex valued functions also. The symmetry immediately yields that the eigenvalues, if any, of $\mathscr{L}$ are all real. Why should they be only discrete is another question that will be addressed later. At this stage, we again stress that the boundary conditions do play an important role in establishing the symmetry of $\mathscr{L}$; see, also, the following remark.

□

Remark 7.2.4.

Looking at the expression on the right side of (7.2.5), we observe that it may be made to vanish for a variety of boundary conditions; for example, we may take the boundary conditions $u(a) = u(b) = 0$ or

$\dot{u}(a) = \dot{u}(b) = 0$. The condition $p(b)W(b) = p(a)W(a)$ is referred to as a *periodic* boundary condition. We should bear in mind that it is not only the form of the operator $\mathscr{L}$ but also the form of the boundary conditions, that play an important role in proving the orthogonality condition. □

Next, we take up the issue of obtaining the eigenvalues and the corresponding eigenfunctions of the operator $\mathscr{L}$. We do this by studying the *oscillations* of a solution u, that is, the zeros of u, of the homogeneous equation $\mathscr{L}u = 0$ in $[a,b]$. The study of oscillations will eventually lead us to the existence of eigenvalues and the eigenfunctions.

7.3   Oscillation Results

Consider the BVP for a second order linear homogeneous equation

$$\frac{d}{dt}\left(P(t)\frac{du}{dt}(t)\right) + Q(t)u(t) = 0, t \in (a,b), \tag{7.3.1}$$

along with the boundary conditions (7.2.3). Here $P > 0$ is differentiable and Q is continuous in the interval $[a,b]$. For the S–L system, we have $P = p$ and $Q = \lambda\rho - q$. Before going to the details of the oscillation results, we have the following result.

Theorem 7.3.1

Any non-trivial solution u ($u \not\equiv 0$) of (7.3.1) can have at most finite number of zeros in $[a,b]$.

Proof: Suppose not. Then, since $[a,b]$ is compact, there is a sequence $\{\xi_n\}$ of zeros of u such that $\xi_n \to \xi$ as $n \to \infty$ for some $\xi \in [a,b]$. By continuity, $u(\xi) = 0$. Put $h_n = \xi_n - \xi$, so that $h_n \to 0$ as $n \to \infty$ and $u(\xi + h_n) = 0$. We have $0 = \dfrac{u(\xi + h_n) - u(\xi)}{h_n}$. Letting $n \to \infty$, it follows that $\dot{u}(\xi) = 0$. Hence, by uniqueness, $u \equiv 0$. This contradiction proves the theorem. □

To study the oscillations of u, we introduce a powerful tool known as *Prüfer substitution*:

$$P(t)\dot{u}(t) = r(t)\cos\theta(t), \quad u(t) = r(t)\sin\theta(t). \tag{7.3.2}$$

We, then, have

$$r^2 = u^2 + P^2\dot{u}^2, \quad \theta = \arctan\left(\frac{u}{P\dot{u}}\right). \tag{7.3.3}$$

Here, r is called the *amplitude* variable and θ, the *phase* variable. When $P \equiv 1$, this gives the usual polar coordinates in the $(\dot{u}, u)$ plane. For $r \neq 0$, the correspondence $(P\dot{u}, u) = (r, \theta)$ as defined here, is smooth (in fact, *analytic*) with non-vanishing Jacobian. Also, for non-trivial solutions u, we have $r > 0$. For, if $r(t_0) = 0$, then $u(t_0) = 0$ and $P(t_0)\dot{u}(t_0) = 0$. Since $P > 0$, it follows that $u(t_0) = 0$ and $\dot{u}(t_0) = 0$. By uniqueness, $u \equiv 0$.

We now derive an equivalent system of ODEs for r and θ. Using $\tan\theta = \dfrac{u}{P\dot{u}}$ or $\cot\theta = \dfrac{P\dot{u}}{u}$, we obtain

$$-\csc^2(\theta)\,\dot{\theta} = \frac{u\dfrac{d}{dt}\left(P\dfrac{du}{dt}\right)}{u^2} - \frac{P\dot{u}^2}{u^2} = -Q - \frac{1}{P}\cot^2\theta.$$

This, after multiplication throughout by $\sin^2\theta$ gives

$$\dot{\theta} = Q\sin^2\theta + \frac{1}{P}\cos^2\theta \equiv F(t, \theta). \tag{7.3.4}$$

Similarly, by differentiating the expression $r^2 = u^2 + P^2\dot{u}^2$, with respect to t, we obtain

$$\dot{r} = \left(\frac{1}{P} - Q\right)r\sin\theta\cos\theta = \frac{1}{2}\left(\frac{1}{P} - Q\right)r\sin 2\theta. \tag{7.3.5}$$

This system is equivalent to the original system (7.3.1) in the sense that every non-trivial solution of (7.3.1) defines a unique solution of the ODEs (7.3.4) and (7.3.5) by the Prüfer substitution. Next, observe that $F(t, \theta)$ is Lipschitz with respect to θ, as

$$\frac{\partial F}{\partial \theta} = Q\sin 2\theta - \frac{1}{P}\sin 2\theta$$

and therefore

$$\left| \frac{\partial F}{\partial \theta} \right| \leq \sup_{t \in [a,b]} |Q(t)| + \sup_{t \in [a,b]} \frac{1}{|P(t)|}.$$

Hence, we obtain a unique solution θ defined on $[a,b]$ for any initial value $\theta(a) = \gamma$. Once θ is known, (7.3.5) for r gives

$$r(t) = r(a) \exp \left[\frac{1}{2} \int_a^t \left\{ \frac{1}{P(s)} - Q(s) \right\} \sin 2\theta(s)\, ds \right],$$

for all $t \in [a,b]$. Each solution of the Prüfer system depends on an initial amplitude $r(a)$ and an initial phase $\gamma = \theta(a)$. Changing $r(a)$ just multiplies the solution u by a constant factor. Thus, the zeros of any solution u can be located by studying only the ODE for the phase θ.

From (7.3.2), we see that the zeros of any non-trivial solution u of (7.3.1) occur where the phase function θ assumes the values $n\pi$, $n \in \mathbb{Z}$. At these points, $\cos^2 \theta = 1$ and $\dot{\theta} > 0$, as follows from (7.3.4). Geometrically, this means that the curve $(P(t)\dot{u}(t), u(t))$, $t \in [a,b]$ in the $(P\dot{u}, u)$ plane, corresponding to a solution u can cross the $P\dot{u}$-axis at $\theta = n\pi$ only counter-clockwise.

The advantage of Prüfer substitution in studying the zeros of the solution u is now evident from (7.3.4) satisfied by the phase variable. It is only a first order equation for θ and *does not* contain r and the solution exists in $[a,b]$ for any given initial condition. We will now make the following observations, which will be useful when we consider S–L systems.

1. If θ is a solution of (7.3.4), so are $-\theta$ and $\theta + n\pi$ for any integer n. We may thus fix the initial condition $\theta(a) = \gamma \in [0, \pi)$.

2. If we are just interested in the location of the zeros of u, then, it is sufficient to solve the first order equation (7.3.4) and find the points where θ takes the values $n\pi$, n a positive integer.

3. Fix a non-negative integer n. If there is a $t_n \in [a,b]$ such that $\theta(t_n) = n\pi$, then, from (7.3.4), it follows that $\dot{\theta}(t_n) = \dfrac{1}{P(t_n)} > 0$. Hence, $\theta(t) > n\pi$ for $t > t_n$, close to t_n.

   We claim that $\theta(t) > n\pi$ for all $t > t_n$. For, if there is a $t > t_n$ such that $\theta(t) = n\pi$, then we would have that $\dot{\theta}(t) \leq 0$. But this contradicts the fact that $\dot{\theta}(t) = \dfrac{1}{P(t)} > 0$, which follows from (7.3.4). Though θ need not be a monotonically increasing function (it is if Q is also non-negative), it remains above the line $\theta = n\pi$, (n any non-negative integer) once it crosses that line, for all future times. In particular, for the chosen initial condition, $\theta > 0$ in $(a,b]$.

The existence of zeros of a non-trivial solution will now be done using comparison theorems.

7.3.1   Comparison theorems

We, now consider the following two equations similar to (7.3.1)

$$\frac{d}{dt}\left(P_i(t)\frac{du_i}{dt}\right) + Q_i(t)u_i(t) = 0, t \in (a,b), \tag{7.3.6}$$

along with the boundary conditions (7.2.3), for $i = 1,2$. Here P_is are positive differentiable functions and Q_is are continuous functions defined on $[a,b]$. If θ_is are the corresponding phase functions given by the Prüfer substitution, then the phase functions satisfy the following equations (see (7.3.4))

$$\dot{\theta}_i = Q_i\sin^2\theta_i + \frac{1}{P_i}\cos^2\theta_i \equiv F_i(t,\theta_i), \tag{7.3.7}$$

for $i = 1,2$. We now wish to compare the solutions θ_1 and θ_2, which will enable us to compare the zeros of the solutions u_1 and u_2.

Theorem 7.3.2

Suppose $F(t,y)$ is a continuous function defined in the strip $D = [a,b] \times \mathbb{R}$. Assume F is Lipschitz continuous with respect to y in D. Let the functions f,g defined on $[a,b]$ satisfy

$$\dot{f}(t) \le F(t,f(t)), \quad \dot{g}(t) = F(t,g(t)),$$

for all $t \in [a,b]$. If $f(a) = g(a)$, then, $f(t) \le g(t)$ for all $t \in [a,b]$.

Proof: Put $\sigma(t) = f(t) - g(t)$, $t \in [a,b]$. Suppose the conclusion of the lemma is false. Then, we can find, by continuity, a subinterval $[a_1,b_1]$ of $[a,b]$ such that $\sigma(a_1) = 0$, $\sigma(t) > 0$ for $t \in (a_1,b_1)$. For $t \in [a_1,b_1]$, we have

$$\dot{\sigma}(t) = \dot{f}(t) - \dot{g}(t) \leq F(t,f(t)) - F(t,g(t)) \leq L\sigma(t).$$

Here, L is the Lipschitz constant of F and we have tacitly used the condition that σ is non-negative in $[a_1,b_1]$. Integrating this inequality, we obtain the estimate $\sigma(t) \leq \sigma(a_1)e^{L(t-a_1)}$ for $t \in [a_1,b_1]$. Since $\sigma(a_1) = 0$, this gives a contradiction to the assumption that $\sigma > 0$ in (a_1,b_1) and completes the proof. $\qquad\square$

Theorem 7.3.3

[Comparison Theorem]   Suppose F,G are continuous functions defined in the strip $D = [a,b] \times \mathbb{R}$. Assume, either F or G is Lipschitz continuous with respect to the second variable in D and that $F(t,y) \leq G(t,y)$ for all $(t,y) \in D$. Let f,g be solutions of the DEs

$$\dot{f}(t) = F(t,f(t)), \dot{g}(t) = G(t,g(t)),$$

respectively, for $t \in [a,b]$. If $f(a) = g(a)$, then $f(t) \leq g(t)$ for all $t \in [a,b]$.

Proof: First assume that G satisfies a Lipschitz condition. Then, we have $\dot{f}(t) = F(t, f(t)) \leq G(t,f(t))$, so that f,g satisfy the hypothesis in Theorem 7.3.2 and the result follows.

Assume now that F satisfies a Lipschitz condition. Put $h(t) = -f(t)$ and $k(t) = -g(t)$ for $t \in [a,b]$. Then,

$$\dot{h}(t) = -\dot{f}(t) = -F(t,f(t)) = -F(t,-h(t)).$$

Similarly, for k, we have

$$\dot{k}(t) = -G(t,-k(t)) \leq -F(t,-k(t)).$$

Applying Theorem 7.3.2 again to functions h,k and $-F(t,-h(t))$, we obtain that $k(t) \leq h(t)$ for all $t \in [a,b]$. This completes the proof. $\qquad\square$

Carefully observing the aforementioned proofs, we obtain the following strict inequalities in the conclusion.

Corollary 7.3.4

[Corollary to Theorem 7.3.2] For any $t_1 \in (a,b]$, either $f(t_1) < g(t_1)$ or $f \equiv g$ in $[a,t_1]$.

Proof: Either $f \equiv g$ in $[a,t_1]$, or there is a $t_0 \in [a,t_1]$ such that $f(t_0) < g(t_0)$. Then, the function $\tilde{\sigma}$ defined by $\tilde{\sigma}(t) = g(t) - f(t)$ for $t \in [a,t_1]$ is non-negative and $\tilde{\sigma}(t_0) > 0$. We then have

$$\dot{\tilde{\sigma}}(t) = \dot{g}(t) - \dot{f}(t) \geq F(t,g(t)) - F(t,f(t)) \geq -L\tilde{\sigma}(t).$$

Here, L is the Lipschitz constant of F and we have tacitly used the condition that $\tilde{\sigma}$ is non-negative in $[a,t_1]$. Integrating this inequality, we obtain the estimate $\tilde{\sigma}(t) \geq \tilde{\sigma}(t_0)e^{-L(t-t_0)}$ for $t \in [a,t_1]$. This gives strict inequality $\tilde{\sigma}(t) > 0$ as $\tilde{\sigma}(t_0) > 0$, and completes the proof. $\qquad\square$

Corollary 7.3.5

[Corollary to Theorem 7.3.3] Suppose both F and G satisfy Lipschitz conditions. If $f(a) < g(a)$, then $f(t) < g(t)$ for all $t \in [a,b]$.

Proof: Suppose the conclusion is false. Then, we can find a $t_1 > a$ such that $f(t_1) = g(t_1)$. Now, consider the functions ϕ and ψ defined by

$$\phi(t) = f(-t), \ \psi(t) = g(-t), t \in [-t_1, -a].$$

Then, ϕ and ψ satisfy the DEs

$$\dot{\phi}(t) = -F(-t,\phi(t)), \dot{\psi}(t) = -G(-t,\psi(t)),$$

for $t \in [-t_1, \ a]$ and satisfy the condition $\phi(-t_1) = \psi(-t_1)$. Since $-F(-t,y) \geq -G(-t,y)$, we can apply Theorem 7.3.3 in the interval $[-t_1,-a]$. We conclude that $\phi(-a) \geq \psi(-a)$. This implies $f(a) \geq g(a)$, a contradiction. The proof is complete. $\qquad\square$

We are now going to apply these comparison results, especially the corollaries 7.3.4 and 7.3.5, to compare the phases θ_1 and θ_2 satisfying (7.3.7). Suppose $P_1 \geq P_2 > 0$ and $Q_1 \leq Q_2$ in $[a,b]$. If $\theta_1(a) \leq \theta_2(a)$, then $\theta_1(t) \leq \theta_2(t)$ for all $t \in [a,b]$. Furthermore, $\theta_1(b) = \theta_2(b)$ only if $\theta_1 \equiv \theta_2$ in $[a,b]$.

We will discuss the equality case in more detail. Suppose $\theta_1 \equiv \theta_2 = \theta$, say, in $[a,b]$. Since $\dot{\theta}(t) = \dfrac{1}{P(t)} > 0$, when $\sin\theta(t) = 0$, the zeros of $\sin\theta(t)$ are only isolated. Now, from (7.3.7), we obtain

$$(Q_2 - Q_1)\sin^2\theta + \left(\frac{1}{P_2} - \frac{1}{P_1}\right)\cos^2\theta = 0.$$

Since, by assumption, both the terms are non-negative, they must separately vanish in $[a,b]$. This gives that $Q_1 \equiv Q_2$. Therefore, we then have, $P_1 \equiv P_2$, except on intervals where $\cos\theta$ vanishes. On such intervals, using (7.3.4), we obtain $\theta \equiv$ constant and therefore, using (7.3.7), we get $Q_1 \equiv Q_2 \equiv 0$. In this situation, we also show that u_1 and u_2 are linearly dependent. On the intervals where $\cos\theta$ vanishes, both u_1 and u_2 are constants, hence they are multiples of each other. On the intervals where $\cos\theta$ does not vanish, we have, $P_1 \equiv P_2$ and therefore u_1 and u_2 satisfy the same second order linear homogeneous equation. Now looking at their Wronskian, we see that they are linearly dependent. By continuity, it follows that $u_2 = cu_1$ on $[a,b]$ for some constant c. We have thus proved the following.

Theorem 7.3.6

[Sturm Comparison Theorem]   Assume P_1, P_2 are in $C^1[a,b]$, $P_1 \geq P_2 > 0$ and $Q_1 \leq Q_2$ are continuous functions in $[a,b]$. Then, between any two zeros of a non-trivial solution u_1 of (7.3.6) with $i = 1$, there lies at least one zero of every real solution u_2 of (7.3.6) with $i = 2$, except when $u_2 \equiv cu_1$, for some constant c. In the latter case, we have $P_1 \equiv P_2$ and $Q_1 \equiv Q_2$, except possibly on intervals where $Q_1 \equiv Q_2 \equiv 0$.

We are now in a position to discuss the oscillation results for the solutions of the Sturm–Liouville system

$$\mathscr{L}u(t) \equiv -\frac{d}{dt}\left(p(t)\frac{du}{dt}\right) + q(t)u(t) = \lambda\rho(t)u(t), \qquad (7.3.8)$$

satisfying the boundary conditions (7.2.3). Comparing (7.3.8) with the equation (7.3.1), we find that $P = p$ and $Q = \lambda\rho - q$. We would like to study the number of zeros of a non-trivial solution of (7.3.8) as the real parameter λ varies. Denote by $\theta(t,\lambda)$, the corresponding phase variable. Then, by (7.3.4), we have

$$\dot{\theta}(t,\lambda) = [\lambda\rho(t) - q(t)]\sin^2\theta(t,\lambda) + \frac{1}{p(t)}\cos^2\theta(t,\lambda), \qquad (7.3.9)$$

for $t \in [a,b]$. Recall that p is a positive C^1 function and $\rho > 0$, q are continuous functions defined on $[a,b]$. Now, fix a real number γ and consider the solution $\theta(t,\lambda)$ of (7.3.9) satisfying $\theta(a,\lambda) = \gamma$, for all λ. Here, γ is determined by the boundary condition (7.2.3) at a as

$$\alpha_1 \sin \gamma + \frac{\alpha_2}{p(a)} \cos \gamma = 0. \tag{7.3.10}$$

There is a unique solution $\gamma \in [0, \pi)$ of (7.3.10). If $\alpha_1 \neq 0$, we have, $\tan \gamma = -\frac{\alpha_2}{\alpha_1}$; if $\alpha_1 = 0$, then, put $\gamma = \frac{\pi}{2}$ and $\tan \gamma = \infty$.

We are now going to obtain the following results as a direct consequence of the comparison theorems and their corollaries proved earlier. See also the observations made on the phase function in the previous section.

Lemma 7.3.7

For any fixed $t > a$, the phase variable $\theta(t, \lambda)$ is a strictly increasing function of λ.

This is a direct consequence of Corollary 7.3.5.

Lemma 7.3.8

Suppose for some $t_n > a$, $\theta(t_n, \lambda) = n\pi$, where n is a non-negative integer. Then, $\theta(t, \lambda) > n\pi$ for all $t > t_n$.

Proof: This follows from the third observation made earlier in the previous section. □

Lemma 7.3.8, combined with the condition $0 \leq \gamma = \theta(a, \lambda) < \pi$, gives the first zero of u in (a, b) when $\theta = \pi$ and the nth zero when $\theta = n\pi$. We now analyze the behavior of $\theta(t, \lambda)$ for fixed $t > a$ and as $\lambda \to \infty$. Denote by $t_n(\lambda)$, the *smallest* $t > a$ for which $\theta(t, \lambda) = n\pi$; here n is a positive integer.

Lemma 7.3.9

Fix a positive integer n. Then, for sufficiently large λ, $t_n(\lambda)$ is defined and is a continuous function. It is a decreasing function of λ, and $\lim_{\lambda \to \infty} t_n(\lambda) = a$.

Proof: From the continuous dependence results in Chapter 4, it follows that the solution $\theta(t, \lambda)$ is a continuous function of $t \in [a, b]$ and $\lambda \in \mathbb{R}$.

Assuming that $t_n(\lambda)$ is well-defined, we first observe that it is a decreasing function of λ. Suppose $\lambda_1 < \lambda_2$. Then,

$$\theta(t_n(\lambda_1), \lambda_1) = n\pi = \theta(t_n(\lambda_2), \lambda_2).$$

By Lemma 7.3.7, we have

$$n\pi = \theta(t_n(\lambda_1), \lambda_1) < \theta(t_n(\lambda_1), \lambda_2),$$

and since $t_n(\lambda_2)$ is the smallest number such that $\theta(t_n(\lambda_2), \lambda_2) = n\pi$, it follows from Lemma 7.3.8 that $t_n(\lambda_1) \geq t_n(\lambda_2)$, proving the assertion of monotonicity.

We now show that $t_n(\lambda)$ is well-defined for large enough λ and prove the limit stated in the theorem. We need to show that for large enough λ, there is $t > a$ such that $\theta(t, \lambda) = n\pi$. We will do this by comparing any non-trivial solution of (7.3.8) and its phase variable given by the Prüfer substitution and satisfying (7.3.9). Thus, it suffices to show that any non-trivial solution of (7.3.8) has at least n zeros in (a, b), which will imply that $\theta(t, \lambda)$ being a continuous function of t, must assume all the values between $\theta(a, \lambda) = \gamma < \pi$ and $n\pi$.

Let q_M and p_M be the maxima of the functions q and p over $[a, b]$ respectively; and ρ_m be the minimum of the function ρ over $[a, b]$. Consider the following S–L system with constant coefficients:

$$p_M \ddot{v}(t) + (\lambda \rho_m - q_M) v(t) = 0, \tag{7.3.11}$$

for $t \in [a, b]$. If we choose λ such that $\lambda > \dfrac{q_M}{\rho_m}$, then (7.3.11) has a non-trivial solution given by $v(t) = \sin k(t - a)$, where $k^2 = (\lambda \rho_m - q_M)/p_M$. Observe that $v(a) = 0$ and $p \leq p_M$ and $(\lambda \rho_m - q_M) \leq (\lambda \rho - q)$ in $[a, b]$ and hence, by the Sturm comparison theorem (Theorem 7.3.6), any non-trivial solution u of (7.3.8) must have at least one zero between any two zeros of v. The zeros of v are given by $a + \frac{n\pi}{k}$, n an integer; the distance between any two consecutive zeros is therefore $\frac{\pi}{k}$. Thus, v has n zeros in (a, b) for large enough λ. This proves that u also has at least n zeros and therefore, $\theta(t, \lambda)$ takes the value $n\pi$ when λ is large enough. Thus, $t_n(\lambda)$ is well-defined.

Note that $t_n(\lambda)$ lies between $a + \dfrac{(n-1)\pi}{k}$ and $a + \dfrac{n\pi}{k}$. This immediately gives the continuity of t_n with respect to λ and also proves the required limit. This completes the proof of the lemma. $\square$

We now come to the main oscillation result.

Theorem 7.3.10

[Oscillation Theorem]

Let $\theta(t,\lambda)$ be the solution of (7.3.9) satisfying the initial condition $\theta(a,\lambda) = \gamma, 0 \leq \gamma < \pi$ for each λ. Then, for fixed $t > a$, $\theta(t,\lambda)$ is a continuous and strictly increasing function of λ. Moreover, for $t \in (a,b]$,

$$\lim_{\lambda \to \infty} \theta(t,\lambda) = \infty, \quad \lim_{\lambda \to -\infty} \theta(t,\lambda) = 0. \tag{7.3.12}$$

Proof: Lemmas 7.3.7–7.3.9 contain all the conclusions of the theorem, except the statement regarding the second limit. We now proceed to show the second assertion in (7.3.12). Given any $\varepsilon > 0$, we need to show that there is a Λ, depending on ε and perhaps on $t \in (a,b]$, such that $\theta(t,\lambda) < \varepsilon$ for any $\lambda < \Lambda$. This proves the assertion, as we have already seen that $\theta > 0$ in $(a,b]$.

Put $|q|_M = \sup_{t \in [a,b]} |q(t)|$ and $p_m = \inf_{t \in [a,b]} p(t)$. Then, it follows from (7.3.9) that

$$\dot{\theta}(t,\lambda) \leq \lambda p(t) \sin^2 \theta(t,\lambda) + K, \tag{7.3.13}$$

for $t \in [a,b]$, where $K = |q|_M + \dfrac{1}{p_m}$. Choose γ_1 such that $\gamma < \gamma_1 < \pi$ and let $\varepsilon > 0, t_1 \in (a,b]$. We may assume that, by taking a smaller ε if necessary,

$$\varepsilon < \gamma_1 \leq \pi - \varepsilon.$$

Now consider the straight line $\theta = s(t)$ in the $t\theta$-plane joining the points (a,γ_1) and (t_1,ε). Its slope is

$$m = -\frac{\gamma_1 - \varepsilon}{t_1 - a} < 0.$$

Observe that $\theta(a,\lambda) = \gamma < \gamma_1 = s(a)$. If we show that the graph of the solution of (7.3.9), for $\lambda < 0$ with large absolute value, lies below this line on the interval $[a,t_1]$, then it follows that $\theta(t_1,\lambda) < s(t_1) = \varepsilon$ and we are done.

From (7.3.13), it follows that $\dot{\theta} \leq K$ for $\lambda < 0$, and so we first obtain that $\theta(t,\lambda) \leq \gamma + K(t-a)$. Therefore, $\theta(t,\lambda) < s(t)$ provided that $t \geq a$

and $t - a < \dfrac{\gamma_1 - \gamma}{K - m}$. Thus, the solution curve lies below the straight line, in $[a, a_1]$ for some $a_1 > a$. Now suppose, if possible, that $\theta(t, \lambda) > s(t)$ for some $t \in [a, t_1]$. We will obtain a contradiction for a choice of λ.

By continuity, we can find the smallest $t_* \in [a, t_1]$ such that $\theta(t_*, \lambda) = s(t_*)$ with $\dot{\theta}(t_*, \lambda) \geq m$, since the solution curve can cross the straight line only from below. Then, observe that $\theta(t_*, \lambda) = s(t_*) = \gamma_1 + m(t_* - a)$. Substituting the expression for m and the upper bound for γ_1, we see that $\theta(t_*, \lambda) \in [\varepsilon, \pi - \varepsilon]$. Therefore, $\sin \theta(t_*, \lambda) \geq \varepsilon$.

Let $\rho_* = \inf\limits_{t \in [a, t_1]} \rho(t)$ and choose $\Lambda = \dfrac{m - K}{\rho_* \sin^2 \varepsilon}$. Then, remembering that $\lambda < 0$, we obtain the following contradiction

$$m \leq \dot{\theta}(t_*, \lambda) \leq \lambda \rho(t_*) \sin^2 \theta(t_*, \lambda) + K \leq \lambda \rho_* \sin^2 \varepsilon + K < m,$$

provided that $\lambda < \Lambda$. This shows that $\theta(t_1, \lambda) < \varepsilon$ if $\lambda < \Lambda$ and completes the proof.   □

7.3.2   Location of zeros

We now obtain some estimates on the location of zeros of a non-trivial solution of (7.3.8) satisfying the condition (7.2.3) at a. To this end, we consider the following S–L system with constant coefficients, as in the proof of Lemma 7.3.9:

$$p_m \ddot{u} + (\lambda \rho_M - q_m) u = 0, \tag{7.3.14}$$

where p_m and q_m are the minima of p and q over $[a, b]$, respectively; ρ_M is the maximum of ρ over $[a, b]$. For the solutions of (7.3.11) and (7.3.14), we impose the initial condition, $\tan \gamma = -\dfrac{\alpha_2}{\alpha_1}$ if $\alpha_1 \neq 0$. The solutions of these equations can be found explicitly and hence, their zeros, for suitable values of λ. The nth zero of the solutions of (7.3.11) and (7.3.14) are, respectively, given by

$$a + \dfrac{(n\pi - \gamma)\sqrt{p_M}}{\sqrt{\lambda \rho_m - q_M}} \quad \text{and} \quad a + \dfrac{(n\pi - \gamma)\sqrt{p_m}}{\sqrt{\lambda \rho_M - q_m}}.$$

By Sturm comparison theorem, we therefore, have the following result:

Lemma 7.3.11

Let u be a non-trivial solution of the S–L system (7.3.8) satisfying the condition $\tan\gamma = \dfrac{u(a)}{p(a)\dot{u}(a)}$. If t_n denotes the nth zero of u in (a,b), then

$$\frac{\sqrt{p_m}}{\sqrt{\lambda\rho_M - q_m}} \leq \frac{t_n - a}{n\pi - \gamma} \leq \frac{\sqrt{p_M}}{\sqrt{\lambda\rho_m - q_M}}. \tag{7.3.15}$$

The preceding estimates were proved under the assumption that $\alpha_1 \neq 0$. The same argument works when $\beta_1 \neq 0$ by changing variable t to $a+b-t$. If both α_1 and β_1 are zero, we can still prove the forgoing estimates by taking $\gamma = \pi/2$.

7.4   Existence of Eigenfunctions

We will now show the existence of eigenvalues and the corresponding eigenfunctions to the S–L system (7.3.8) with boundary conditions in (7.2.3), using the results developed in the previous sections.

Theorem 7.4.1

[Existence of Eigenfunctions] The regular S–L system (7.3.8) has an infinite sequence of real eigenvalues $\lambda_0 < \lambda_1 < \cdots$ with $\lim_{n\to\infty} \lambda_n - \infty$.

The eigenfunction u_n corresponding to eigenvalue λ_n has exactly n zeros in (a,b). Also, u_n is unique up to a constant factor.

Proof: The initial condition on the phase variable θ ensures that the boundary condition (7.2.3) is satisfied at a for the corresponding solution u of (7.3.8). This u will be an eigenfunction if the boundary condition at b in (7.2.3) is also satisfied. Translating this condition at b on θ, we see that $\theta(b,\lambda) = \delta + n\pi$ for $n = 0,1,2,\cdots$, provided that δ satisfies

$$\beta_1 \sin\delta + \beta_2(p(b))^{-1}\cos\delta = 0. \tag{7.4.1}$$

There is a unique value $\delta \in (0,\pi]$ satisfying (7.4.1). For this value of δ, we now ask whether there is a λ satisfying $\theta(b,\lambda) = \delta$. Theorem 7.3.10 gives the positive answer. Call this value λ_0 and the corresponding eigenfunction u_0. Since $\delta \leq \pi$, it follows that $\theta < \pi$ in (a,b) and therefore, u_0 does not vanish in this interval.

Next, we ask whether there is a λ satisfying $\theta(b,\lambda) = \delta + \pi$. The answer is again yes, because of Theorem 7.3.10. Call this value λ_1 and the corresponding eigenfunction u_1. This eigenfunction has a zero in (a,b) since θ takes the value π there. Continuing in this fashion and repeatedly invoking Theorem 7.3.10, we obtain all the eigenvalues and the corresponding eigenfunctions.

For the last assertion in the theorem, let u and v be two eigenfunctions corresponding to the same eigenvalue λ and W be their Wronskian. From the boundary condition (7.2.3) at a, we see that $W(a) = 0$ and hence $W \equiv 0$. Therefore, u and v are linearly dependent and the proof is complete. $\square$

7.5 Exercises

1. Let u, v satisfy the following equations

$$\frac{d}{dt}\left(P_1(t)\frac{du}{dt}(t)\right) - Q_1(t)u(t) = 0,$$

$$\frac{d}{dt}\left(P_2(t)\frac{dv}{dt}(t)\right) - Q_2(t)v(t) = 0,$$

on some interval in $\mathbb{R}$, where $P_1 \geq P_2 > 0$ are differentiable functions and $Q_1 \geq Q_2$ are continuous functions. If v does not vanish at any point in a closed interval $[a,b]$, show that

$$\left[\frac{u}{v}(P_1\dot{u}v - P_2u\dot{v})\right]_a^b = \int_a^b (Q_1 - Q_2)u^2\, dt + \int_a^b (P_1 - P_2)\dot{u}^2\, dt$$

$$+ \int_a^b P_2 \frac{(\dot{u}v - u\dot{v})^2}{v^2}\, dt$$

where $[\chi]_a^b = \chi(b) - \chi(a)$. This formula is known as the *Picone formula*. Deduce Sturm comparison theorem from this formula.

2. Suppose u satisfies the following equations

$$\frac{d}{dt}\left(P_1(t)\frac{du}{dt}(t)\right) - Q_1(t)u(t) = 0,$$

$$\frac{d}{dt}\left(P_2(t)\frac{du}{dt}(t)\right) - Q_2(t)u(t) = 0,$$

on some interval in $\mathbb{R}$, where $P_1 \geq P_2 > 0$ are differentiable functions and $Q_1 \geq Q_2$ are continuous functions. If a and b are two consecutive zeros of u, show that $P_1 \equiv P_2$ and $Q_1 \equiv Q_2$ in $[a,b]$.

3. Consider the constant coefficient equation

$$P\ddot{u} + Qu = 0,$$

where P, Q are positive constants. Write down the corresponding equation of the phase variable given by the Prüfer substitution and solve it.

4. Verify the estimates in (7.3.15).

5. Suppose P is a positive C^1 function on $[a,b]$. Solve explicitly, the equations

$$P\ddot{u} + \frac{u}{P} = 0 \text{ and } P\ddot{u} - \frac{u}{P} = 0.$$

6. In the following exercises, a second order differential operator $\mathscr{L}$ is given on an interval I along with the boundary conditions. Find the eigenvalues and eigenfunctions of the resulting Sturm–Liouville problem.

   (a) $\mathscr{L}u = \ddot{u}, u(0) = \dot{u}(1) = 0, I = [0,1]$.

   (b) $\mathscr{L}u = \alpha\ddot{u}, u(0) = u(l) = 0, I = [0,l] \, (\alpha > 0)$.

   (c) $\mathscr{L}u = \ddot{u}, \dot{u}(0) = \dot{u}(1) = 0, I = [0,1]$.

7. In the text, we have only considered boundary conditions which are separated at the end points of the given interval $[a,b]$. Now consider the following boundary conditions for $\mathscr{L}$:

$$\mathbf{A} \begin{bmatrix} u(a) \\ \dot{u}(a) \end{bmatrix} + \mathbf{B} \begin{bmatrix} u(b) \\ \dot{u}(b) \end{bmatrix} = \begin{bmatrix} 0 \\ 0 \end{bmatrix}$$

with

$$p(a)\det(\mathbf{B}) = p(b)\det(\mathbf{A}).$$

Here $\mathbf{A}, \mathbf{B}$ are 2×2 real matrices such that the block matrix $[\mathbf{A} \quad \mathbf{B}]$ has rank 2. Verify that the boundary conditions given in (7.2.3) satisfy the aforementioned conditions.

8. Find the eigenvalues and eigenfunctions of the following Sturm–Liouville problems:

(a) $\mathscr{L}u = \ddot{u}, I = [0,1], u(0) = \dot{u}(1), \dot{u}(0) = \dot{u}(1).$

(b) $\mathscr{L}u = \ddot{u} - u, I = [0,1], u(0) - \dot{u}(0) + u(1) = 0,$
$u(0) + \dot{u}(0) + 2\dot{u}(1) = 0.$

7.6 Notes

We have studied regular Sturm–Liouville boundary value problems in this chapter, mainly concentrating on the existence of eigenvalues and the corresponding eigenfunctions. We have followed the approach in [BR03] by the consideration of the Prüfer substitution. For other approaches to this problem, the reader is referred to [Inc26, CL72, Sim91, SK07] among others. We have not done the expansion in terms of eigenfunctions, as this requires tools from Hilbert space. We also have not considered the more difficult topic of singular S–L systems. Representation of solutions of BVP through Green's function will be taken up in Chapter 9. We may also use the integral operator defined through Green's function to show the existence of eigenvalues and eigenfunctions; but this also requires tools from functional analysis (compact operators).

8

Qualitative Theory

8.1 Introduction

Nonlinear dynamics, essentially concentrated around the study of planetary motions, has some claim to be the most ancient of scientific problems, perhaps as old as geometry. It, therefore, seems surprising that until the twentieth century, geometric methods in nonlinear dynamics were not much pursued. Henri Poincarè is universally acknowledged as the founder of geometric dynamics, followed by G. D. Birkoff. But apart from a few instances such as the stability analysis of Liapunov,[1] Poincarè's ideas seemed to have had little impact on applied dynamics for almost half a century. A reason perhaps could be that Poincarè and Birkoff concentrated on conservative systems motivated by problems in celestial mechanics. Dissipative systems, on the other hand, have the property that an evolving ensemble of states occupies a region of phase space whose volume decreases with time. Over a long period of time, this volume has the tendency to simplify the topological structure of the orbits in the phase space; this may be true even in an infinite dimensional phase space, for example, governed by a partial differential equation.

In this chapter, we study the qualitative behavior of solutions to nonlinear ODE. We wish to do this by plotting the phase portrait of these systems similar to the one that was done for 2×2 linear systems in Chapter 5. The material in this chapter will be developed through important examples described in Chapter 1, which will be recalled in the sequel frequently.

---

[1] also spelled Lyapunov in some text books

We close this section with a few remarks on the phase portrait. There is a similarity between the plotting of a phase portrait of a system and plotting of a plane or space curve given by a parametric representation. In both the situations, we suppress the independent variable t while plotting the curve in question. We have already observed this in great detail while analyzing 2×2 linear systems. In general, it will be more difficult to have a complete phase portrait for nonlinear systems. Physically, the position vector $\mathbf{x}(t)$ and its velocity vector $\dot{\mathbf{x}}(t)$ are called *phases* of the system, hence the name *phase portrait*.

8.2   General Definitions and Results

In this section, we will introduce several concepts that will be used throughout this chapter. Consider a system of n first order equations:

$$\dot{\mathbf{x}} = \mathbf{f}(\mathbf{x}), \tag{8.2.1}$$

or, as explicitly written

$$\dot{x}_j = f_j(x_1, x_2, \cdots, x_n), \quad j = 1, 2, \cdots, n.$$

Here x_js, the unknowns, the components of $\mathbf{x}$, and f_j are real valued functions defined on $\mathbb{R}^n$, the components of the vector valued function $\mathbf{f}$. The positive integer n is referred to as the *dimension* of the system. Since an ODE of any given order can be written in the form of a system of first order equations, the aforementioned consideration is more general. System (8.2.1) is called an *autonomous* system as the right side function $\mathbf{f}$ does not depend on t explicitly. If $\mathbf{f}$ depends on t explicitly as well, the system is referred to as *non-autonomous*. For example, the equation $\dot{x} = x + t$ (the 1D or one-dimensional equation) is non-autonomous. A non-autonomous system may be converted into an autonomous system by increasing the dimension of the system by 1 as follows: Introduce a new unknown variable x_{n+1} satisfying $\dot{x}_{n+1} = 1$. Then, an nth order non-autonomous system $\dot{\mathbf{x}} = \mathbf{f}(\mathbf{x}, t)$ can be written as an $(n+1)^{\text{th}}$ order autonomous system $\dot{\mathbf{X}} = \mathbf{F}(\mathbf{X})$, where $\mathbf{X} = (\mathbf{x}, x_{n+1})$ and $\mathbf{F} = (\mathbf{f}, 1)$. However, even after this reduction, the analysis of a non-autonomous system does not become easier. One reason for this is that, since $\dot{x}_{n+1} = 1$, the reduced system does not have any equilibrium points (defined later in this section). For example, if we consider the linear system $\dot{\mathbf{x}} = \mathbf{A}(t)\mathbf{x}(t)$, then $\mathbf{x} = \mathbf{0} \in \mathbb{R}^n$ is an equilibrium point for the unreduced system, but $\mathbf{0} \in \mathbb{R}^{n+1}$ is not an equilibrium point for the enlarged system, namely $\dot{\mathbf{x}}(t) = \mathbf{A}(t)\mathbf{x}(t), \dot{x}_{n+1}(t) = 1$.

Usual assumptions on $\mathbf{f}$ are made so that the system (8.2.1) has a unique, global solution for a prescribed initial condition; global means existence for all t; see Chapter 4. Even when a solution does not exist for all t, we can still plot the phase portrait by considering the maximum interval of existence of the solution in question; we will consider an example later. However, uniqueness plays a crucial role.

The solution $\mathbf{x}(t)$ has the geometrical meaning of a curve in $\mathbb{R}^n$ and (8.2.1) gives its tangent vector at every $\mathbf{x}(t)$. For this reason, $\mathbf{f}$ in (8.2.1) is also referred to as a *vector field* and the corresponding solution as an *integral curve* of the vector field.

If $\mathbf{x}$ is a solution of (8.2.1), we say that $\mathbf{x}$ passes through $\mathbf{x}^0 \in \mathbb{R}^n$ if $\mathbf{x}(t_0) = \mathbf{x}^0$, for some $t_0 \in \mathbb{R}$.

Definition 8.2.1

Given a solution $\mathbf{x}$ of (8.2.1) passing through $\mathbf{x}^0 \in \mathbb{R}^n$ with $\mathbf{x}(t_0) = \mathbf{x}^0$, for some $t_0 \in \mathbb{R}$, the **orbit** through $\mathbf{x}^0$, is the set $\mathcal{O}(\mathbf{x}^0)$ defined by

$$\mathcal{O}(\mathbf{x}^0) = \{\mathbf{x}(t) \in \mathbb{R}^n : t \in \mathbb{R}\},$$

and the **positive orbit** through $\mathbf{x}^0$ is the set $\mathcal{O}^+(\mathbf{x}^0)$ defined by

$$\mathcal{O}^+(\mathbf{x}^0) = \{\mathbf{x}(t) \in \mathbb{R}^n : t \geq t_0\}.$$

Lemma 8.2.3 shows that any solution passing through $\mathbf{x}^0$ may be used to define $\mathcal{O}(\mathbf{x}^0)$ or $\mathcal{O}^+(\mathbf{x}^0)$ unambiguously. Generally speaking, the phase space (plane) analysis is about describing all the (positive) orbits of (8.2.1). The other terminologies used for orbit are **trajectory** and **path.**

We will now discuss some important properties of solutions of autonomous systems. In the following results, statements regarding t refer to all $t \in \mathbb{R}$.

Lemma 8.2.2

If $\mathbf{x}$ is a solution of (8.2.1), define $\mathbf{x}_c$ by $\mathbf{x}_c(t) = \mathbf{x}(t+c)$ for any fixed c and for all t. Then, $\mathbf{x}_c$ is also a solution of (8.2.1).

Proof: Direct differentiation.   □

We remark that this lemma is not true for a non-autonomous system.

Lemma 8.2.3

If $\mathbf{x}$ and $\mathbf{y}$ are solutions of (8.2.1) passing through $\mathbf{x}^0 \in \mathbb{R}^n$ with $\mathbf{x}(t_0) = \mathbf{y}(t_1) = \mathbf{x}^0$, for some $t_0, t_1 \in \mathbb{R}$, then $\mathbf{y}(t) = \mathbf{x}(t + t_0 - t_1)$ and $\mathbf{x}(t) = \mathbf{y}(t + t_1 - t_0)$ for all t.

proof: Define $\mathbf{z}(t) = \mathbf{x}(t + t_0 - t_1)$ for $t \in \mathbb{R}$. By Lemma 8.2, $\mathbf{z}$ is a solution of (8.2.1) and $\mathbf{z}(t_1) = \mathbf{x}(t_0) = \mathbf{y}(t_1)$. By uniqueness, $\mathbf{z} \equiv \mathbf{y}$. This completes one part of the proof and the other part is similar.   □

This lemma shows that $\mathcal{O}(\mathbf{x}^0)$ or $\mathcal{O}^+(\mathbf{x}^0)$ is the same set whether $\mathbf{x}$ or $\mathbf{y}$ is used in its definition.

Corollary 8.2.4

If $\mathbf{x}^0, \mathbf{x}^1 \in \mathbb{R}^n$ and $\mathbf{x}^1 \in \mathcal{O}(\mathbf{x}^0)$ (respectively, $\mathbf{x}^1 \in \mathcal{O}^+(\mathbf{x}^0)$), then, $\mathcal{O}(\mathbf{x}^0) = \mathcal{O}(\mathbf{x}^1)$ (respectively, $\mathcal{O}^+(\mathbf{x}^0) \supset \mathcal{O}^+(\mathbf{x}^1)$).

Lemma 8.2.5.

Let, $\mathbf{x}^0, \mathbf{x}^1 \in \mathbb{R}^n$, then, either $\mathcal{O}(\mathbf{x}^0) = \mathcal{O}(\mathbf{x}^1)$ or $\mathcal{O}(\mathbf{x}^0) \cap \mathcal{O}(\mathbf{x}^1) = \phi$, the empty set.

Proof: If $\tilde{\mathbf{x}} \in \mathcal{O}(\mathbf{x}^0) \cap \mathcal{O}(\mathbf{x}^1)$, then by Corollary 8.2.4, it follows that $\mathcal{O}(\mathbf{x}^0) = \mathcal{O}(\tilde{\mathbf{x}}) = \mathcal{O}(\mathbf{x}^1)$ and the proof is complete.   □

Similar statements may be made regarding the positive orbits.

Lemma 8.2.6

Suppose $\mathbf{x}$ is a solution of (8.2.1) and there exist t_0 and $T > 0$ such that $\mathbf{x}(t_0 + T) = \mathbf{x}(t_0)$. Then, $\mathbf{x}(t + T) = \mathbf{x}(t)$ for all t.

Proof: Define $\mathbf{x}_T$ by $\mathbf{x}_T(t) = \mathbf{x}(t + T)$. Then, $\mathbf{x}_T$ is a solution and by hypothesis, $\mathbf{x}_T(t_0) = \mathbf{x}(t_0)$. The proof is complete by uniqueness.   □

Remark 8.2.7

The solution in Lemma 8.2.6 is termed as a *periodic solution*. The smallest such $T > 0$ is called the *period* of **x**. The orbit of a periodic solution is called a **periodic orbit** or **closed orbit**. If a periodic orbit is *isolated* in the sense that there is no other periodic orbit in its immediate vicinity, then the periodic orbit is called a **limit cycle**. For example, the orbits of $\ddot{x} + x = 0$ are all periodic orbits but, none of them is a limit cycle. Limit cycles can only occur in nonlinear systems. This is clear from the detailed study of linear systems that has been carried out in Chapter 5. The existence of periodic solutions to (8.2.1) is an important aspect of qualitative theory and two important results, namely the Poincarè–Bendixon theorem and Leinard's theorem give sufficient conditions for the existence of periodic solutions in $2D$ systems. These will be discussed later in Section 8.7.

We now discuss an important class of solutions to (8.2.1).

Definition 8.2.8

A point $\bar{\mathbf{x}} \in \mathbb{R}^n$ is called an **equilibrium point** or **equilibrium solution** of (8.2.1) if $\mathbf{f}(\bar{\mathbf{x}}) = 0$. An equilibrium point $\bar{\mathbf{x}}$ is **isolated** if there is a neighborhood of $\bar{\mathbf{x}}$ not containing any other equilibrium point of (8.2.1). Otherwise, the equilibrium point is **non-isolated**.

Thus, an equilibrium point is precisely a constant solution of (8.2.1) and for this reason, it is also called a *fixed point, critical point, steady state solution, stationary point,* or a *singularity*. The equilibrium solutions of (8.2.1) are obtained by solving the system of algebraic equations $\mathbf{f}(\mathbf{x}) = \mathbf{0}$; the equilibrium solutions are the common zeros (roots) of the functions f_j. For example, for a 2×2 linear system with a nonsingular coefficient matrix, the origin $(0,0) \in \mathbb{R}^2$ is the only equilibrium point. Here is an important property of equilibrium points.

Lemma 8.2.9

Suppose **x** is a solution of (8.2.1) and $\lim_{t \to \infty} \mathbf{x}(t) = \xi$ exists. Then, ξ is an equilibrium point.

Proof: For any fixed $h > 0$, $\mathbf{x}(t+h)$ is also a solution and converges to ξ as $t \to \infty$. By mean value theorem, we have

$$\mathbf{x}(t+h) - \mathbf{x}(t) = h\dot{\mathbf{x}}(\tilde{t}) = h\mathbf{f}(\mathbf{x}(\tilde{t}))$$

for some $\tilde{t}$ between t and $t+h$. Hence $\tilde{t} \to \infty$ and $\mathbf{x}(t+h) - \mathbf{x}(t) \to 0$ as $t \to \infty$. By continuity, we therefore get $h\mathbf{f}(\xi) = \mathbf{0}$ and conclude that $\mathbf{f}(\xi) = \mathbf{0}$ as required.   □

Thus, if a solution $\mathbf{x}(t)$ has a finite limit as $t \to \pm\infty$, then, the limit is an equilibrium point. In one dimension, it is easy to see that in the absence of equilibrium points, all orbits will be unbounded and will not have finite limits as $t \to \pm\infty$.

8.2.1 Examples

Example 8.2.10

Consider the $1D$ equation $\dot{x} = x^2$.

Here $x = 0$ is the only equilibrium point. Hence, it is isolated.

Example 8.2.11

For the $1D$ equation

$$\dot{x} = \sin x,$$

the equilibrium points are $n\pi, n \in \mathbb{Z}$. All these equilibrium points are isolated.

Example 8.2.12

Let us write Duffing's equations (1.2.32) and (1.2.33) as a $2D$ system:

$$\dot{x} = y, \quad \dot{y} = \pm x - x^3 - \delta y. \tag{8.2.2}$$

If we take the negative sign in the second equation, then the origin $(0,0)$ is the only equilibrium point. On the other hand, for the case of the positive sign, $(0,0)$ and $(\pm 1, 0)$ are the equilibrium points. In either case, they are isolated.

Example 8.2.13

Writing the van der Pol equation (1.2.34) as the following $2D$ system:

$$\dot{x} = y, \quad \dot{y} = \mu(x^2 - 1)y - x, \tag{8.2.3}$$

we immediately see that $(0,0)$ is the only equilibrium point; hence, it is isolated.

Example 8.2.14

Write the pendulum equation (1.2.35) as a $2D$ system:

$$\dot{x} = y, \quad \dot{y} = -k\sin x. \tag{8.2.4}$$

We find that $(n\pi, 0), n \in \mathbb{Z}$ are the equilibrium points and each one of them is isolated.

Example 8.2.15

Consider the following $2D$ system:

$$\dot{x} = -y + x(x^2 + y^2), \quad \dot{y} = x + y(x^2 + y^2).$$

The origin $(0,0)$ is the only equilibrium point of this system (Why?).

Example 8.2.16

Consider the second order equation $\ddot{x} + x\dot{x} = 0$.

Writing this as a first order system in $x, \dot{x}$, we see that each point on the line $\dot{x} = 0$ is an equilibrium point. Hence, *none* of the equilibrium points is isolated.

8.3   Liapunov Stability, Liapunov Function

We confine ourselves to a discussion of stability of equilibrium solutions of (8.2.1) in the sense of Liapunov. Further, we consider only isolated equilibrium points. For a discussion of stability of arbitrary solutions, see [Per01, Wig90]. There are also important notions of *orbital stability* and *structural stability*; see, for example, [Wig90]

Definition 8.3.1

[Liapunov stability] An isolated equilibrium point $\bar{\mathbf{x}} \in \mathbb{R}^n$ of (8.2.1) is said to be **stable (Liapunov stable)** if given $\varepsilon > 0$, there is a $\delta = \delta(\varepsilon) > 0$ such that for any solution $\mathbf{x}$ of (8.2.1) satisfying $|\mathbf{x}(t_0) - \bar{\mathbf{x}}| < \delta$, we have $|\mathbf{x}(t) - \bar{\mathbf{x}}| < \varepsilon$ for all $t > t_0$. Otherwise, $\bar{\mathbf{x}}$ is said to be **unstable**. $\quad\square$

Usually, we take $t_0 = 0$. Here $|\mathbf{x}|$ denotes the Euclidean norm in $\mathbb{R}^n$ defined by

$$|\mathbf{x}|^2 = x_1^2 + x_2^2 + \cdots + x_n^2, \text{ for } \mathbf{x} = (x_1, x_2, \cdots, x_n) \in \mathbb{R}^n.$$

Definition 8.3.2

[Asymptotic stability] An isolated equilibrium point $\bar{\mathbf{x}}$ of (8.2.1) is said to be **asymptotically stable** if it is stable and for any solution $\mathbf{x}$ of (8.2.1), there exists $b > 0$ such that if $|\mathbf{x}(t_0) - \bar{\mathbf{x}}| < b$, then $\lim\limits_{t \to \infty} |\mathbf{x}(t) - \bar{\mathbf{x}}| = 0$. $\quad\square$

Before proceeding further, we again recall from 2D linear theory that, in the case of a non-singular coefficient matrix, the only equilibrium point $(0,0)$ is asymptotically stable if all the eigenvalues have negative real parts (complex eigenvalues occur in conjugate pairs); it is stable, but not asymptotically stable if eigenvalues have zero real parts; and unstable in all the other cases. In higher dimensions, the possibilities are more.

8.3.1   Linearization

We now discuss the linearization around an equilibrium point of (8.2.1). We assume that $\mathbf{f}$ in (8.2.1) is a C^2 function. If $\bar{\mathbf{x}}$ is an equilibrium point, then by Taylor's formula (see Chapter 2), we have

$$\mathbf{f}(\bar{\mathbf{x}} + \mathbf{y}) = \mathbf{f}(\bar{\mathbf{x}}) + \mathbf{A}\mathbf{y} + O(|\mathbf{y}|^2) = \mathbf{A}\mathbf{y} + O(|\mathbf{y}|^2), \qquad (8.3.1)$$

where $\mathbf{A} = D\mathbf{f}(\bar{\mathbf{x}}) \equiv \left[\dfrac{\partial f_i}{\partial x_j}(\bar{\mathbf{x}}) \right]$ denotes the Jacobian matrix of $\mathbf{f}$ at $\bar{\mathbf{x}}$. Writing $\mathbf{x} = \bar{\mathbf{x}} + \mathbf{y}$ and ignoring quadratic and higher order terms in $\mathbf{y}$, we obtain from (8.2.1) and (8.3.1), the following linear system:

$$\dot{\mathbf{y}} = \mathbf{A}\mathbf{y}. \qquad (8.3.2)$$

Equation (8.3.2) is referred to as *linearized system* of (8.2.1) around the equilibrium point $\bar{\mathbf{x}}$. It is also referred to as the *variation equation* of (8.2.1). Since we are concerned about the stability of $\bar{\mathbf{x}}$, it is quite reasonable to examine the stability of $\mathbf{0}$ for the linearized system (8.3.2). This is termed as *linear stability analysis*. However, we see through some examples that the stability of $\mathbf{0}$ of the linearized system (8.3.2) may or may not imply the stability of $\bar{\mathbf{x}}$ for (8.2.1). We do have a result in the positive direction, namely:

Theorem 8.3.3

Suppose the eigenvalues of $\mathbf{A}$ in (8.3.2) all have negative real parts. Then, the equilibrium point $\bar{\mathbf{x}}$ is asymptotically stable for (8.2.1).

This result is a special case of *Perron's theorem*, discussed next.

Theorem 8.3.4

[Perron's Theorem] [CL72] Let $\mathbf{A}$ be a real $n \times n$ matrix whose eigenvalues all have negative real parts and consider the n-dimensional system

$$\dot{\mathbf{x}} = \mathbf{A}\mathbf{x} + \mathbf{f}(t, \mathbf{x}), \tag{8.3.3}$$

where $\mathbf{f}$ is a continuous function satisfying

$$\mathbf{f}(t, \mathbf{x}) = o(|\mathbf{x}|), \tag{8.3.4}$$

as $\mathbf{x} \to \mathbf{0}$, *uniformly* in t. Then, a solution $\mathbf{x}$ of (8.3.3) with sufficiently small $\mathbf{x}(0)$ exists for all $t \geq 0$ and $\mathbf{x}(t)$ tends to $\mathbf{0}$ as $t \to \infty$.

Thus, the identically zero solution of (8.3.3) is asymptotically stable. The assumption on $\mathbf{A}$ implies that the zero solution, which is the only equilibrium point of the linear system $\dot{\mathbf{x}} = \mathbf{A}\mathbf{x}$ is asymptotically stable. Therefore, the theorem asserts that the asymptotic stability persists under small nonlinear perturbations. The hypothesis may be modified so that a solution exists for small positive time; uniqueness is not an issue here. For related issues, see [CL72].

We also remark that the existence of the solution for all $t \geq 0$ is not trivial; the reader may wish to compare in this regard the 1D equations

$$\dot{x} = x^2, \quad \dot{x} = -\mu x + x^2, \; \mu > 0.$$

Proof of Theorem 8.3.4:   The hypothesis on $\mathbf{A}$ implies that there exist positive constants K, σ such that

$$|\exp(t\mathbf{A})| \le K e^{-\sigma t}, \qquad \text{for all } t \ge 0. \tag{8.3.5}$$

(see, Chapter 2, Theorem 2.5.6). By the local existence result, given any $\mathbf{x}_0 \in \mathbb{R}^n$, there exists a solution $\mathbf{x}$ of (8.3.3) for small positive times, say $t \in [0, t^*]$, with $\mathbf{x}(0) = \mathbf{x}_0$. Further, $\mathbf{x}$ satisfies the integral relation

$$\mathbf{x}(t) = \exp(t\mathbf{A})\mathbf{x}_0 + \int_0^t \exp((t-s)\mathbf{A})\mathbf{f}(s, \mathbf{x}(s))\, ds,$$

for $t \in [0, t^*]$. The hypothesis (8.3.4) on $\mathbf{f}$ means that given any $\varepsilon > 0$, there is a $\delta > 0$, depending *only* on ε such that

$$|\mathbf{f}(t, \mathbf{x})| \le \frac{\varepsilon}{K}|\mathbf{x}|, \tag{8.3.6}$$

for all $\mathbf{x}$ satisfying $|\mathbf{x}| \le \delta$ and for all t. Therefore, we have from (8.3.5) and (8.3.6), using (8.3.4),

$$|\mathbf{x}(t)| \le K e^{-\sigma t}|\mathbf{x}_0| + \varepsilon \int_0^t e^{-\sigma(t-s)}|\mathbf{x}(s)|\, ds \tag{8.3.7}$$

as long as the solution $\mathbf{x}(t)$ satisfies the condition $|\mathbf{x}(t)| \le \delta$, for $t \in [0, t^*]$; this may be achieved by choosing small $|\mathbf{x}_0|$ and t^* if necessary. Next, by multiplying the inequality (8.3.7) throughout by $e^{\sigma t}$ and using Gronwall's inequality, we obtain

$$e^{\sigma t}|\mathbf{x}(t)| \le K|\mathbf{x}_0|e^{\varepsilon t}.$$

Choosing $\varepsilon = \dfrac{\sigma}{2}$, we obtain the *a priori* estimate

$$|\mathbf{x}(t)| \le K|\mathbf{x}_0|e^{-\frac{\sigma}{2}t}, \tag{8.3.8}$$

provided that $|\mathbf{x}(t)| \le \delta$. If we choose $\mathbf{x}_0$ such that $K|\mathbf{x}_0| \le \delta$, then, from (8.3.8), we see that $|\mathbf{x}(t)| \le \delta$, for all t, where the local solution exists. Therefore, for the chosen initial data $\mathbf{x}_0$, all the aforementioned arguments are justified and the solution $\mathbf{x}(t)$ satisfies (8.3.8) in its interval of existence. This allows us to extend the solution for all $t \ge 0$ and (8.3.8) holds, from which the theorem follows, using results similar to Proposition 4.5.5 and Theorem 4.5.6 when applied to systems. □

An equilibrium point $\bar{x}$ of (8.2.1) is called a **hyperbolic equilibrium point** if all the eigenvalues of $Df(\bar{x})$ have non-zero real parts.

8.3.2   Examples

Consider the $1D$ equation $\dot{x} = \sin x$.

The equilibrium points are $n\pi, n \in \mathbb{Z}$. We now linearize around an equilibrium point $n\pi$. For this, we write $y = x - n\pi$ and use the trigonometric formula $\sin x = \sin(y + n\pi) = (-1)^n \sin y$. Thus, the linearized equation is $\dot{y} = (-1)^n y$ and therefore, the equilibrium point is unstable if n is even and asymptotically stable if n is odd.

Let us analyze the nonlinear equation directly. See Fig. 8.1. Consider two consecutive equilibrium points $n\pi$ and $(n + 1)\pi$. These are themselves (trivial) orbits of the equation. Now consider a solution $x(t)$ starting at $n\pi + \alpha$ with $0 < \alpha < \pi$, that is, $x(0) = n\pi + \alpha$. Since, $\sin x$ is globally Lipschitz, the solution exists for all time and it can approach $n\pi$ or $(n + 1)\pi$ only as $t \to \infty$. Since $\sin x$ is positive or negative in $(n\pi, (n + 1)\pi)$ according as n is even or odd, x is, respectively, either increasing or decreasing in $(n\pi, (n + 1)\pi)$. Therefore, we see that $x(t) \to (n + 1)\pi$ or $n\pi$ as $t \to \infty$ according as n is even or odd respectively. Thus, the phase portrait in this example consists of an infinite number of orbits consisting of open intervals $(n\pi, (n + 1)\pi)$ along with the equilibrium points $n\pi$ with $n \in \mathbb{Z}$. The equilibrium point $n\pi$ is unstable if n is even and asymptotically stable if n is odd, a conclusion same as in linearization.

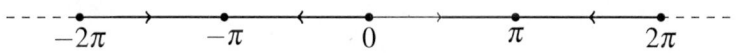

$$-2\pi \qquad -\pi \qquad 0 \qquad \pi \qquad 2\pi$$

Fig. 8.1   Phase line for $\dot{x} = \sin x$

In this case, we also have an explicit formula for the solution, from which these conclusions may be drawn. We have, with α as earlier

$$x(t) = \begin{cases} n\pi + 2\arctan(Ce^{-t}) & \text{if } n \text{ is odd} \\ n\pi + 2\arctan(Ce^{t}) & \text{if } n \text{ is even,} \end{cases}$$

where the constant C satisfies $\alpha = 2\arctan(C)$.

Example 8.3.7

Let us consider one more $1D$ equation $\dot{x} = x^2$.

In this case, we know that the solutions do not exist for all time; but there is a maximum interval of existence depending on the initial condition. Here 0 is the only equilibrium point and linearization gives the equation $\dot{y} = 0$. Thus, in the linearization, 0 is stable but not asymptotically stable. For the nonlinear equation, we have the solution in explicit form:

$$x(t) = \frac{x_0}{1 - x_0 t}$$

with $x(0) = x_0$. The solution is defined in the interval $(\frac{1}{x_0}, \infty)$ if $x_0 < 0$ and in the interval $(-\infty, \frac{1}{x_0})$ if $x_0 > 0$. Since any solution is always increasing, we see that 0 is an unstable equilibrium point and its behavior is more like a saddle point: if $x_0 < 0$, then $x(t) \to 0$ as $t \to \infty$ and when $x_0 > 0$, then $x(t) \to \infty$ as $t \to \frac{1}{x_0}$.

Example 8.3.8

Consider the $2D$ system

$$\dot{x} = -y + x(x^2 + y^2), \quad \dot{y} = x + y(x^2 + y^2).$$

The only equilibrium point is $(0,0)$. The corresponding linearized system is

$$\dot{x} = -y, \quad \dot{y} = x$$

and $(0,0)$ is stable, but not asymptotically stable for the linearized system. However, by considering the original equations in polar coordinates, we see that $\dot{r} = r^3$, where $r^2 = x^2 + y^2$. Thus, r is increasing and the orbits starting near the origin spiral away from the origin as t

increases and therefore the origin is unstable. On the other hand, if we change the signs of the nonlinear terms in the equations, we find that the origin is asymptotically stable as r now satisfies $\dot{r} = -r^3$.

Example 8.3.9

Consider the Duffing system (8.2.2), with negative sign

$$\dot{x} = y, \quad \dot{y} = x - x^3 - \delta y.$$

The equilibrium points are $(0,0)$ and $(\pm 1, 0)$. We now discuss the linearization of the system around each of these equilibrium points. At any point $(x,y) \in \mathbb{R}^2$, the Jacobian of the right side functions is given by

$$\begin{bmatrix} 0 & 1 \\ 1 - 3x^2 & -\delta \end{bmatrix}.$$

At $(0,0)$, this becomes $\begin{bmatrix} 0 & 1 \\ 1 & -\delta \end{bmatrix}$, whose eigenvalues are $\frac{1}{2}(-\delta \pm \sqrt{\delta^2 + 4})$. Hence, for $\delta \geq 0$, there is always one positive eigenvalue and the equilibrium point $(0,0)$ is linearly unstable.

At $(\pm 1, 0)$, the Jacobian matrix is given by

$$\begin{bmatrix} 0 & 1 \\ -2 & -\delta \end{bmatrix}.$$

Here, the eigenvalues are $\frac{1}{2}(-\delta \pm \sqrt{\delta^2 - 8})$. Hence, the equilibrium points $(\pm 1, 0)$ are asymptotically stable when $\delta > 0$. If $\delta = 0$, the eigenvalues are $\pm \sqrt{2}i$ and the equilibrium points are now stable, but not asymptotically stable in the linear approximation.

Example 8.3.10

For the van der Pol system (8.2.3)

$$\dot{x} = y, \quad \dot{y} = \mu(x^2 - 1)y - x,$$

the origin $(0,0)$ is the only equilibrium point.

The linearized system at $(0,0)$ is given by

$$\dot{x} = y, \quad \dot{y} = -x - \mu y.$$

The corresponding Jacobian matrix is $\begin{bmatrix} 0 & 1 \\ -1 & -\mu \end{bmatrix}$, whose eigenvalues are $\frac{1}{2}(\mu \pm \sqrt{\mu^2 - 4})$. Thus, $(0,0)$ is asymptotically stable for $\mu < 0$ and unstable for $\mu > 0$. For $\mu = 0$, it is stable in the linear approximation; the original system itself is linear when $\mu = 0$.

Example 8.3.11

Consider the Lorenz system (1.2.36)

$$\dot{x} = -\sigma x + \sigma y, \quad \dot{y} = Rx - y - xz, \quad \dot{z} = -bz + xy,$$

where R, σ and b are constants. We will now find the equilibrium points of this system and do their linear analysis. We assume that all the constants are positive.

The origin $(0,0,0)$ is always an equilibrium point. It will be the only equilibrium point for $R \leq 1$. But, if $R > 1$, there are two more equilibrium points given by $(\pm\sqrt{b(R-1)}, \pm\sqrt{b(R-1)}, R - 1)$. These are symmetrically situated in the $z = R - 1$ plane, in the (x, y, z) space.

Next, the Jacobian of the right hand side at a general point (x, y, z) is given by

$$\begin{bmatrix} -\sigma & \sigma & 0 \\ R - z & -1 & -x \\ y & x & -b \end{bmatrix}.$$

At the origin, it becomes

$$\begin{bmatrix} -\sigma & \sigma & 0 \\ R & -1 & 0 \\ 0 & 0 & -b \end{bmatrix},$$

whose eigenvalues are $-b$ and the roots of the quadratic equation

$$\lambda^2 + (1 + \sigma)\lambda + \sigma(1 - R) = 0.$$

Therefore, for $R < 1$, all the eigenvalues are negative and when $R > 1$, one eigenvalue is positive. Hence, in the linear analysis, the equilibrium point at the origin is asymptotically stable for $R < 1$ and unstable for $R > 1$. For $R = 1$, the eigenvalues are $-b, 1 + \sigma, 0$.

We, next consider the equilibrium points $(\pm\sqrt{b(R-1)}, \pm\sqrt{b(R-1)},$ $R-1)$, which exist when $R > 1$. The corresponding Jacobian matrix is given by

$$
\begin{bmatrix}
-\sigma & \sigma & 0 \\
1 & -1 & \mp\sqrt{b(R-1)} \\
\pm\sqrt{b(R-1)} & \pm\sqrt{b(R-1)} & -b
\end{bmatrix},
$$

whose eigenvalues are the roots of the cubic equation

$$
\lambda^3 + (\sigma + b + 1)\lambda^2 + (R + \sigma)b\lambda + 2\sigma b(R - 1) = 0.
$$

The analysis of this cubic equation becomes more difficult, as there are three parameters. Being a cubic equation with real coefficients, there is always a real root, which can be shown to be negative. The Hurewitz criterion (see, for instance, [Mer97]) shows that the eigenvalues all have negative real parts if and only if

$$
R > \sigma(\sigma + b + 1)(\sigma - b - 3)^{-1}.
$$

We will not discuss this further and the interested reader can refer to many works on the subject. For example, see [GH83, Hao84, Wig90]. However, we wish to make the following remark on the value $R = 1$, which is special as observed earlier. As the value of R moves from the region $R < 1$ to the region $R > 1$, we have observed either a change in the stability of an equilibrium point or increase in the number of equilibrium points. For this reason, $R = 1$ is referred to as a *bifurcation point*. The topic of *bifurcation* is an important and difficult part of the qualitative analysis. The interested reader may refer to the works cited earlier.

Example 8.3.12

(See [Wig90]) We now consider a time dependent, that is, a non-autonomous system exhibiting a somewhat unexpected behavior. Consider the following 2D linear non-autonomous system

$$
\begin{bmatrix} \dot{x}_1 \\ \dot{x}_2 \end{bmatrix} = A(t) \begin{bmatrix} x_1 \\ x_2 \end{bmatrix},
$$

where

$$A(t) = \begin{bmatrix} -1 + \dfrac{3}{2}\cos^2 t & 1 - \dfrac{3}{2}\cos t \sin t \\ -1 - \dfrac{3}{2}\cos t \sin t & -1 + \dfrac{3}{2}\sin^2 t \end{bmatrix}.$$

The eigenvalues of this system are given by $\frac{1}{4}(-1 \pm \sqrt{7}i)$ for all t and thus, they have negative real parts. However, this system has the following two linearly independent solutions:

$$v_1(t) = e^{t/2}\begin{bmatrix} -\cos t \\ \sin t \end{bmatrix}, \quad v_2(t) = e^{-t}\begin{bmatrix} \sin t \\ \cos t \end{bmatrix}.$$

Hence, the only equilibrium point $(0,0)$ is unstable of saddle type, a conclusion that does not follow from the nature of the eigenvalues of $A(t)$.

We end the section with a mention of an important theorem, namely, the Hartman–Grobman theorem, which states that *in a neighborhood of a hyperbolic equilibrium point, the orbits of a nonlinear system and those of its linearized system are linked via a homeomorphism.* For more details, see [Per01]. We have already seen (Perron's theorem) that if the linearized system around an equilibrium point has eigenvalues, all with negative real parts, then, the equilibrium point is also asymptotically stable for the nonlinear system. Thus, the only case when the orbits of the nonlinear system and those of the linearized system around an isolated equilibrium point may not be comparable is when the linearized system has at least one eigenvalue with zero real part, that is, when the equilibrium point is non-hyperbolic. This case can be handled by the construction of an appropriate *Liapunov function*, which is the topic of the next section. The *center manifold theorem*, see [Wig90], also deals in detail with the case of a non-hyperbolic equilibrium point.

8.4 Liapunov Function

We, now discuss the stability of the equilibrium points of (8.2.1) using a Liapunov function. This, indeed, is a very useful and powerful tool, quite well suited to the situation of non-hyperbolic equilibrium points. Recall that these equilibrium points are not covered by Perron's theorem or the Hartmann–Grobman theorem.

Assume that **0** is an isolated equilibrium point of (8.2.1). Thus, there is an open neighborhood Ω containing **0** which does not contain any other equilibrium point of (8.2.1). The case of a non-zero equilibrium point can be brought to the case of a zero equilibrium point by some suitable translation.

Definition 8.4.1

A C^1 function $V : \Omega \to \mathbb{R}$ satisfying

(1) $V(\mathbf{0}) = 0,\quad V(\mathbf{x}) > 0$ for all $\mathbf{x} \in \Omega \setminus \{\mathbf{0}\}$,

(2) $\nabla V \cdot \mathbf{f} \leq 0$ in Ω

is called a **Liapunov function** for (8.2.1).

Here, ∇V denotes the gradient of V and $\mathbf{a} \cdot \mathbf{b}$ for vectors $\mathbf{a}, \mathbf{b} \in \mathbb{R}^n$ denotes the standard scalar or dot (inner) product.

Theorem 8.4.2

Assume that **0** is an isolated equilibrium point of (8.2.1) and the system (8.2.1) possesses a Liapunov function V. Then, the origin **0** is stable. If, in addition, $\nabla V \cdot \mathbf{f} < 0$ in $\Omega \setminus \{\mathbf{0}\}$, then **0** is asymptotically stable.

Proof: We begin with a discussion on *level surfaces* of V. For $c > 0$, define the *level surface* V_c by

$$V_c = \{\mathbf{x} \in \Omega : V(\mathbf{x}) = c\}.$$

If $\mathbf{x} \in V_c$, we say $\mathbf{x}$ is at an energy level[2] c. By the continuity of V, the set V_c is a closed subset. Also in a neighborhood of a point $\mathbf{x}^0$ where $\nabla V(\mathbf{x}^0) \neq \mathbf{0}$, V_c represents a surface in $\mathbb{R}^n$.

If $\mathbf{x}(t)$ is a solution of (8.2.1), then, by condition (2) and the chain rule, we have

$$\frac{d}{dt} V(\mathbf{x}(t)) = \nabla V(\mathbf{x}(t)) \cdot \dot{\mathbf{x}}(t) = \nabla V(\mathbf{x}(t)) \cdot \mathbf{f}(\mathbf{x}(t)) \leq 0.$$

---

[2]Later, while discussing conservative equations we will see that V may be taken as the sum of kinetic energy and potential energy. Thus, V_c may be thought of as the surface at energy level c.

Thus, the positive function V is non-increasing along the orbits of (8.2.1). Therefore, if an orbit begins on V_c, then for all future times $t \geq t_0$, the orbit either stays on V_c or moves to a level surface $V_{c'}$ with $c' < c$.

To prove stability, let $\varepsilon > 0$ be such that the sphere C_1 with radius ε, centered at the origin and its interior (namely, the closed ball of radius ε) are inside Ω. See Fig. 8.2. Let $m > 0$ be the minimum of V on C_1. Since $V(\mathbf{0}) = 0$, by continuity, we can find a positive $\delta \leq \varepsilon$ such that $V(\mathbf{x}) \leq \dfrac{m}{2} < m$ for all $|\mathbf{x}| \leq \delta$. Let C_2 be the circle with radius δ, centered at the origin. Thus, if $|\mathbf{x}(t_0)| \leq \delta$, then, $V(\mathbf{x}(t_0)) \leq \dfrac{m}{2}$; that is, $\mathbf{x}(t_0)$ is at an energy level $c \leq \dfrac{m}{2}$. Hence, by our observation, for $t \geq t_0$, $\mathbf{x}(t)$ cannot intersect C_1 as all the points on C_1 have energy level $\geq m$. Thus, $\mathbf{x}(t)$ lies inside C_1, proving stability.

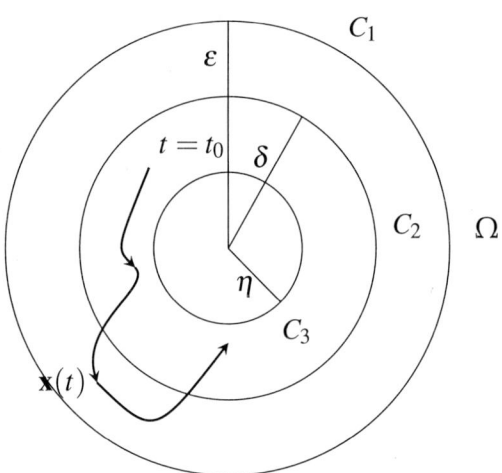

Fig. 8.2   Proof of Liapunov's theorem

Next, suppose, in addition, that $\nabla V \cdot \mathbf{f} < 0$ in $\Omega \setminus \{\mathbf{0}\}$. To prove asymptotic stability, we first show that $V(\mathbf{x}(t)) \to 0$ as $t \to \infty$, if the orbit $\mathbf{x}(t)$ starts at $t = t_0$ in a suitable neighborhood around $\mathbf{0}$. Let $\mathbf{x}(t_0)$ lie inside C_2. Since $\dfrac{d}{dt} V(\mathbf{x}(t)) \leq 0$, the function $V(\mathbf{x}(t))$ is non-increasing and bounded below by 0 as $V(\mathbf{0}) = 0$. Hence, it follows that $\lim\limits_{t \to \infty} V(\mathbf{x}(t))$ exists. Call the limit L; clearly, $L \geq 0$. We claim that $L = 0$.

Suppose $L > 0$. Let C_3 be the sphere of radius $\eta < \delta$, centered at the origin such that $V(\mathbf{x}) \leq \dfrac{L}{2} < L$ on and inside C_3; that is, for all $|\mathbf{x}| \leq \eta$. The existence of η again follows from the continuity of V and $V(\mathbf{0}) = 0$. Now $\nabla V \cdot \mathbf{f}$ is continuous and negative in the closed annular region between C_1 and C_3. Hence, $\nabla V \cdot \mathbf{f}$ has a negative minimum denoted by $-k$ with $k > 0$. Indeed, we have $V(\mathbf{x}(t_0)) \geq V(\mathbf{x}(t)) \geq L$ and hence, the trajectory $\mathbf{x}(t)$ cannot intersect the circle C_3. Therefore, the trajectory lies in the annular region for all $t \geq t_0$. Now for $t > t_0$,

$$V(\mathbf{x}(t)) = V(\mathbf{x}(t_0)) + \int_{t_0}^{t} \frac{d}{ds}(V(\mathbf{x}(s))\,ds \leq V(\mathbf{x}(t_0)) - k(t - t_0).$$

Since $k > 0$, $V(\mathbf{x}(t)) < 0$ for large t, which contradicts the positivity of V. Thus, $L = 0$, that is, $\lim_{t \to \infty} V(\mathbf{x}(t)) = 0$. We leave it as an exercise to the reader to show that $\lim_{t \to \infty} \mathbf{x}(t)$ exists and the limit is $\mathbf{0}$ as $V(\mathbf{x}) > 0$ for $\mathbf{x} \neq \mathbf{0}$. Thus, $\mathbf{0}$ is asymptotically stable. The proof is complete. $\qquad\square$

On similar lines, we have the following instability result.

Theorem 8.4.3

[Instability Result] Suppose there is a C^1 function $V : \Omega \to \mathbb{R}$ satisfying:

(1) $V(\mathbf{0}) = 0$.

(2) Every sphere centered at $\mathbf{0}$ contains at least one point where V is positive.

(3) $\nabla V \cdot \mathbf{f} > 0$ in $\Omega \backslash \{\mathbf{0}\}$.

Then, $\mathbf{0}$ is unstable for (8.2.1).

Actually, one does not need to assume the condition (2) that every sphere around $\mathbf{0}$ contains a point where $V > 0$; it may be replaced by a weaker assumption. Theorem 8.4.3 follows from the following theorem, due to Chetaev.

Theorem 8.4.4

[Chetaev] Suppose in any neighborhood of **0**, there is a non-empty set where $V > 0$ and $\nabla V \cdot \mathbf{f} > 0$ in the region $\{V > 0\}$. Then **0** is unstable for (8.2.1).

Proof: Choose an $\varepsilon > 0$ so that the sphere C_1 centered at **0** with radius ε is such that the closed ball B_ε of radius ε is inside Ω. See Figure 8.3. By hypothesis, there is an $\mathbf{x}^0 \in B_\varepsilon$ satisfying $V(\mathbf{x}^0) > 0$. Let $\mathbf{x}(t)$ be a solution of (8.2.1) with $\mathbf{x}(t_0) = \mathbf{x}^0$. Now, $V > 0$ in a neighborhood of $\mathbf{x}^0$ and therefore, by hypothesis, $\nabla V \cdot \mathbf{f} > 0$ in the same neighborhood. By a similar integration as in Liapunov's theorem 8.4.2, we see that $V(\mathbf{x}(t)) \geq V(\mathbf{x}(t_0)) > 0,$ for all $t \geq t_0$. In fact, the function $t \to V(\mathbf{x}(t))$ is non-decreasing for $t \geq t_0$. This shows that the positive orbit $\mathcal{O}^+(\mathbf{x}^0)$ through $\mathbf{x}^0$ is confined to $\{V > 0\}$.

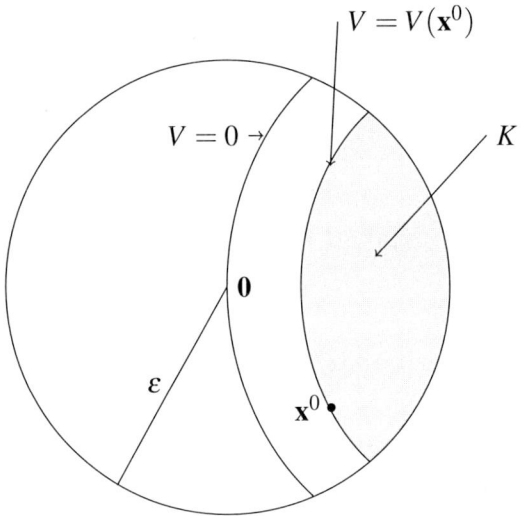

Fig. 8.3   Domain for the proof of Chetaev's theorem

Claim: This positive orbit crosses C_1 after a finite time greater than t_0.

Suppose not. Then, $\mathcal{O}^+(\mathbf{x}^0) \subset K \equiv \{\mathbf{x} \in \Omega : |\mathbf{x}| \leq \varepsilon, V(\mathbf{x}) \geq V(\mathbf{x}^0) > 0\}$. Observe that **0** does not belong to K.

The continuous function V is bounded in the compact set K. Also, by hypothesis, $\nabla V \cdot \mathbf{f} > 0$ in K and therefore has a minimum $m > 0$ in K.

Hence, one more integration shows that $V(\mathbf{x}(t)) \geq V(\mathbf{x}^0) + m(t - t_0)$ for all $t > t_0$, which contradicts the boundedness of V in K. This completes the proof. □

We now consider a few examples to illustrate stability and instability results.

Example 8.4.5

Consider the 2D system

$$\dot{x}_1 = -x_2^3 \quad \text{and} \quad \dot{x}_2 = x_1^3.$$

By considering $V(x_1, x_2) = x_1^4 + x_2^4$, we see that the orbits $(x_1(t), x_2(t))$ of the given system are given by

$$x_1^4(t) + x_2^4(t) = c^2,$$

where c is a constant. Therefore, all the orbits are closed and surround the origin, the only equilibrium point. Hence, $(0,0)$ is stable, but not asymptotically stable. Indeed, V is a Liapunov function.

Example 8.4.6

Consider the 3D system

$$\dot{x}_1 = -2x_2 + x_2 x_3, \quad \dot{x}_2 = x_1 - x_1 x_3, \quad \dot{x}_3 = x_1 x_2.$$

The Jacobian matrix of this system at the equilibrium point $(0,0,0)$ is given by

$$\begin{bmatrix} 0 & -2 & 0 \\ 1 & 0 & 0 \\ 0 & 0 & 0 \end{bmatrix},$$

whose eigenvalues are $0, \pm 2i$. Consider the function $V(x_1, x_2, x_3) = c_1 x_1^2 + c_2 x_2^2 + c_3 x_3^2$. Then, we find that

$$\frac{1}{2} \nabla V \cdot \mathbf{f} = (c_1 - c_2 + c_3) x_1 x_2 x_3 + (-2c_1 + c_2) x_1 x_2.$$

Choosing $c_1 = c_3 > 0$ and $c_2 = 2c_1$, we find that $V(x) > 0$ for $x \neq 0$ and $\nabla V \cdot \mathbf{f} \equiv 0$. Thus, the orbits of the system lie on the ellipsoids

$$x_1^2 + 2x_2^2 + x_3^2 = c^2,$$

where c is a constant. These are ellipsoids surrounding the origin and therefore $(0,0,0)$ is stable, but not asymptotically stable. This conclusion follows directly from the nature of the trajectories. The Liapunov theorem is not applicable here as the equilibrium point $(0,0,0)$ is not isolated.

Note that this system has many more equilibrium points. In fact, the points $(0,0,c), (0,b,2)$ and $(a,0,1)$, where $a,b,c \in \mathbb{R}$, are all equilibrium points and none of them is isolated! The reader should be able to construct suitable Liapunov functions and do the stability analysis.

Example 8.4.7

[Slight modification of Example 8.4.6] Consider

$$\dot{x}_1 = -2x_1 + x_2 + x_2 x_3 - x_1^3$$
$$\dot{x}_2 = x_1 - x_1 x_3 - x_2^3$$
$$\dot{x}_3 = x_1 x_3 - x_3^3$$

The reader can verify that the origin $(0,0,0)$ is the only equilibrium point. The Jacobian matrix is same as in Example 8.4.6. If we take $V(x) = x_1^2 + 2x_2^2 + x_3^2$, we find that $\nabla V \cdot f = -2(x_1^4 + 2x_2^4 + x_3^4) < 0$ for $\mathbf{x} \neq \mathbf{0}$. Thus, $(0,0,0)$ is asymptotically stable.

Example 8.4.8

Consider the $2D$ system

$$\dot{x} = x^2 + 2y^5, \quad \dot{y} = xy^2.$$

The origin $(0,0)$ is the only equilibrium point. It is easy to see that the linearization does not reveal much regarding the nature of the equilibrium point. Consider the Liapunov function $V(x,y) = x^2 - y^4$. This is not a positive definite function, but has subsets in any neighborhood around the origin, where it is positive. These subsets are bounded by the parabolas $x = y^2$ and $x = -y^2$. Next, along a trajectory $(x(t), y(t))$ of the given system, we find that

$$\dot{V}(x(t),y(t)) = 2x\dot{x} - 4y^3\dot{y} = 2x^3,$$

using the given equations. Thus, in the region $x > y^2$ bounded by the parabola $x = y^2$, we see that both V and $\dot{V}$ are positive. By Chetaev's theorem, we therefore conclude that the origin is unstable for the given system.

8.5   Invariant Subspaces and Manifolds

Consider the autonomous system (8.2.1), namely, $\dot{x} = f(x)$. Denote by $x(t, t_0, x^0)$, the orbit passing through $x^0 \in \mathbb{R}^n$ at time t_0, that is, $x(t_0, t_0, x^0) = x^0$.

Definition 8.5.1

A set $S \subset \mathbb{R}^n$ is said to be **invariant** under (8.2.1) if for any $x^0 \in S$, we have $x(t, t_0, x^0) \in S$ for all $t \in \mathbb{R}$. If we restrict this to $t \geq t_0$, then, we say that S is **positively invariant**, that is, $x(t, t_0, x^0) \in S$ for all $t \geq t_0$.

$\square$

If an invariant set S possesses a smooth manifold structure, like a smooth curve in $\mathbb{R}^2$ or an open set or a subspace in $\mathbb{R}^n$, then S is said to be an *invariant manifold*. Obviously, a positive orbit is a positively invariant set; so are the unions of positive orbits. Conversely, any positively invariant set is a union of positive orbits. Similar statements hold for invariant sets.

Invariant sets play an important role in many applications. For example, when (8.2.1) represents a population dynamics model, we are interested in positive solutions of (8.2.1). Thus, it is necessary to ensure that a solution of such a model remains positive for all future times when started positive initially.

The invariant subspaces associated with a linear system was discussed in detail in Chapter 5. We now discuss a nonlinear example. See [Wig90].

Example 8.5.2

Consider the 2D system $\dot{x} = x$, $\quad \dot{y} = -y + x^2$.

The linearized system at the only equilibrium point $(0,0)$ is given by $\dot{x} = x$ and $\dot{y} = -y$, and has stable and unstable subspaces given by

$$E^s = \{(x,y) \in \mathbb{R}^2 : x = 0\}, \quad E^u = \{(x,y) \in \mathbb{R}^2 : y = 0\}.$$

Turning to the nonlinear system, we eliminate the independent variable t and obtain

$$\frac{dy}{dx} = \frac{\dot{y}}{\dot{x}} = -\frac{y}{x} + x.$$

Upon integration, we find that $y(x) = \frac{x^2}{3} + \frac{c}{x}$, with c a constant of integration. Therefore, any solution (x,y) of the given system satisfies

$$x\left(y - \frac{x^2}{3}\right) = c,$$

c a constant; $c = x_0\left(y_0 - \frac{x_0^2}{3}\right)$, x_0, y_0 being the initial values of x and y respectively.

We choose $c = 0$. If $x_0 = 0$, then $x(t) = 0$ for all t. This in turn gives the y-axis, which is an invariant set for the given system. It is a stable manifold, denoted by $W^s(0,0)$, which is the same as E^s, as any solution of the given system starting on the y-axis remains there for all t and converges to 0 as $t \to \infty$. Now suppose $x_0 \neq 0$. Then, as c is assumed to be 0, we have $y_0 - \frac{x_0^2}{3} = 0$. It is not difficult to see that the parabola given by,

$$W^u(0,0) = \{(x,y) \in \mathbb{R}^2 : y = \frac{x^2}{3}\},$$

is an invariant manifold of the given nonlinear system. It is unstable as any non-trivial solution starting on this parabola remains there and moves away from the origin as t increases.

Example 8.5.3

For the system in Example 8.4.6, it is easily verified that

$$\frac{d}{dt}(x_1^2(t) + 2x_2^2(t) + x_3^2(t)) = 0 \text{ and } \frac{d}{dt}(x_1^2(t) + x_2^2(t) + 2x_3(t)) = 0,$$

for any solution $(x_1(t), x_2(t), x_3(t))$ of the system. Therefore, the ellipsoids $x_1^2 + 2x_2^2 + x_3^2 = a^2$, $a \geq 0$ is a constant and the paraboloids $x_1^2 + x_2^2 + 2x_3 = b$, b is a constant, are invariant manifolds for the system.

Analogous to the linear theory concerning stability of an equilibrium point, we have the following:

Theorem 8.5.4

[The Stable Manifold Theorem]

Consider the autonomous system (8.2.1), namely $\dot{\mathbf{x}} = \mathbf{f}(\mathbf{x})$. Assume that $\mathbf{f} \in C^1(\Omega)$, where Ω is an open set in $\mathbb{R}^n$ containing the origin and $\mathbf{f}(\mathbf{0}) = \mathbf{0}$. Let the Jacobian matrix of $\mathbf{f}$ at $\mathbf{0}$, $D\mathbf{f}(\mathbf{0})$, have k eigenvalues with negative real parts and $n - k$ eigenvalues with positive real parts. Then, the following hold:

- There exists a k-dimensional differentiable manifold S tangent to the stable subspace E^s of the linear system (8.3.2) at $\mathbf{0}$ such that S is invariant under (8.2.1) and for all $\mathbf{x}^0 \in S$,

$$\lim_{t \to \infty} \mathbf{x}(t, t_0, \mathbf{x}^0) = \mathbf{0}.$$

- There exists an $n - k$ dimensional differentiable manifold U tangent to the unstable subspace E^u of (8.3.2) at $\mathbf{0}$ such that $\mathbf{x}^0 \in U$ implies $\mathbf{x}(t, t_0, \mathbf{x}^0) \in U$ for all $t \leq t_0$ and

$$\lim_{t \to -\infty} \mathbf{x}(t, t_0, \mathbf{x}^0) = \mathbf{0}. \qquad \square$$

For the description of the subspaces E^s and E^u, see Theorem 5.6.3. A more delicate and detailed analysis is contained in the following:

Theorem 8.5.5

[Hartman–Grobman Theorem]

Consider the autonomous system (8.2.1), namely $\dot{\mathbf{x}} = \mathbf{f}(\mathbf{x})$. Assume that $\mathbf{f} \in C^1(\Omega)$ for some open set $\Omega \subset \mathbb{R}^n$ containing the origin, $\mathbf{0}$ is a hyperbolic equilibrium point of (8.2.1) and put $\mathbf{A} = D\mathbf{f}(\mathbf{0})$, the Jacobian matrix of $\mathbf{f}$ at $\mathbf{0}$. Then, there exists a homeomorphism H of an open set U containing the origin onto an open set V containing the origin such that for each $\mathbf{x}^0 \in U$, there is an open interval $I(t_0) \in \mathbb{R}$ containing t_0 such that

$$H(\mathbf{x}(t, t_0, \mathbf{x}^0)) = e^{(t-t_0)\mathbf{A}} H(\mathbf{x}^0), \text{ for all } t \in I(t_0);$$

that is, H maps orbits of (8.2.1) near the origin onto the orbits of (8.3.2) near the origin and preserves the parametrization by time. $\qquad \square$

For the description of the terminologies in the aforementioned two theorems and for their proofs, the interested reader is referred to [Per01].

8.6 Phase Plane Analysis

We begin by analyzing a nonlinear equation in the conservative form:

$$\ddot{x} + \frac{d}{dx}V(x) = 0, \tag{8.6.1}$$

where V is a given function of the position x, called *potential function*. A simple computation shows that the total energy (kinetic energy + potential energy) is conserved; that is,

$$\frac{1}{2}(\dot{x}(t))^2 + V(x(t)) = E, \tag{8.6.2}$$

where E is a constant, for any solution of (8.6.1) and for all t; hence the name *conservative system*. Though it may not be possible, in general, to obtain $x(t)$ explicitly from (8.6.2), it is however possible to obtain valuable information regarding $x(t)$ and $\dot{x}(t)$ by the use of (8.6.2). This is the content of the *phase plane analysis*.

   If we assume that the potential function V is bounded below, then, using (8.6.2), it is not very difficult to show that a solution of (8.6.1), with given initial conditions, exists for all $t \in \mathbb{R}$; see [Arn98]. The pendulum equation and Duffing's equation with no damping, discussed in the next section, are examples of conservative equations with the corresponding potential functions bounded below.

8.6.1   Examples

Example 8.6.1

   [Pendulum equation for small oscillations] This is the well-known equation $\ddot{x} + kx = 0, k > 0$.

Solving the equation to obtain the solution explicitly, we see that every solution is periodic. The same conclusion may be reached by analyzing (8.6.2), which turns out to be

$$(\dot{x}(t))^2 + k(x(t))^2 = 2E$$

and E is obtained from the initial conditions. The orbits in this case are ellipses (circles if $k = 1$) surrounding the origin in the phase plane.

Example 8.6.2

[Pendulum equation] This is the equation $\ddot{x} + k \sin x = 0, k > 0$,

and the corresponding relation (8.6.2) is given by

$$(\dot{x}(t))^2 + 2k(1 - \cos x) = 2E,$$

with E again obtained from the initial conditions.

We now describe the phase portrait of solutions using this relation for different values of E. The potential function may be taken as $V(x) = k(1 - \cos x)$. Since $V(x) \geq 0$, we also have $E \geq 0$. Also note that $V(x) = 0$ at $x = 2n\pi$ and $V(x) = 2k$ at $x = (2n - 1)\pi, n \in \mathbb{Z}$. We consider the following different cases: (See Fig. 8.4).

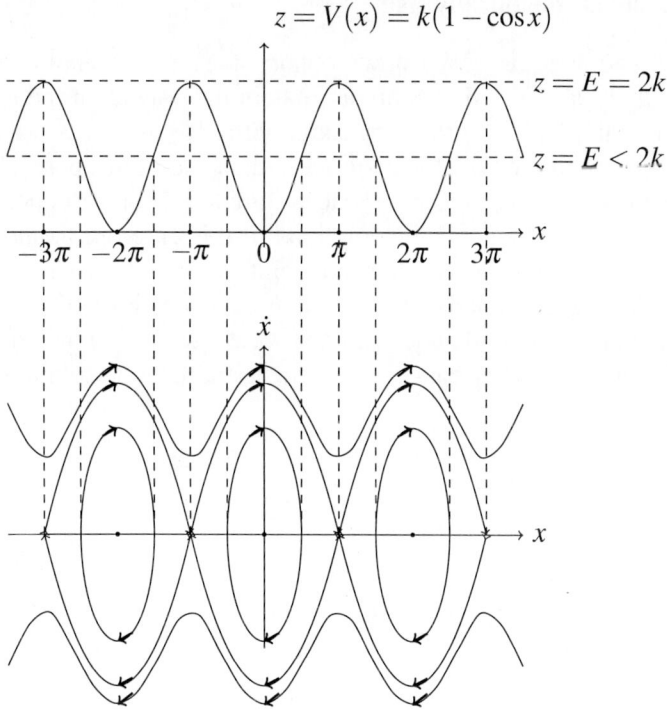

Fig. 8.4   Potential function and phase plane for the pendulum equation

Case 1: $E = 0$:   Here, we obtain the equilibrium solutions $(2n\pi,0)$, $n \in \mathbb{Z}$.

Case 2: $0 < E < 2k$:   In this case, the values of x will be restricted to a symmetric interval around each equilibrium point $2n\pi$, $n \in \mathbb{Z}$ of length $2b$, where $0 < b < \pi$ and $V(2n\pi \pm b) = E$; see the graph of the potential function V in Figure 8.4. To see the nature of the orbits, consider now an orbit $x(t)$, around $(0,0)$ with $x(0) \in (-b,b)$. If $\dot{x}(0) > 0$, then the orbit starts moving towards b and when it reaches b, at that time $\dot{x}$ is 0, because of energy conservation. Then, $\dot{x}$ must become negative, for otherwise, x will exceed b, which is not allowed. Now the orbit starts moving towards $-b$ and when it reaches $-b$, similar observations can be made. Thus, the orbit is a periodic orbit traveling between $-b$ and b with corresponding values of $\dot{x}$ given by the energy conservation equation. Similar periodic orbits exist around each of the equilibrium points $(2n\pi,0)$, $n \in \mathbb{Z}$. We need to just watch the orientation of the orbits: whether they rotate in clockwise or counter-clockwise direction. Therefore, these equilibrium points are all stable but not asymptotically stable.

Case 3: $E = 2k$:   We now obtain the equilibrium solutions $((2n-1)\pi,0)$, $n \in \mathbb{Z}$. We can argue in a similar way as in the previous case. For example, if an orbit $x(t)$ starts with $x(0) \in (-\pi,\pi)$ (the other cases of odd multiples of π are similar), then the orbit approaches the equilibrium point $(\pm\pi,0)$ as $t \to \infty$ according to whether $\dot{x}$ is positive or negative at $t = 0$. Thus, all the solutions in this case are bounded, but none is periodic and the equilibrium points $((2n-1)\pi,0)$, $n \in \mathbb{Z}$ are unstable. In this case, it is also possible to write down the solutions explicitly. For example, if $x(0) = x_0 \in (-\pi,\pi)$ with corresponding $\dot{x}(0)$ obtained from the energy conservation equation, then the solution $x(t)$ is given by

$$x(t) = 2\arctan\left(\frac{y - y^{-1}}{2}\right),$$

where $y = y_0 \exp(\pm k^{1/2}t)$ with $y_0 = \sec\left(\frac{x_0}{2}\right) + \tan\left(\frac{x_0}{2}\right)$ and the sign is chosen as the sign of $\dot{x}(0)$. We also observe that, if $\dot{x}(0) > 0$, then $\lim_{t \to \infty} x(t) = \pi$ and $\lim_{t \to -\infty} x(t) = -\pi$. Such an orbit is called a *heteroclinic orbit*.

Case 4: $E > 2k$:   Here, we have $(\dot{x})^2 \geq E - 2k > 0$ and therefore, $\dot{x}$ is bounded away from 0. Thus, the solution is either strictly increasing or decreasing for all time according to whether $\dot{x}$ is positive or negative initially. Therefore, the solutions in this case are unbounded, though $\dot{x}$ remains bounded.

The orbit of Case 3 is referred to as a *separatrix* as it separates the periodic orbits (Case 2) from the non-periodic (Case 4) orbits.

Example 8.6.3

[Duffing's equation with no damping] This is the equation given by

$$\ddot{x} - x + x^3 = 0$$

and is in the conservative form, with the potential function $V(x) = -\frac{x^2}{2} + \frac{x^4}{4}$. The corresponding energy equation (8.6.2) is given by

$$\frac{1}{2}\dot{x}^2 - \frac{x^2}{2} + \frac{x^4}{4} = E.$$

We will now do a similar phase portrait analysis as in the previous example.   Note that V is symmetric around the origin, that is, $V(x) = V(-x)$ and attains its minimum at ± 1 with $V(\pm 1) = -\frac{1}{4}$ and $V(\pm\sqrt{2}) = 0$. Thus, $E \geq -\frac{1}{4}$ and we consider the following cases. See Fig. 8.5.

Case 1: $E = -\frac{1}{4}$: Here, we obtain only the equilibrium solutions $(\pm 1, 0)$.

Case 2: $-\frac{1}{4} < E < 0$:   In this case, the values of x are restricted to the symmetric intervals around the equilibrium points $(\pm 1, 0)$ of length $2b$ where $b > 0$ satisfies $V(\pm 1 \pm b) = E$; see the graph of the potential function in Fig. 8.5. We, then, obtain periodic solutions surrounding each of these equilibrium points *separately*.

Case 3: $E = 0$:   Now, we first obtain the equilibrium solution $(0,0)$. Thus, any other orbit can reach this equilibrium point only as $t \to \pm\infty$. Any such solution x therefore lies either in the interval $(0, \sqrt{2}]$ or $[-\sqrt{2}, 0)$ and is thus bounded. The direction of the orbit for increasing t can easily be determined and is shown in Fig. 8.5. In this case, we obtain 2 orbits, one with $x(0) > 0$ and the other with $x(0) < 0$. Each one of these orbits approaches the equilibrium point $(0,0)$ *both* as $t \to \pm\infty$. Such orbits are termed *homoclinic orbits*. Note that the equilibrium point $(0,0)$ is unstable.

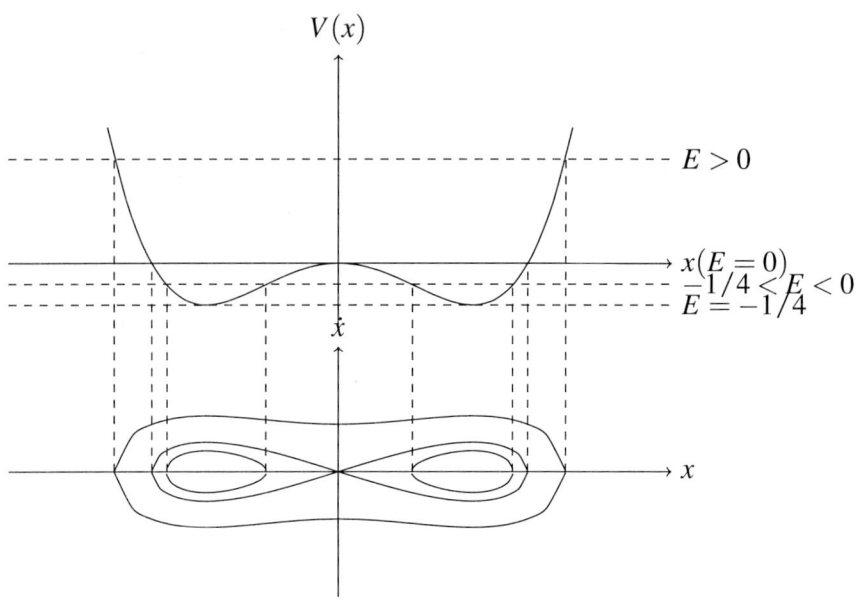

Fig. 8.5   Potential function and phase portrait of a Duffing oscillator

We also have an explicit formula for a solution when $E = 0$. For example, if $x(0) = x_0 > 0$ and the corresponding $\dot{x}(0)$, given by energy conservation equation (8.6.2) is non-negative, then, the solution $x(t)$ is given by

$$x(t) = 2\sqrt{2}y_0^{1/2}\frac{e^t}{1 + y_0 e^t} \equiv 2\sqrt{2}(z + z^{-1})^{-1},$$

where $y_0 = \dfrac{\sqrt{2} - \sqrt{2 - x_0^2}}{\sqrt{2} + \sqrt{2 - x_0^2}}$ and $z = y_0^{1/2}e^t$. (Replace x by $-x$ if $x_0 < 0$ and z by y_0/z if $\dot{x}(0) \le 0$.)

Case 4: E > 0:   The values of a solution x are now restricted to an interval $[-b, b]$, where $b > \sqrt{2}$ satisfies $V(\pm b) = E$. We again obtain a periodic orbit, now surrounding all the three equilibrium points.

8.7 Periodic Orbits

So far we have discussed the equilibrium points (*trivial solutions*) of an autonomous system $\dot{\mathbf{x}} = \mathbf{f}(\mathbf{x})$, and their linear and nonlinear stability. Next in simplicity are periodic solutions (also called, *closed paths* or *orbits*), which are 'simpler' than the general solutions. Recall that from Lemma 8.2.6, it follows that if $x(t)$ is a solution such that $x(t_0 + T) = x(t_0)$ for some t_0 and $T > 0$, then $x(t + T) = x(t)$ for all t, so that x is periodic. The smallest such $T > 0$ is the *period* of x. The equilibrium points may be considered as periodic solutions with period 0. Unlike equilibrium points, it is, in general, very difficult to say when a given autonomous system has a periodic solution. The class of periodic solutions is interesting and important in applications as many systems exhibit them; for example, planetary motion and the motion of a simple pendulum. We have also seen in the previous section that the pendulum equation and Duffing's equation having periodic orbits in certain, but not all, situations.

We now restrict our discussion to 2D systems as results are more complete in this case. For higher dimensional systems, the reader is referred to advanced texts cited in the references. Now, we change the notation a bit and consider the following 2D autonomous system:

$$\dot{x} = f(x,y), \quad \dot{y} = g(x,y), \tag{8.7.1}$$

where x, y are real valued functions and f, g are real valued, C^k $(k \geq 2)$ functions defined in a domain of $\mathbb{R}^2$. We shall now discuss the notion of *Poincarè index* and its implications to the periodic orbits of 2D systems.

We begin with a discussion on smooth curves in the plane. We will not go into much detail and the reader should refer good text books on one variable complex analysis and multivariable calculus.

Let Γ be a *smooth curve* in $\mathbb{R}^2$. This means that as a set, Γ is the image of a smooth mapping from a closed interval $[a,b]$ in $\mathbb{R}$ into $\mathbb{R}^2$. Thus, there are functions $x, y\colon [a,b] \to \mathbb{R}$ such that

$$\Gamma = \{(x(t), y(t)) \in \mathbb{R}^2 : a \leq t \leq b\}.$$

This is referred to as a parametric representation of a curve in $\mathbb{R}^2$, t is referred to as the *parameter* and the smoothness refers to the smoothness of the functions x, y defining Γ. The points $(x(a), y(a))$ and $(x(b), y(b))$ are, respectively, referred to as initial and end points of Γ. A point on

Γ is called a *multiple point* if it corresponds to more than one parameter value in $[a,b)$. A curve with no multiple points is called a *Jordan curve*. When initial and end points coincide, Γ is said to be a *closed curve*. In what follows, we are mainly interested in closed Jordan curves.

For example, $\{(\cos t, \sin t): 0 \leq t \leq 2\pi\}$ represents the unit circle centered at the origin, and it is a closed Jordan curve. On the other hand, the curve $\{(\cos t, \sin t): 0 \leq t \leq 4\pi\}$, though same as the unit circle centered at the origin, is *not* a Jordan curve.

Let $\mathbf{v} = (v_1, v_2)$ be a smooth vector field, that is, $\mathbf{v}$ is a mapping from a bounded open set of $\mathbb{R}^2$ into $\mathbb{R}^2$. Thus, v_1 and v_2 are smooth real valued functions defined on a bounded open set in $\mathbb{R}^2$. Eventually, we will be interested in taking $v_1 = f$ and $v_2 = g$, where f, g are as in (8.7.1). Let Γ be a closed Jordan curve, contained in the domain of $\mathbf{v}$ such that $\mathbf{v}$ does not vanish on Γ. The points where $\mathbf{v}$ vanishes may be called the equilibrium points of the vector field $\mathbf{v}$; the assumption means that Γ does not pass through any equilibrium points of $\mathbf{v}$.

In this set up, at each point (x, y) of Γ, the vector field $\mathbf{v}(x, y)$ defines a unique direction making an angle ϕ with some fixed direction, which we take as the x-axis. Starting with a point on Γ, if we now move in the counter-clockwise direction along Γ, these vectors on Γ rotate, (See Fig. 8.6) and when we get back to the point where we started, these vectors would have rotated through an angle $2\pi k$ for some integer k. This integer is called the *Poincarè index* or simply *index* of Γ and is denoted by $I_{\mathbf{v}}(\Gamma)$.[3]

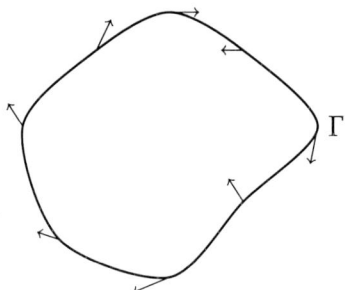

Fig. 8.6 A typical closed curve Γ with a vector field

A good way to visualize the variation of the angle the vector $\mathbf{v}$ makes with the x-axis as it moves in the positive direction along Γ, is to place all these

---

[3]A reader familiar with one variable complex analysis would realize that the Poincarè index is similar to the notion of the **winding number** of a closed curve in the plane.

vectors at one point and see how these vectors rotate. The reader should try this with the simple examples mentioned a little later and more.

Since the angle $\phi = \arctan(v_2/v_1)$, it is not difficult to see that the analytic expression for the index is as follows. Suppose the vector field is given by $\mathbf{v} = (v_1, v_2)$, where v_1, v_2 are smooth real valued functions. Then,

$$I_{\mathbf{v}}(\Gamma) = \frac{1}{2\pi} \int_{\Gamma} d\phi = \frac{1}{2\pi} \int_{\Gamma} d\left(\arctan \frac{v_2(x,y)}{v_1(x,y)}\right) = \frac{1}{2\pi} \int_{\Gamma} \frac{v_1 \, dv_2 - v_2 \, dv_1}{v_1^2 + v_2^2}.$$
(8.7.2)

It is important to keep the direction right as far as the line integral is concerned; it is always in the counter-clockwise direction, which we may call *positive* direction. Using the parametric representation of Γ, the line integral in (8.7.2) may be expressed as the following one-dimensional integral:

$$I_{\mathbf{v}}(\Gamma) = \frac{1}{2\pi} \int_a^b \left(v_1^2 + v_2^2\right)^{-1} \left(v_1 \frac{dv_2}{ds} - v_2 \frac{dv_1}{ds}\right) ds,$$
(8.7.3)

where, in the integrand, $v_i = v_i(x(s), y(s))$, $i = 1, 2$. The derivatives with respect to s may be evaluated, using the chain rule, in terms of the partial derivatives of v_1, v_2 and the derivatives of x, y. Before proceeding to see the relevance of the index with the periodic orbits of (8.7.1), we will see some examples.

Example 8.7.1

Let Γ be the unit circle centered at the origin.

1. Let the vector field be given by $v_1(x,y) = x$ and $v_2(x,y) = y$. If we use the parametric representation $x(s) = \cos(s)$ and $y(s) = \sin(s)$, $0 \le s \le 2\pi$, for Γ, we obtain $I_{\mathbf{v}}(\Gamma) = 1$.

2. Now take the vector field given by $v_1(x,y) = y$ and $v_2(x,y) = -x$. We again find that $I_{\mathbf{v}}(\Gamma) = 1$.

3. We consider the vector field $v_1(x,y) = x$, $v_2(x,y) = -y$. This time, we find that $I_{\mathbf{v}}(\Gamma) = -1$.

Example 8.7.2

Let Γ be the unit circle centered at $(-2, 0)$ and the vector field be $v_1(x,y) = x$, $v_2(x,y) = y$. We now find that $I_{\mathbf{v}}(\Gamma) = 0$.

Example 8.7.3

Let Γ be again the unit circle centered at the origin and the vector field be given by $v_1(x,y) = x^2$, $v_2(x,y) = y^2$.

The reader should try to visualize the rotation of the given vector field along the positive direction on Γ as described earlier. We leave it to the reader as an exercise to show that $I_{\mathbf{v}}(\Gamma) = 2$. On the other hand, if we now take $v_1(x,y) = -x^2$, $v_2(x,y) = y^2$, the index will be -2.

These are typical examples covering many vector fields with isolated equilibrium points.

Theorem 8.7.4

Let Γ be a closed Jordan curve such that Γ and its interior D_Γ do not contain any equilibrium points of a smooth vector field $\mathbf{v}$. Then, $I_{\mathbf{v}}(\Gamma) = 0$.

The theorem is true with minimal assumption of continuity. See [CL72, Lef77]. However, the following proof is simpler using in addition, the smoothness assumption (see [JS03]).

Proof: By Green's theorem,[4] we have

$$\int_\Gamma P\,dy - Q\,dx = \iint_{D_\Gamma} \left(\frac{\partial P}{\partial x} + \frac{\partial Q}{\partial y} \right) dx\,dy \qquad (8.7.4)$$

for any continuously differentiable functions P, Q. The line integral on the left side of (8.7.4) may be expressed as a one-dimensional integral for any given parametric representation of Γ and on the right side, we have a double integral. In (8.7.2), we write, using chain rule,

$$\frac{dv_1}{ds} = \frac{\partial v_1}{\partial x}\frac{dx}{ds} + \frac{\partial v_1}{\partial y}\frac{dy}{ds}, \quad \frac{dv_2}{ds} = \frac{\partial v_2}{\partial x}\frac{dx}{ds} + \frac{\partial v_2}{\partial y}\frac{dy}{ds}.$$

Therefore, (8.7.3) becomes the line integral

$$I_{\mathbf{v}}(\Gamma) = \frac{1}{2\pi} \int_\Gamma P(x,y)\,dx - Q(x,y)\,dy,$$

---

[4]Green's theorem is more generally valid for any bounded open set with a smooth boundary.

where,

$$Q = -\left(v_1^2 + v_2^2\right)^{-1}\left(v_1\frac{\partial v_2}{\partial x} - v_2\frac{\partial v_1}{\partial x}\right), P = \left(v_1^2 + v_2^2\right)^{-1}\left(v_1\frac{\partial v_2}{\partial y} - v_2\frac{\partial v_1}{\partial y}\right)$$

and they satisfy the conditions of Green's theorem, as the denominator is never zero on Γ and in its interior D_Γ. Hence, using (8.7.4), we obtain

$$I_{\mathbf{v}}(\Gamma) = \iint_{D_\Gamma}\left[\frac{\partial P}{\partial x} + \frac{\partial Q}{\partial y}\right]dxdy.$$

It is easy to check that the integrand, after the evaluation of the partial derivatives, is identically zero. Thus, $I_{\mathbf{v}}(\Gamma) = 0$ and the proof is complete.

□

Example 8.7.2 illustrates this theorem. If we examine the proof carefully, we see that the theorem is valid for any closed curve, not necessarily a Jordan curve, for which the hypothesis is satisfied and Green's theorem is valid. For example, the closed curve may be taken as the union of two concentric circles connected by a line. This is illustrated in the next interesting corollary.

Corollary 8.7.5

Suppose Γ_1 and Γ_2 are two simple closed curves in the plane, one lying in the interior of the other, such that the vector field $\mathbf{v}$ has no equilibrium points on Γ_1, Γ_2 and the 'annular' region between them. Then, $I_{\mathbf{v}}(\Gamma_1) = I_{\mathbf{v}}(\Gamma_2)$.

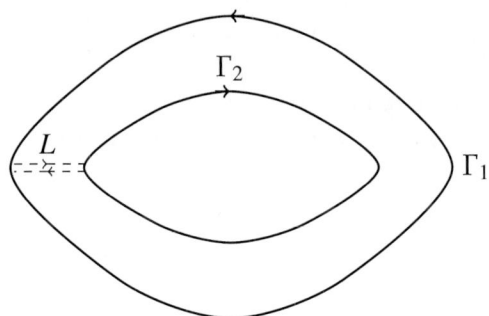

Fig. 8.7   The closed curve to prove $I_{\mathbf{v}}(\Gamma_1) = I_{\mathbf{v}}(\Gamma_2)$

Proof: Suppose Γ_2 lies in the interior of Γ_1. Let L be a line joining them. Consider the closed curve Γ shown in Fig. 8.7. By the theorem, $I_\mathbf{v}(\Gamma) = 0$. Performing the line integral, we see that $I_\mathbf{v}(\Gamma) = I_\mathbf{v}(\Gamma_1) - I_\mathbf{v}(\Gamma_2)$. This completes the proof.   □

The conclusion of the corollary is usually stated as: If a closed curve not containing any equilibrium points of $\mathbf{v}$, is continuously deformed without crossing any equilibrium points of $\mathbf{v}$, then the index is unchanged. Thus, the index is in a way independent of the closed curve in question and is associated with some special points in the plane. This enables us to talk of the *index of an isolated equilibrium point* $\mathbf{x}_0$ of the vector field $\mathbf{v}$ as the index of any Jordan curve containing $\mathbf{x}_0$ in its interior, and no other equilibrium point of $\mathbf{v}$. This will be denoted by $I_\mathbf{v}(\mathbf{x}_0)$. The reader is advised to look at the examples discussed earlier keeping this discussion in mind. For isolated equilibrium points, it can be shown that the index of an equilibrium point, computed from the linearized system, is the same for the nonlinear system as well. For more details, see [CL72, JS03]. For linear systems, we can do the computations easily and find the following. Suppose the vector field $\mathbf{v}$ is linear and is given by

$$v_1(x,y) = ax + by, \; v_2(x,y) = cx + dy, \; ad - bc \neq 0.$$

Then, the origin is the only equilibrium point of $\mathbf{v}$ and one finds, using (8.7.4), that $I_\mathbf{v}(\mathbf{0}) =$ the sign of $ad - bc$. This is left as an exercise to the reader.

The idea of the proof of the corollary can also be extended to prove the following general result.

Theorem 8.7.6

If Γ is a closed Jordan curve surrounding a finite number of equilibrium points $\mathbf{a}_1, \cdots, \mathbf{a}_n$ of a vector field $\mathbf{v}$ (See Fig. 8.8) which does not vanish on Γ, then, $I_\mathbf{v}(\Gamma) = \displaystyle\sum_{i=1}^{n} I_\mathbf{v}(\mathbf{a}_i)$.

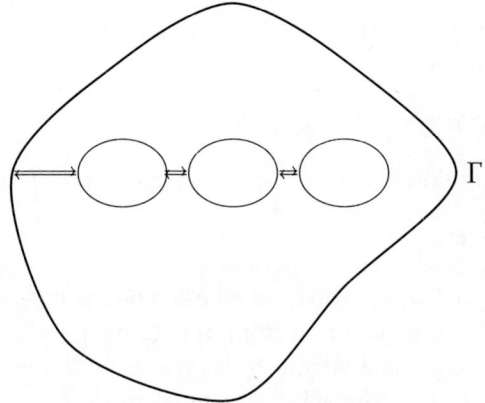

Fig. 8.8   The closed curve to prove Theorem 8.7.7

Theorem 8.7.7

Let Γ be a closed Jordan curve with a continuous tangent vector $\mathbf{v}$ at each point of Γ, which has no equilibrium points on Γ. Then, $I_{\mathbf{v}}(\Gamma)=1$.

Proof: If $\mathbf{u}(\mathbf{p})$ is the unit tangent vector to Γ at $\mathbf{p}$, then, clearly $I_{\mathbf{v}}(\Gamma) = I_{\mathbf{u}}(\Gamma)$. Thus, it suffices to prove the theorem for $\mathbf{u}$. As rotation and translation of Γ will not affect its index, we may assume that Γ lies in the region $y \geq 0$ in the (x, y)-plane, and that Γ is parameterized by $(x(s), y(s))$, $a \leq s \leq b$. Thus, $\mathbf{v}(s) = \left(\dfrac{dx}{ds}(s), \dfrac{dy}{ds}(s) \right)$. We will assume that $\mathbf{v}(a)$ is in the direction of the positive x-axis. See Fig. 8.9.

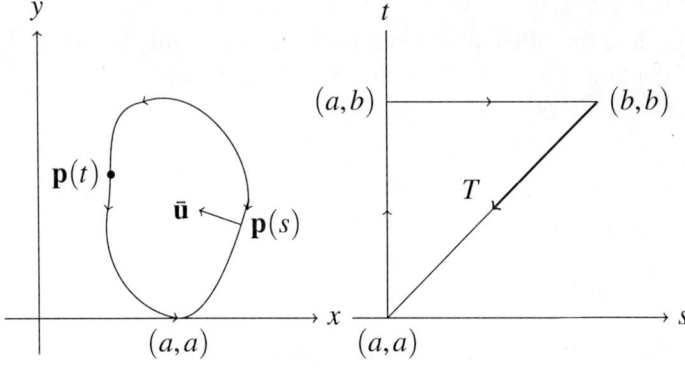

Fig. 8.9   Proof of Theorem 8.7

We will now construct an auxiliary vector field $\bar{\mathbf{u}}$, which will be used to prove the theorem. Let

$$T = \{(s,t) : a \leq s \leq b, s \leq t \leq b\},$$

be the triangular region in the (s,t)-plane. Define $\bar{\mathbf{u}}$ on T by

$$\bar{\mathbf{u}}(s,s) = \mathbf{u}(s),$$

for $a \leq s \leq b$ and $\bar{\mathbf{u}}(a,b) = -\mathbf{u}(a)$; at all other points in T, define $\bar{\mathbf{u}}(s,t)$ to be the unit vector in the direction from $\mathbf{p}(s)$ to $\mathbf{p}(t)$ on Γ. See Figure 8.7. Let $\theta(s,t)$ be the angle the vector $\bar{\mathbf{u}}(s,t)$ makes with the positive x-axis. Therefore, $\theta(a,a) = 0$. Since Γ is assumed to lie in the region $y \geq 0$, $\theta(a,t)$ varies from 0 to π as t varies from a to b. Similarly, $\theta(s,b)$ varies from π to 2π as s varies from a to b. Also, by the definition, $\bar{\mathbf{u}}$ does not vanish on the boundary $\bar{\Gamma}$ of T. Therefore, by Theorem 8.7.4, we conclude that $I_{\bar{\mathbf{u}}}(\bar{\Gamma}) = 0$. This means that the variation of $\theta(s,s)$ as s varies from a to b is 2π. But this is precisely the same as saying that the variation of the angle that $\mathbf{u}$ makes with the positive x-axis as Γ is traversed once in the positive direction is 2π. Hence, $I_{\mathbf{u}}(\Gamma) = 1$ and the proof is complete.   □

The following theorem and corollaries are immediate consequences of these results.

Theorem 8.7.8

Suppose C is a periodic orbit of (8.7.1). Put $\mathbf{v} = (f,g)$. Then,

1. $I_{\mathbf{v}}(C) = 1$.

2. Assume that the equilibrium points of $\mathbf{v}$ are only isolated. Then, C contains only finitely many equilibrium points of $\mathbf{v}$, the sum of whose indices is 1.

From these two properties of the index, the following theorem is an immediate consequence.

Theorem 8.7.9

A periodic orbit of (8.7.1) necessarily surrounds an equilibrium point.

Theorem 8.7.9, thus, asserts that equilibrium points are necessary for the existence of a periodic orbit in two-dimensions. However, it is interesting to note that the presence of equilibrium points is not a necessary condition for the existence of a periodic orbit in higher dimensions.

Example 8.7.10

Consider the following 3D system

$$\dot{x}_1 = x_2, \quad \dot{x}_2 = -x_1, \quad \dot{x}_3 = 1 - x_1^2 - x_2^2.$$

This system does not have any equilibrium points, but has periodic orbits given by $(\cos t, \sin t, c)$, for arbitrary constants c.

Theorem 8.7.11

[Bendixon's criterion] If in a region Ω of $\mathbb{R}^2$ (the phase plane), the term $\dfrac{\partial f}{\partial x} + \dfrac{\partial g}{\partial y}$ has a definite sign, then Ω contains no periodic orbit of (8.7.1).

Proof: If Ω contains a periodic orbit $C = \{(x(t), y(t)) : 0 \le t \le T\}$, let D be the region interior to C. Then, by Green's theorem, we have

$$\int_C f\, dy - g\, dx = \iint_D \left(\frac{\partial f}{\partial x} + \frac{\partial g}{\partial y} \right) dx\, dy. \tag{8.7.5}$$

The line integral on the left side of (8.7.5) is

$$\int_0^T (f\dot{y} - g\dot{x})\, dt,$$

which vanishes using (8.7.1) and, by hypothesis, the right side is either positive or negative. This contradiction proves the theorem. □

Now, we state the important Poincarè–Bendixon theorem, whose proof will be given in the Appendix.

Theorem 8.7.12

[Poincarè–Bendixon Theorem] Let Ω be a bounded region in the phase plane together with its boundary and assume that Ω does not contain any equilibrium point of (8.7.1). If $C = \{(x(t), y(t))\}$ is an orbit which lies in Ω at $t = t_0$ and remains there for all $t > t_0$, then, C itself is a periodic orbit or it spirals towards a periodic orbit as $t \to \infty$.

$\square$

Thus, in the case of two-dimensional systems, a non-trivial orbit is either unbounded or bounded. In the latter case, either it eventually enters an equilibrium point or the Poincarè–Bendixon theorem applies.

While working on a particular situation, Ω will be taken as an annular region centered at an equilibrium point of (8.7.1), but not containing any equilibrium points of (8.7.1). We, then, have the existence of a periodic orbit once we show that a positive orbit gets trapped in Ω. We illustrate this in the following example.

Example 8.7.13

Consider the following $2D$ system

$$\dot{x} = -y + x(1 - x^2 - y^2), \quad \dot{y} = x + y(1 - x^2 - y^2).$$

Using polar coordinates $x = r\cos\theta$ and $y = r\sin\theta$, we find that $r^2 = x^2 + y^2$ and $\tan\theta = y/x$. Therefore, $r\dot{r} = x\dot{x} + y\dot{y}$ and $\dot{\theta} = \dfrac{1}{r^2}(x\dot{y} - y\dot{x})$. Thus, using the given equations satisfied by x and y, we obtain

$$\dot{r} = r(1 - r^2) \text{ and } \dot{\theta} = 1.$$

Observe that for any orbit $(x(t), y(t))$, $r(t)$ is increasing in the region $r < 1$ and decreasing in the region $r > 1$. Therefore, if we take, for example, Ω to be the annulus $\{1/2 < r < 3/2\}$, we see that any orbit that starts in Ω remains there for all future times; this incidentally proves that the solution exists in $[t_0, \infty)$ for any initial time t_0. Also, Ω does not contain the origin, the only equilibrium point of the given system. Therefore, by the Poincarè–Bendixon theorem, we conclude that the given system possesses a periodic orbit in Ω.

Looking at the system satisfied by r and θ, we see that it is decoupled and therefore, it is straightforward to integrate this system. We ask the reader to show that

$$r(t) = (1 + c\exp(-2(t - t_0)))^{-1/2}, \quad \theta(t) = \theta_0 + (t - t_0),$$

where t_0 is the initial time and $c = \frac{1-r_0^2}{r_0^2}$ is given in terms of the initial condition for $r(t_0) = r_0 > 0$. Hence, every orbit that either begins inside or outside the circle $r = 1$, spirals towards the circle $r = 1$ as $t \to \infty$. Thus, $r = 1$ is the only periodic orbit for the given system and it is a stable limit cycle. ☐

The following theorem, due to Leinard, (see [Sim91, SK07]) for the existence of periodic orbits of second order equations has particularly verifiable hypothesis compared to the Poincarè–Bendixon theorem.

Theorem 8.7.14

[Leinard's Theorem] Consider the second order equation

$$\ddot{x} + f(x)\dot{x} + g(x) = 0, \tag{8.7.6}$$

where f, g satisfy the following conditions.

(1) f, g are C^1 functions.
(2) g is odd, that is, $g(-x) = -g(x)$ and $g(x) > 0$ for $x > 0$; f is even, that is, $f(-x) = f(x)$.
(3) The odd function $F(x) = \int_0^x f(s)\,ds$ has exactly one positive zero a; F is negative in $(0,a)$ and is positive and non-decreasing in (a,∞) with $F(x) \to \infty$ as $x \to \infty$.

Then, (8.7.6) has a unique periodic orbit surrounding the only equilibrium point $(0,0)$ in the phase plane. Further, this periodic orbit is spirally approached by every other (non-trivial) orbit as $t \to \infty$.

Thus, the periodic orbit is a **stable limit cycle**. The proof will be given in the Appendix. We, illustrate the applicability of Leinard's theorem by an example.

Example 8.7.15

(van der Pol Equation) Applying Leinard's theorem to the van der Pol equation,

$$\ddot{x} + \mu(x^2 - 1)\dot{x} + x = 0, \mu > 0,$$

we see that it possesses a unique periodic orbit, which is a stable limit cycle.

8.8 Exercises

1. In the following systems, find the equilibrium points, draw the phase portraits and find explicit solutions wherever possible.

   (a) Consider the Lotka–Volterra prey–predator model discussed in Example 1.2.6. This system is given by $\dot{x} = ax - bxy$, $\dot{y} = -cy + dxy$, where a, b, c, d are all positive real numbers.

   (b) Consider the system $\dot{x} = y + \sin x$, $\dot{y} = x - \cos x$.

   (c) Show that all solutions of the system

   $$\dot{x} = x^2 + y \sin x,$$

   $$\dot{y} = 1 + xy \cos y,$$

   which start in the first quadrant must remain in the first quadrant for all future time.

   (d) (**Competition between two species**) Consider the system

   $$\dot{x} = ax - bxy - ex^2,$$

   $$\dot{y} = -cy + dxy - fy^2,$$

   where the constants $a, b, \ldots$ are all positive. If $c/d > a/e$, show that every orbit that starts in the first quadrant approaches the equilibrium point $(a/e, 0)$ as $t \to \infty$. (Note that this system is a generalization of the logistic model with two competing species, that is, the Lotka–Volterra prey–predator model).

2. Consider the equation $\ddot{x} = e^x$. There are no equilibrium points! Hence, all solutions are unbounded. Find the solution explicitly using the equation of conservation of energy (8.6.2).

3. Do the same as in the previous exercise for the equation $\ddot{x} = -e^x$. Is there any difference?

4. Work out all the details in Example 8.6.2 of the pendulum equation when $E = 2k$.

5. Work out all the details in Example 8.6.3 of Duffing's equation with no damping when $E = 0$.

6. Consider the equation $-\ddot{x} + x\dot{x} = 0$. For this equation, all the equilibrium points are **non-isolated**! Solve this equation explicitly.

7. The viscous **Burgers' equation** is the following nonlinear partial differential equation

$$u_\tau + u u_\xi = \varepsilon u_{\xi\xi},$$

where $u = u(\tau, \xi)$ is the unknown function of the real variables τ and ξ and $\varepsilon > 0$. By rescaling the variables, we may take $\varepsilon = 1$. A *traveling wave solution* is a solution of the form $u(\tau, \xi) = x(t)$, $t = \xi - c\tau$ for a constant c. If we substitute this into Burgers' equation, we see that x satisfies, after some adjustments, the equation in Exercise 6 above.

Now it is required to find a solution x of the aforementioned equation satisfying

$$\lim_{t \to -\infty} x(t) = a \text{ and } \lim_{t \to \infty} x(t) = -a.$$

for appropriate $a \neq 0$. Using the knowledge of the solutions, show that this is possible if and only if $a > 0$ and write down the solution in explicit form.

8. Consider a $2n$-dimensional linear system

$$\dot{x} = \mathbf{A}x + \mathbf{B}y,$$

$$\dot{y} = \mathbf{C}x + \mathbf{D}y.$$

Here, $\mathbf{x}, \mathbf{y} \in \mathbb{R}^n$ are functions of t and $\mathbf{A}, \mathbf{B}, \mathbf{C}$ and $\mathbf{D}$ are real $n \times n$ (constant) matrices. If $\mathbf{B}, \mathbf{C}$ are symmetric and $\mathbf{D} = -\mathbf{A}^t$ (superscript t denotes the transpose of the matrix), show that the given system is a Hamiltonian system and write down a corresponding Hamiltonian function.

9. Work out all the details in Example 8.7.13. Do similar analysis for the following:

   (a) $\dot{x} = x + y - x\exp(x^2 + y^2 - 4), \dot{y} = -x + y - y\exp(x^2 + y^2 - 4)$.
   (b) $\dot{x} = 2x + 3y - x(x^2 + y^2), \dot{y} = -3x + 2y - y(x^2 + y^2)$.

10. Draw the phase portrait of $\ddot{x} = \frac{1}{2}(x^2 - 1)$. Derive the formulas for the solution x in Example 8.2.6, using the method of separable variables.

11. Work out all the details in Example 8.6.11.

12. Let (x, y) be a solution of the system in Example 8.5.2 with initial data (x_0, y_0) at $t = 0$. Show that

(a) $x \left(y - \dfrac{x^2}{3} \right)$ is a constant.

(b) $\left(y(t) - \dfrac{x(t)^2}{3} \right) = e^{-t} \left(y_0 - \dfrac{x_0^2}{3} \right)$ for all $t \in \mathbb{R}$.

8.9 Notes

In this chapter, we have studied the very basic notion of stability of an equilibrium point of an autonomous system. This is an important aspect in many physical systems such as mechanical systems, motion of a satellite, etc. The stability analysis of a hyperbolic equilibrium point of a general autonomous system follows from the analysis of the linearized system. This follows from the theorems of Perron and Hartman–Grobman. However, the case of a non-hyperbolic equilibrium point is more delicate. We have discussed a powerful tool, namely the Liapunov function, to deal with this situation. Though the stability results are easy to state and prove using the method of Liapunov, it is not trivial to construct a Liapunov function for a general system. For polynomial vector fields, one may try to use the quadratic forms to generate a Liapunov function.

In the same way, one can consider the stability of an orbit of a solution. However, a notion called structural stability is required, which is not considered here, and which is more involved and complicated. Another aspect we have completely left out is the study of stability of a periodic orbit. We have briefly mentioned this during our discussion on Floquet theory, which concerns linear systems with periodic orbits. An interested reader, after thoroughly going through the present chapter, may look into more advanced texts regarding the other aspects of stability theory. A good list of references is [CL72, HSD04, Wig90, Per01, JS03].

9

Two Point Boundary Value Problems

9.1 Introduction

In this chapter, we discuss some boundary value problems (BVPs) for linear and nonlinear second order equations. These problems arise in a vast number of practical situations ranging from physics, engineering to biology. For a very good collection of such problems and their detailed descriptions, refer to [AMR95].

The analysis, in the linear case, makes use of the existence of two linearly independent solutions, discussed thoroughly in Chapter 3. These are, then, used to construct the so-called *Green's function* of the given BVP, which in turn will generate the required solution.

The nonlinear case is more delicate. We describe a well-known method, *shooting method*, to prove the existence and uniqueness of a solution to BVP. Several examples will be discussed in detail as illustrations of the theory developed. Doing a phase plane analysis, at least in the case of autonomous equations, may help decide whether a solution to the given BVP is possible or not. However, in most situations one needs to use a suitable numerical scheme to obtain a solution.

The study of the existence and uniqueness of solutions to a BVP is more difficult than that of an IVP, even in the linear case. We first look at some examples to see these difficulties.

Example 9.1.1

Consider the equation $\ddot{u} + u = 0$, in $(0, b)$ with boundary conditions $u(0) = 0$, $u(b) = \beta$.

Here, $b > 0$. The general solution to this equation is given by

$$u(t) = c_1 \sin t + c_2 \cos t,$$

for arbitrary constants c_1, c_2. If this solution were to satisfy the given boundary conditions, then, we have $c_2 = 0$ and $c_1 \sin b = \beta$. From this, we observe the following:

- If $b = \pi$ and $\beta \neq 0$, then there is no solution.
- If $b = \pi$ and $\beta = 0$, then there are infinitely many solutions.
- If $b < \pi$, then there is a unique solution $u(t) = c \sin t$ provided that $|\beta| \leq |c|$.

Example 9.1.2

Consider the equation $-\ddot{u} = f(t)$ in $(0, 1)$, with the boundary conditions $\dot{u}(0) = \gamma_1$ and $-\dot{u}(1) = \gamma_2$.

This represents a steady state heat flow in a rod; the boundary conditions represent the heat fluxes at the ends of the rod. The given function f, represents the external heating or cooling of the rod.

Assuming that a solution exists, we obtain after an integration that

$$\int_0^1 f(t)\, dt = -\dot{u}(1) + \dot{u}(0) = \gamma_2 + \gamma_1.$$

The left hand side represents total heat (or cooling) supply to the rod and the term on the right side represents the total heat flux at the ends.

Therefore, we immediately conclude that no solution exists if, for example, $f \equiv 1$ and $\gamma_1 = \gamma_2 = 0$. On the other hand, if $f(t) = \sin(2\pi t)$, $0 \leq t \leq 1$ and $\gamma_1 = \gamma_2 = 0$, then there are infinitely many solutions given by

$$u(t) = a - \frac{t}{2\pi} + \frac{1}{4\pi^2} \sin(2\pi t),$$

with an arbitrary constant a.

Example 9.1.3

Consider the equation $\ddot{u} + a_0 u = 0$ in $(0, 1)$ with boundary conditions $u(0) = a_1$, $u(1) = a_2$.

Here a_0, a_1, a_2 are real constants. The reader should work out the details to find out various cases of existence and non-existence of solutions.

9.2 Linear Problems

We consider the following BVP for a general second order linear equation

$$\ddot{y} + \alpha(t)\dot{y}(t) + \beta(t)y(t) = g(t), \quad t \in (a,b), \tag{9.2.1}$$

$$\mathbf{A}\begin{bmatrix} y(a) \\ \dot{y}(a) \end{bmatrix} + \mathbf{B}\begin{bmatrix} y(b) \\ \dot{y}(b) \end{bmatrix} = \begin{bmatrix} \gamma_1 \\ \gamma_2 \end{bmatrix}. \tag{9.2.2}$$

Here, α, β and g are given continuous functions defined on a compact interval $[a,b]$ in $\mathbb{R}$ and $\mathbf{A}, \mathbf{B}$ are given 2×2 non-zero real matrices; γ_1, γ_2 are arbitrarily prescribed real numbers. The problem is to find a C^2 function y defined on $[a,b]$ that satisfies (9.2.1) and (9.2.2). The linear boundary conditions given by (9.2.2) are most general. If $\gamma_1 = \gamma_2 = 0$, the boundary conditions are called *homogeneous*; otherwise, they are called *non-homogeneous*. For example, the homogeneous boundary conditions $y(a) = 0$ and $y(b) = 0$ result if we take $\mathbf{A} = \begin{bmatrix} 1 & 0 \\ 0 & 0 \end{bmatrix}$, $\mathbf{B} = \begin{bmatrix} 0 & 0 \\ 1 & 0 \end{bmatrix}$ and $\gamma_1 = \gamma_2 = 0$.

Corresponding to (9.2.1) and (9.2.2), we now consider the following homogeneous problem

$$\ddot{y} + \alpha(t)\dot{y}(t) + \beta(t)y(t) = 0, \quad t \in (a,b), \tag{9.2.3}$$

$$\mathbf{A}\begin{bmatrix} y(a) \\ \dot{y}(a) \end{bmatrix} + \mathbf{B}\begin{bmatrix} y(b) \\ \dot{y}(b) \end{bmatrix} = \begin{bmatrix} 0 \\ 0 \end{bmatrix}. \tag{9.2.4}$$

We analyze the question of existence and uniqueness of solutions by relating problems (9.2.1), (9.2.2) and (9.2.3), (9.2.4). Suppose y_1, y_2 are any two linearly independent solutions of the homogeneous equation (9.2.3). Then, any general solution of (9.2.3) is given by $y = c_1 y_1 + c_2 y_2$ for arbitrary real constants c_1 and c_2. If y were to satisfy the boundary conditions (9.2.4), then c_1, c_2 should satisfy

$$(\mathbf{A}\mathbf{W}(a) + \mathbf{B}\mathbf{W}(b)) \begin{bmatrix} c_1 \\ c_2 \end{bmatrix} = \begin{bmatrix} 0 \\ 0 \end{bmatrix}. \tag{9.2.5}$$

Here, $\mathbf{W}(t) = \begin{bmatrix} y_1(t) & y_2(t) \\ \dot{y}_1(t) & \dot{y}_2(t) \end{bmatrix}$ denotes the Wronskian matrix of the solutions y_1 and y_2 at t. The aforementioned system of linear algebraic equations has a non-trivial solution if and only if the matrix $\mathbf{A}\mathbf{W}(a) + \mathbf{B}\mathbf{W}(b)$ is singular, that is, its rank is either 0 or 1; notice that $y \equiv 0$ is always a solution of (9.2.3) and (9.2.4) (the trivial solution). Also, note that $\mathbf{W}(t)$ is non-singular and the rank of $\mathbf{A}\mathbf{W}(a) + \mathbf{B}\mathbf{W}(b)$ does not depend on any particular choice of a pair of linearly independent solutions.

Now, let y_0 be any particular solution of the non-homogeneous equation (9.2.1). Then, any general solution of (9.2.1) is given by $y = y_0 + c_1 y_1 + c_2 y_2$ for arbitrary real constants c_1 and c_2. If this y were to satisfy the boundary conditions (9.2.2), then c_1, c_2 should satisfy

$$(\mathbf{A}\mathbf{W}(a) + \mathbf{B}\mathbf{W}(b)) \begin{bmatrix} c_1 \\ c_2 \end{bmatrix} = \gamma - \xi \equiv \begin{bmatrix} \gamma_1 - \xi_1 \\ \gamma_2 - \xi_2 \end{bmatrix}, \tag{9.2.6}$$

where $\begin{bmatrix} \xi_1 \\ \xi_2 \end{bmatrix} = \mathbf{A} \begin{bmatrix} y_0(a) \\ \dot{y}_0(a) \end{bmatrix} + \mathbf{B} \begin{bmatrix} y_0(b) \\ \dot{y}_0(b) \end{bmatrix}$. The system of algebraic equations (9.2.6) has a solution if and only if

$$\operatorname{rank}(\mathbf{A}\mathbf{W}(a) + \mathbf{B}\mathbf{W}(b)) = \operatorname{rank}[\mathbf{A}\mathbf{W}(a) + \mathbf{B}\mathbf{W}(b) \quad \gamma - \xi]. \tag{9.2.7}$$

Note that this condition on ranks is automatically satisfied if $\mathbf{A}\mathbf{W}(a) + \mathbf{B}\mathbf{W}(b)$ is non-singular. We now state the foregoing discussion in the following theorem.

Theorem 9.2.1

The following hold:

1. If $\mathbf{A}\mathbf{W}(a) + \mathbf{B}\mathbf{W}(b)$ is non-singular, then, the BVP (9.2.1) and (9.2.2) has a unique solution for arbitrary γ_1, γ_2. In this case, the BVP (9.2.3) and (9.2.4) has only the trivial solution.

2. If $\mathbf{A}\mathbf{W}(a) + \mathbf{B}\mathbf{W}(b)$ is singular, then, the BVP (9.2.1) and (9.2.2) has a solution if and only if the rank condition (9.2.7) is fulfilled; the

solution is not unique. In this case, the BVP (9.2.3) and (9.2.4) has non-trivial solutions.

For simplicity of presentation, we now consider the BVP (9.2.1) with the following homogeneous boundary conditions, replacing the general boundary conditions in (9.2.2):

$$y(a) = 0 \text{ and } y(b) = 0. \tag{9.2.8}$$

Fix α, β. We wish to derive a formula for the solution y, when it exists, in the form of an integral involving the 'input' function g, much similar to the case of first order linear equations. We rewrite (9.2.1) in the equivalent form as

$$\frac{d}{dt}[p(t)\dot{y}(t)] + q(t)y(t) = f(t), \tag{9.2.9}$$

where p is a positive C^1 function on $[a,b]$. Given (9.2.9), it is obvious that it can be written in the form (9.2.1) by taking $\alpha(t) = \dot{p}(t)/p(t)$, $\beta(t) = q(t)/p(t)$ and $g(t) = f(t)/p(t)$. Conversely, given (9.2.1), we define $p(t) = \exp\left(\int_a^t \alpha(s)\,ds\right)$ and find that (9.2.1) can be put in the form (9.2.9) with $q(t) = p(t)\beta(t)$ and $f(t) = p(t)g(t)$.

We begin with a heuristic description of the method to obtain a solution to the problem (9.2.9) satisfying the boundary conditions (9.2.8). Let u_1, u_2 be two linearly independent solutions of (9.2.9) with $f = 0$, that is, the homogeneous equation corresponding to (9.2.9). By the method of variation of parameters, we find a general solution of (9.2.9) as

$$y(t) = Au_1(t) + Bu_2(t) + \int_a^t \frac{f(s)}{W(s)}[u_1(s)u_2(t) - u_1(t)u_2(s)]\,ds, \tag{9.2.10}$$

where A, B are constants and W is the Wronskian of u_1 and u_2 defined by $W(t) = u_1(t)\dot{u}_2(t) - \dot{u}_1(t)u_2(t)$. By the linear independence of u_1, u_2, it follows that W is never zero. If we now require that y given by (9.2.10) satisfy the boundary conditions (9.2.8), then we must have, using (9.2.10),

$$Au_1(a) + Bu_2(a) = 0$$

$$Au_1(b) + Bu_2(b) = \int_a^b \frac{f(s)}{W(s)}[u_1(b)u_2(s) - u_1(s)u_2(b)]\,ds.$$

$$\tag{9.2.11}$$

Solving for A, B and substituting in the expression (9.2.10) for y, we obtain

$$y(t) = I_1(t) + I_2(t),$$

where

$$I_1(t) = \int_a^b \frac{f(s)}{W(s)} \frac{[u_1(b)u_2(s) - u_1(s)u_2(b)] \cdot [u_1(a)u_2(t) - u_1(t)u_2(a)]}{u_1(a)u_2(b) - u_1(b)u_2(a)} \, ds,$$

$$I_2(t) = \int_a^t \frac{f(s)}{W(s)} [u_1(s)u_2(t) - u_1(t)u_2(s)] \, ds.$$

Looking at these expressions, we find it convenient to introduce the following:

$$w_1(t) = u_1(a)u_2(t) - u_1(t)u_2(a)$$

$$w_2(t) = u_1(b)u_2(t) - u_1(t)u_2(b). \tag{9.2.12}$$

Then, w_1, w_2 being linear combinations of u_1, u_2, are themselves solutions of the homogeneous equation (9.2.9) ($f \equiv 0$). They are also linearly independent satisfying $w_1(a) = 0, w_2(b) = 0$, provided that $u_1(a)u_2(b) - u_1(b)u_2(a)$ is not zero. Looking at the aforementioned expression for I_1, I_2, we see that the required solution y may be written as

$$y(t) = \int_a^t [\cdots] f(s) \, ds + \int_t^b [\cdots] f(s) ds.$$

Having obtained the knowledge of the form of the solution, we now proceed directly to obtain a representation of a solution to the BVP (9.2.9) and (9.2.8). Choose *any* linearly independent solutions w_1, w_2 of the homogeneous equation corresponding to (9.2.9) satisfying $w_1(a) = 0$, $w_2(b) = 0$.[1] Then, the required solution is given by

$$y(t) = \int_a^b G(t,s) f(s) \, ds \tag{9.2.13}$$

where the function G, called *Green's function* of (9.2.9) and (9.2.8), is given by

---

[1]Note that the existence of w_1 and w_2 is *not* automatic and the given boundary conditions play an important role in their existence.

$$G(t,s) = \begin{cases} w_1(s)w_2(t) & \text{if } a \leq s \leq t \\ w_1(t)w_2(s) & \text{if } t < s \leq b. \end{cases}$$

We now directly verify that y given by (9.2.13) is a solution of (9.2.9) satisfying the boundary conditions (9.2.8), after suitably normalizing w_1, w_2. Clearly y satisfies (9.2.8) as $w_1(a) = 0$ and $w_2(b) = 0$.

Note that

$$\lim_{s \to t-} \frac{\partial}{\partial t} G(t,s) - \lim_{s \to t+} \frac{\partial}{\partial t} G(t,s) = w_1(t)\dot{w}_2(t) - \dot{w}_1(t)w_2(t).$$

The right side expression being the Wronskian of w_1, w_2, it follows from (9.2.9) that this limit equals $C/p(t)$, where C is a constant. This follows from the fact that the Wronskian satisfies a first order linear equation; see Chapter 3. We may normalize w_1, w_2 so that $C = 1$. With this normalization, we have[2]

$$\dot{y}(t) = \int_a^t \frac{\partial}{\partial t} G(t,s)f(s)\,ds + G(t,t)f(t) + \int_t^b \frac{\partial}{\partial t} G(t,s)f(s)\,ds$$

$$-G(t,t)f(t)$$

and therefore,

$$\dot{y}(t) = \dot{w}_2(t) \int_a^t w_1(s)f(s)\,ds + \dot{w}_1(t) \int_t^b w_2(s)f(s)\,ds.$$

Now multiply this expression by $p(t)$ and differentiate once again with respect to t to obtain

$$\frac{d}{dt}[p(t)\dot{y}(t)] = \int_a^t w_1(s) \frac{d}{dt}[p(t)\dot{w}_2(t)]f(s)\,ds + p(t)w_1(t)\dot{w}_2(t)f(t)$$

$$+ \int_t^b \frac{d}{dt}[p(t)\dot{w}_1(t)w_2(s)f(s)\,ds - p(t)\dot{w}_1(t)w_2(t)f(t).$$

Using the normalization and that w_1, w_2 satisfy the homogeneous equation (9.2.9), we see that the expression on the right equals $-q(t)y(t) + f(t)$. This completes the verification that y is indeed a solution of the BVP. We take up the uniqueness question in the next section.

---

[2]See differentiation under the integral sign in Chapter 2.

If we consider more general boundary conditions in (9.2.2), homogeneous or not, instead of (9.2.8), the computation of Green's function becomes more tedious even in the case of constant coefficient equations. The case of non-homogeneous boundary conditions can always be reduced to the case of homogeneous boundary conditions by subtracting a solution of the homogeneous equation satisfying the non-homogeneous boundary conditions. The details will be given in the following discussion on BVP of linear systems. Some BVP with non-homogeneous boundary conditions may lead to 'non-self-adjoint' problems.

9.2.1   BVP for linear systems

The earlier discussion on BVP can be easily generalized to linear systems with the aid of a fundamental matrix. Consider the following BVP:

$$\dot{\mathbf{x}}(t) = \mathbf{A}(t)\mathbf{x}(t) + \mathbf{f}(t), \quad t \in [a,b], \tag{9.2.14}$$

$$\mathbf{M}\mathbf{x}(a) + \mathbf{N}\mathbf{x}(b) = \xi. \tag{9.2.15}$$

Here, $\mathbf{A}$ is an $n \times n$ matrix valued continuous function defined on $[a,b]$, a given interval on the real line, $\mathbf{f}: [a,b] \to \mathbb{R}^n$ is a given continuous vector function and $\mathbf{M}$, $\mathbf{N}$ are (constant) $n \times n$ matrices and ξ is a given vector in $\mathbb{R}^n$. The problem is to find a solution $\mathbf{x}(t)$ satisfying (9.2.14) in $[a,b]$ and the boundary conditions (9.2.15).

Let $\mathbf{\Phi}$ be a fundamental matrix of the homogeneous equation in (9.2.14) satisfying $\mathbf{\Phi}(a) = \mathbf{I}$. Then, any solution of (9.2.14) is given by $\mathbf{x}(t) = \mathbf{\Phi}(t)\mathbf{c} + \mathbf{\Phi}(t)\int_a^t \mathbf{\Phi}^{-1}(s)\mathbf{f}(s)\,ds$ for some constant vector $\mathbf{c}$ in $\mathbb{R}^n$. If $\mathbf{x}(t)$ were to satisfy the boundary conditions (9.2.15), then, the following relation must hold:

$$\mathbf{M}\mathbf{c} + N\mathbf{\Phi}(b)\mathbf{c} + \mathbf{x}_b = \xi,$$

where $\mathbf{x}_b = N\mathbf{\Phi}(b)\displaystyle\int_a^b \mathbf{\Phi}^{-1}(s)\mathbf{f}(s)\,ds$. Therefore, the BVP (9.2.14), (9.2.15) has a unique solution if and only if the matrix $\mathbf{Y} \equiv \mathbf{M} + N\mathbf{\Phi}(b)$ is non-singular. These arguments enable us to construct *(matrix) Green's function* for the homogeneous BVP (9.2.14) and (9.2.15) with $\xi = 0$, as follows. Define

$$
G(t,s) = \begin{cases} \boldsymbol{\Phi}(t)\boldsymbol{\Phi}^{-1}(s) - \boldsymbol{\Phi}(t)\mathbf{Y}^{-1}N\boldsymbol{\Phi}(b)\boldsymbol{\Phi}^{-1}(s), & \text{if } a \le s \le t; \\[2mm] -\boldsymbol{\Phi}(t)\mathbf{Y}^{-1}N\boldsymbol{\Phi}(b)\boldsymbol{\Phi}^{-1}(s), & \text{if } t < s \le b. \end{cases}
$$

Then, the required solution is given by

$$
\mathbf{x}(t) = \int_a^b \mathbf{G}(t,s)\mathbf{f}(s)\,ds.
$$

The two variable (matrix) function G is called *(matrix) Green's function* and this representation is *Green's formula* to obtain a solution to the given homogeneous BVP. A solution to the non-homogeneous BVP (9.2.14) and (9.2.15) can be found by adding a solution of the homogeneous system in (9.2.14) satisfying (9.2.15) (for example, $\boldsymbol{\Phi}(t)\mathbf{Y}^{-1}\xi$) to the solution obtained from Green's formula.

9.2.2 Examples

Example 9.2.2

Consider the following simple BVP

$$
\ddot{y} = f(t), \; y(0) = y(1) = 0.
$$

Choose $w_1(t) = t$ and $w_2(t) = t - 1$ for $t \in [0,1]$. These are linearly independent solutions of the given equation. Moreover, $w_1(0) = 0$ and $w_2(1) = 0$ and their Wronskian is 1, the required normalization constant. Here $p \equiv 1$. Therefore Green's function of the problem is given by

$$
G(t,s) = \begin{cases} s(t-1), & \text{if } 0 \le s \le t, \\[2mm] t(s-1), & \text{if } t < s \le 1. \end{cases}
$$

and the solution to the given BVP is given by

$$
y(t) = \int_0^1 G(t,s)f(s)\,ds.
$$

In particular, if $f \equiv 1$, we find that $y(t) = \dfrac{1}{2}t(t-1)$.

Example 9.2.3

Consider the following BVP

$$\ddot{y} + 4y = f(t), \, y(0) = y(\pi/4) = 0.$$

Since, $\sin(2t)$ and $\cos(2t)$ form a basis for the solution space of the homogeneous equation, we choose $w_1(t) = \sin(2t)$ and $w_2(t) = -\frac{1}{2}\cos(2t)$ to satisfy the required boundary conditions and normalization condition. Thus, Green's function is given by

$$G(t,s) = \begin{cases} -\frac{1}{2}\sin(2s)\cos(2t), & \text{if } 0 \le s \le t, \\ -\frac{1}{2}\sin(2t)\cos(2s), & \text{if } t < s \le 1. \end{cases}$$

If, again, $f \equiv 1$, then the solution is given by $y(t) = \frac{1}{4}(1 - \cos(2t))$.

Example 9.2.4

In this example, we consider the BVP

$$\ddot{y} - 4\dot{y} + 4y = f(t), \quad y(0) + 2\dot{y}(0) = 0, \quad y(1) - \dot{y}(1) = 0.$$

Here $p(t) = e^{-4t}$ and we need to normalize the linearly independent solutions we choose. Since e^{2t} and te^{2t} are basis functions for the solution space of the homogeneous equation, we choose $w_1(t) = c_1 e^{2t}(2 - 5t)$ and $w_2(t) = c_2 e^{2t}(2 - t)$. Then, w_1 and w_2 satisfy the homogeneous equation and $w_1(0) + 2\dot{w}_1(0) = 0$; $w_2(1) - \dot{w}_2(1) = 0$. The constants c_1, c_2 are chosen to satisfy $8c_1c_2 = 1$ for the normalization requirement. We leave further construction details of Green's function and the solution to the reader.

9.3 General Second Order Equations

Consider the following BVP

$$\ddot{y} = f(t, y, \dot{y}) \tag{9.3.1}$$

with boundary conditions

$$a_0 y(a) - a_1 \dot{y}(a) = \alpha, \quad |a_0| + |a_1| \neq 0$$

$$b_0 y(b) + b_1 \dot{y}(b) = \beta, \quad |b_0| + |b_1| \neq 0$$

(9.3.2)

The negative sign of a coefficient in the boundary condition is chosen just for later convenience. We look for a solution in the interval $[a, b]$ as in the previous section. A formal approach to the solution of (9.3.1) satisfying the boundary conditions (9.3.2) is as follows: Start with the initial value problem (IVP)

$$\ddot{u} = f(t, u, \dot{u})$$

(9.3.3)

satisfying

$$a_0 u(a) - a_1 \dot{u}(a) = \alpha,$$

$$c_0 u(a) - c_1 \dot{u}(a) = s,$$

(9.3.4)

where, for the purpose of independence, we assume $a_1 c_0 - a_0 c_1 \neq 0$. We may fix c_0, c_1 by requiring $a_1 c_0 - a_0 c_1 = 1$. Denote by $u(t; s)$, the corresponding solution of the IVP, emphasizing the dependence on s. We seek a value of s so that

$$\phi(s) \equiv b_0 u(b; s) + b_1 \dot{u}(b; s) - \beta = 0.$$

(9.3.5)

If $s = s^*$ is a solution of (9.3.5), then $y(t) \equiv u(t; s^*)$ would be a solution of (9.3.1) satisfying the boundary conditions (9.3.2). This method is called the *shooting method* and is quite extensively used in solving (9.3.1) and (9.3.2) numerically. The following theorem is a consequence of the standard existence, uniqueness and continuous dependence of solutions of IVP.

Theorem 9.3.1

Let the function $f(t, u_1, u_2)$ be continuous in the region $R = \{(t, u_1, u_2) : t \in [a, b], \ u_1, u_2 \in \mathbb{R}\}$ and satisfy the uniform Lipschitz condition in u_1 and u_2. Then, the BVP (9.3.1), (9.3.2) has as many solutions as there are distinct roots, $s = s_j$ of (9.3.5). The solutions of (9.3.1), (9.3.2) are given by

$$y(t) = y_j(t) \equiv u(t; s_j),$$

where u solves IVP (9.3.3).   □

Theorem 9.3.2

[Existence and Uniqueness] In addition to the hypotheses in Theorem 9.3.1, suppose that

$$\frac{\partial f}{\partial u_1} > 0 \;\; \text{and} \;\; \left| \frac{\partial f}{\partial u_2} \right| \leq M,$$

in R. Also assume that

$$a_0 a_1 \geq 0, b_0 b_1 \geq 0 \;\; \text{and} \;\; |a_0| + |b_0| \neq 0. \tag{9.3.6}$$

Then, the BVP (9.3.1) and (9.3.2) has a unique solution. The condition $\frac{\partial f}{\partial u_1} > 0$ can be replaced by $\frac{\partial f}{\partial u_1} < 0$.   □

Corollary 9.3.3

[Uniqueness for the Linear Problem] Consider the linear equation

$$\ddot{y} + p(t)\dot{y} + q(t)y = r(t),$$

with the boundary conditions (9.3.2) and the same conditions on the coefficients a_0, a_1, b_0, b_1 as in Theorem 9.3.1 and Theorem 9.3.2. If p, q are continuous in $[a, b]$ and $q > 0$ in $[a, b]$, then, the linear BVP has a unique solution.

Proof: (of Theorem 9.3.2) It suffices to show that (9.3.5) has a unique root. If $u(t; s)$ denotes a solution of (9.3.3), put $\xi(t) = \frac{\partial u}{\partial s}(t; s)$. By differentiating (9.3.3) with respect to s, we obtain

$$\ddot{\xi} = p(t)\dot{\xi} + q(t)\xi, \tag{9.3.7}$$

where

$$p(t) = \frac{\partial f}{\partial \dot{u}}(t, u(t; s), \dot{u}(t; s)) \;\; \text{and} \;\; q(t) = \frac{\partial f}{\partial u}(t, u(t; s), \dot{u}(t; s)).$$

Equation (9.3.7) is referred to as the *variational equation*. By hypothesis, $q > 0$ and p is bounded. Also by differentiating the second equation in (9.3.4) with respect to s, we obtain

$$\xi(a) = a_1, \dot{\xi}(a) = a_0.$$

Since $|a_0| + |a_1| \neq 0, \xi \neq 0$ for $a < t \leq a + \varepsilon$ for some $\varepsilon > 0$.

Claim: $\xi(t) \neq 0$ for $a < t \leq b$: Suppose the claim is not true. We assume that $a_0 \geq 0$ and $a_1 \geq 0$; the other case is similar. With this assumption, we have $\xi(t) > 0$ for $a < t \leq a + \varepsilon$. Suppose there exists $t_* \in (a,b]$ such that $\xi(t_*) \leq 0$. Therefore, ξ has a positive maximum in $[a,t_*)$. If $a_0 > 0$, then $\dot{\xi}(a) = a_0 > 0$, so maximum cannot occur at a; if $a_0 = 0$, then, $\ddot{\xi}(a) = q(a)a_1 > 0$, so, again, maximum cannot occur at a. If $t_0 \in (a,t_*)$ is a point where a positive maximum occurs, then $\xi(t_0) > 0, \dot{\xi}(t_0) = 0$ and $\ddot{\xi}(t_0) \leq 0$. But $\ddot{\xi}(t_0) = q(t_0)\xi(t_0) > 0$. This contradiction proves that $\xi(t) > 0$ for $a < t \leq b$.

   Thus, $q(t)\xi(t) > 0$ for $a < t \leq b$ and from (9.3.7), we have

$$\ddot{\xi}(t) > p(t)\dot{\xi}(t), a < t \leq b.$$

Multiplying this inequality by $\exp\left(-\int_a^t p(t_1)\,dt_1\right)$ and integrating, we obtain

$$\dot{\xi}(t) > a_0 \exp\left(\int_a^t p(t_1)\,dt_1\right), a < t \leq b.$$

One more integration gives,

$$\xi(t) > a_1 + a_0 \int_a^t \exp\left(\int_a^{t_2} p(t_1)\,dt_1\right) dt_2, a < t \leq b.$$

Using $p \geq -M$, we have

$$\exp\left(\int_a^{t_2} p(t_1)\,dt_1\right) \geq \exp(-M(t_2 - a))$$

and thus,

$$\dot{\xi}(t) = \frac{\partial \dot{u}}{\partial s}(t;s) > a_0 \exp(-M(t-a)) \geq 0,$$

for $a < t \le b$. In particular, we see that by taking $t = b, u(b;s)$ is a monotone function of s whose derivative (with respect to s) is bounded away from 0 for any a_0. The same is true with $\dot{u}(b;s)$ if $a_0 \ne 0$; if $a_0 = 0$, its derivative (with respect to s) is not bounded away from 0. But, since b_0, b_1 do not both vanish and have same sign, and $b_0 \ne 0$ if $a_0 = 0$, the function $\phi(s) \equiv b_0 u(b;s) + b_1 \dot{u}(b;s) - \beta$ must have a derivative of one sign which is bounded away from 0 for any a_0. Such a function takes on each real value once and only once; hence $\phi(s) = 0$ has a unique root. This completes the proof.   □

9.3.1 Examples:

Example 9.3.4

Consider the BVP

$$\ddot{u} = 2 + u^2, t \in [0,1]; \; u(0) = u(1) = 0. \tag{9.3.8}$$

This is an equation in the conservative form and its phase portrait is not hard to draw. This is depicted in Fig. 9.1. Note that $E = E(t) = \frac{1}{2}\dot{u}^2(t) - 2u(t) - \frac{1}{3}u^3(t)$ is the total conserved energy.

If we do not insist that $u(1) = 0$, but just require that $u(b) = 0$ for some $b > 0$, then we obtain an infinite number of solutions of (9.3.8) by choosing $u(0) = 0$ and $\dot{u}(0) < 0$, as can be seen from the phase portraits shown in Fig. 9.1.

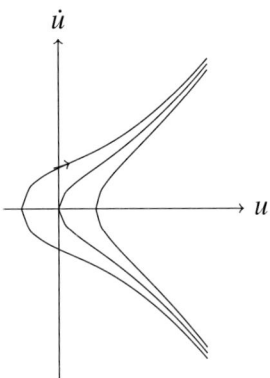

Fig. 9.1   Phase portrait for $\ddot{u} = 2 + u^2$

The requirement of the latter condition that $\dot{u}(0) < 0$ may also be seen as follows. From the equation in (9.3.8), we see that u is convex. Thus, if it satisfies the boundary conditions in (9.3.8), we must have $\dot{u}(0) < 0$ and $\dot{u}(1) > 0$. It is not possible to integrate the given equation explicitly, but the equation may be solved numerically. We find that if we 'shoot' with $\dot{u}(0)$ slightly less than -1, we obtain a solution of (9.3.8).

Example 9.3.5

Consider the BVP

$$\ddot{u} + \lambda e^{u} = 0, t \in [0,1]; u(0) = u(1) = 0. \tag{9.3.9}$$

Here, $\lambda > 0$. A complete phase portrait of the equation is shown in Fig. 9.2. Note that in this case a solution u is concave and therefore, we need to 'shoot' with $\dot{u}(0) > 0$ to possibly obtain a solution of (9.3.9). In the present situation, an explicit solution is possible. Here, the constant conserved energy is given by $E = E(t) = \frac{1}{2}(\dot{u})^{2}(t) + \lambda e^{u(t)}$.

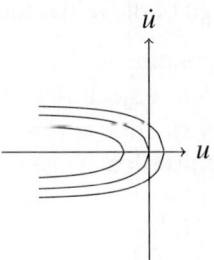

Fig. 9.2   Phase portrait for $\ddot{u} + \lambda e^{u} = 0, \lambda > 0$

A function of the form[3]

$$u(t) = -\log(C^{2}\cosh^{2}(\kappa(1-2t))),$$

where, C and κ are positive constants, may be tested for a solution of (9.3.9). The function u satisfies the boundary conditions in (9.3.9) if $C\cosh(\kappa) = 1$ and it satisfies the equation if $8\kappa^{2} = \dfrac{\lambda}{C^{2}}$. Eliminating C, we obtain the following nonlinear equation connecting κ and λ:

---

[3]The reader should try to obtain such an expression for the solution by integrating the conservation equation $\frac{1}{2}(\dot{u})^{2} + \lambda e^{u} = E \equiv$ constant.

$$\frac{\kappa}{\cosh(\kappa)} = \frac{\sqrt{\lambda}}{2\sqrt{2}}$$

The function of $\kappa > 0$ on the left side, has a unique positive maximum and tends to 0 as $\kappa \to \infty$. It follows, therefore, that there is a critical value λ_{cr} of λ such that (9.3.9) has no solution for $\lambda > \lambda_{cr}$, unique solution for $\lambda = \lambda_{cr}$ and two solutions for $\lambda < \lambda_{cr}$.

For a slightly different representation of the solution, see [AMR95].

Example 9.3.6

We change the sign of the nonlinear term in Example 9.3.5 and consider the BVP

$$\ddot{u} - \lambda e^u = 0, \quad t \in [0,1]; \qquad u(0) = u(1) = 0. \tag{9.3.10}$$

We again take $\lambda > 0$. Surprisingly, the phase portrait of this equation is substantially different from the previous one. The behavior of a solution very much depends on the sign of the initial total energy $E = \frac{1}{2}(\dot{u}(0))^2 - \lambda e^{u(0)}$. The solutions corresponding to $E \geq 0$, all are unbounded and tend to $\pm\infty$ as $t \to \infty$. Some typical trajectories are shown in Fig. 9.3, when $E \geq 0$. On the other hand, any solution corresponding to $E < 0$, is bounded below and there is a possibility that it may satisfy the given boundary conditions. Some trajectories for $E < 0$ are shown in Fig. 9.4.

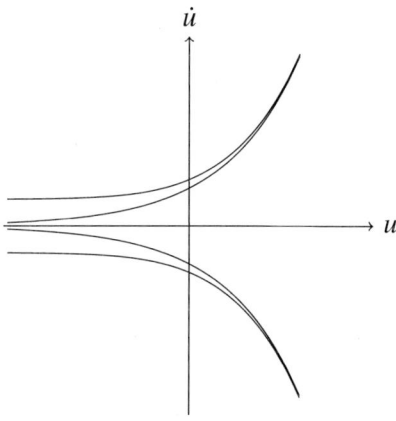

Fig. 9.3   Phase portrait for $\ddot{u} - \lambda e^u = 0, \lambda > 0, E \geq 0$

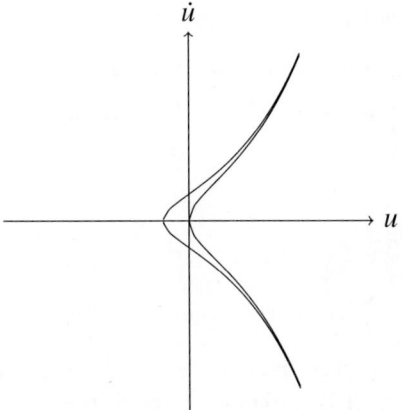

Fig. 9.4   Phase portrait for $\ddot{u} - \lambda e^u = 0, \lambda > 0, E < 0$

We now proceed to obtain explicit expressions for the solutions like in the previous example. Let

$$u(t) = -2\log(C\sinh(\kappa_1 - \kappa_2 t)). \qquad (9.3.11)$$

The constants C, κ_1, κ_2 in (9.3.11) are all positive. We should observe that there is a need to choose κs so that the argument of sinh function in (9.3.11) is never 0 and permits the values of t in an interval containing $[0,1]$, the interval of our interest. This leads to the choice

$$\kappa_1 > 0 \quad \text{and} \quad \kappa_2 + 1 \le \kappa_1, \quad \kappa_2 \ne 0.$$

(We may also use $\kappa_1 + \kappa_2 t$ instead of $\kappa_1 - \kappa_2 t$ in (9.3.11).) It is easily checked that u is a solution of the equation in (9.3.10) if $2\kappa_2^2 = \dfrac{\lambda}{C^2}$. However, for none of these values of C, κ_1, κ_2, are both the boundary conditions satisfied. We leave it to the reader to verify that this solution corresponds to $E \ge 0$.

Next, consider the function u defined by

$$u(t) = -2\log(C\cos(\kappa(2t - 1))) \qquad (9.3.12)$$

We again remark that such an expression arises from the situation when $E < 0$. Note that now the function is defined for only those t for which the argument of cos function lies in $(-\pi/2, \pi/2)$ as we are interested in an interval containing the interval $[0,1]$.

The function given by (9.3.12) is a solution of the equation in (9.3.10) if $8\kappa^2 = \dfrac{\lambda}{C^2}$ and it satisfies the boundary conditions therein if $C\cos(\kappa) = 1$. Since $\kappa > 0$, we must have $C > 1$. Eliminating C, we now obtain the equation

$$\frac{\kappa}{\cos(\kappa)} = \frac{\sqrt{\lambda}}{2\sqrt{2}}. \tag{9.3.13}$$

Since the function $\dfrac{\kappa}{\cos(\kappa)}$ for $\kappa \in (0, \pi/2)$ is an increasing function and $\to \infty$ as $\kappa \to \pi/2$, we see that for each $\lambda > 0$, there is a unique κ in $(0, \pi/2)$ satisfying (9.3.13). This in turn gives the unique solution of BVP (9.3.10) through (9.3.12) for each $\lambda > 0$.

9.4 Exercises

1. Determine the values of λ for which a Green's function can be constructed for the equation $\ddot{y} + \lambda y = f(t)$, with the following prescribed boundary conditions. Construct a Green's function for all such values of λ:

   (a) $y(0) = y(1) = 0$.

   (b) $y(0) + \dot{y}(0) = 0$ and $y(1) = \dot{y}(1)$.

   (c) $y(0) = \dot{y}(0)$ and $y(l) = 0 \, (l > 0)$.

   (d) $y(0) = \dot{y}(0)$ and $y(\pi) = \dot{y}(\pi)$.

2. Determine the values of λ for which the BVP

   $$\ddot{u} + 2\dot{u} + \lambda u = 0, \, u(0) = u(l) = 0, \, (l > 0),$$

   has non-trivial solutions.

3. Work out all the details in the examples of Section 9.3.1.

9.5 Notes

Two point BVP are studied in this chapter for both linear and nonlinear second order equations. For linear equations, we take the advantage of the existence of linearly independent solutions in order to construct Green's function of the given BVP. A brief discussion on the construction of a matrix Green's function for linear system is also included. For nonlinear equations, we no longer have the advantage of the existence of two linearly independent solutions. An existence and uniqueness result is proved under some sufficient conditions on the coefficients. The examples considered, however, show that these conditions may not be necessary. For other results on this topic, the interested reader is referred to [Sim91, SK07, CL72, Kel90, AMR95].

10

First Order Partial Differential Equations: Method of Characteristics

In Chapter 3, we have seen an example of a PDE (see Example 3.4.1) which can be solved by solving two ODEs. In this chapter, we will present the method of characteristics, where the problem of local solvability of any first order PDE, linear or nonlinear, can be reduced to solvability of a system of ODE. However, it must be observed that the geometry involved turns out to be more and more complicated when we go from linear to quasi-linear to nonlinear equations.

10.1 Linear Equations

In Example 3.4.1, the way we arrived at the characteristic curves, which were straight lines, may look artificial. We will now explain the geometry of the problem and see how we can extend these ideas to general first order equations. First, consider the general first order linear PDE in two variables;

$$a(x,y)u_x + b(x,y)u_y = c(x,y)u + d(x,y), \qquad (10.1.1)$$

where $(x,y) \in \Omega$. Here, Ω is a smooth bounded domain in $\mathbb{R}^2$; a,b,c,d are given smooth functions defined on Ω and $u = u(x,y)$ is the unknown function. Let $\Gamma_0 \subset \Omega$ be an initial curve, which is given in the parametric form:

$$\Gamma_0 = \{(x_0(s), y_0(s)) : 0 \leq s \leq 1\}.$$

Let $u_0 = u_0(s)$ be a given function defined on the initial curve. The IVP for the PDE can be defined as follows: Find $u = u(x,y)$ satisfying the PDE (10.1.1) together with the initial condition

$$u(x_0(s), y_0(s)) = u_0(s) \tag{10.1.2}$$

for all $s \in [0,1]$. The problem of local solvability of IVP is to find u in a neighborhood in Ω of the initial curve satisfying (10.1.1) and (10.1.2).

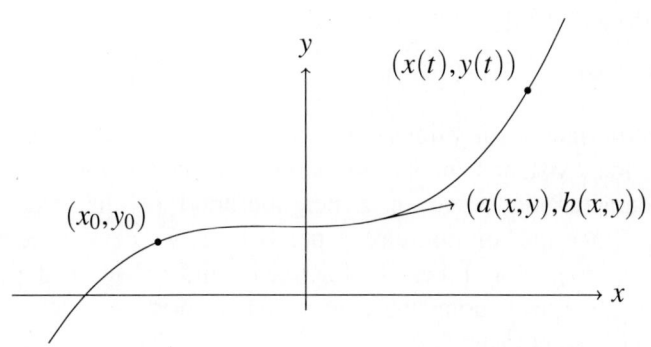

Fig. 10.1 A characteristic curve

We now give the geometric idea behind introducing the characteristics. First observe that for a fixed point (x,y), the term on the left side of (10.1.1) is the directional derivative of u at (x,y) in the direction $(a(x,y), b(x,y))$. Thus, if we consider any curve $(x(t), y(t))$, parameterized by the t variable, in the (x,y) plane whose tangent at each point is the direction $(a(x,y), b(x,y))$ (see Fig. 10.1), then the term on the left side of (10.1.1) will become a total derivative along the curve with the parameter describing the curve. Thus, consider the one parameter family of curves defined by the system of ODE

$$\frac{dx}{a(x,y)} = \frac{dy}{b(x,y)} \text{ or } \frac{dy}{dx} = \frac{b(x,y)}{a(x,y)} \tag{10.1.3}$$

or in the parametric form

$$\frac{dx}{dt}(t) = a(x(t), y(t)), \quad \frac{dy}{dt}(t) = b(x(t), y(t)). \tag{10.1.4}$$

Along these curves, u will satisfy the ODE

$$\frac{d}{dt}u(x(t),y(t)) = c(x(t),y(t))\,u(x(t),y(t)) + d(x(t),y(t)). \quad (10.1.5)$$

Note that t here is a parameter along the curves. The one parameter family of curves defined by (10.1.4) are called the *characteristic curves* of the PDE (10.1.1).

Now fix s and fix a point $(x_0(s),y_0(s))$ on the initial curve and look for the characteristic curve passing through this point; that is, solve the ODE system (10.1.4) together with the initial values

$$x(0) = x_0(s),\ y(0) = y_0(s).$$

Under appropriate assumptions on a and b, for example, a,b are C^1 functions, the existence theory discussed in Chapter 4 will give us a unique characteristic curve in a neighborhood of the origin (local existence). Since the solution also depends on s, we denote the solution as $(x(t,s),y(t,s))$. More precisely, for fixed s and varying t, the solution $(x(t,s),y(t,s))$ moves along the characteristic curve. As s changes, we get different characteristic curves.

Now by restricting u along a characteristic curve (that is, fixing an s), we get the equation for u as

$$\frac{d}{dt}u(x(t,s),y(t,s)) = c(x(t,s),y(t,s))u(x(t,s),y(t,s))$$

$$+d(x(t,s),y(t,s)) \quad (10.1.6)$$

with the initial values

$$u(x(0,s),y(0,s)) = u_0(s). \quad (10.1.7)$$

Note that in this ODE, s is merely a parameter, whereas t is the independent variable. Thus, we have obtained u in the fixed characteristic curve and changing s, we can obtain u along different characteristic curves. However, we still need to answer the question whether u is indeed defined in a neighborhood Ω_1 in Ω of the initial curve Γ_0 satisfying the PDE and the initial conditions. To achieve this, we need to assure ourselves that the family of characteristics $\{(x(t,s),y(t,s))\}_s$ covers such an Ω_1. In other words, we need to solve the following inverse problem:

Given an arbitrary point $(x,y) \in \Omega_1$, find a characteristic curve passing through (x,y) and meeting the initial curve; more precisely, given $(x,y) \in \Omega_1$, find (t,s) and a solution of (10.1.4) such that

$$x(t,s) = x, \ y(t,s) = y. \tag{10.1.8}$$

An answer to this inverse problem imposes certain conditions on the initial curve known as *transversality condition.* We will study this problem in a more general situation, in the next section, while dealing with quasi-linear equations.

Example 10.1.1

Consider the transport equation $u_y(x,y) + ku_x(x,y) = 0$ with the initial condition $u(x,0) = u_0(x)$ on $\Gamma : y = 0$ considered in (3.4.1).

Here $a = k, b = 1, c = d = 0$. Thus, $\dfrac{dy}{dx} = \dfrac{1}{k}$ or $x - ky = $ constant, are the characteristics which we have already seen in Example 3.4.1. Note that we had used t instead of y.

Example 10.1.2

Consider the PDE, $xu_x + yu_y = \alpha u$ and $u = \phi(x)$ on the initial curve $y = 1$.

It is easy to see that $y = cx$, c constant, are the characteristic curves and along any of these curves, u satisfies

$$\frac{d}{dx}u(x,cx) \ = \ u_x(x,cx) + u_y(x,cx).c = u_x + \frac{y}{x}u_y = \frac{\alpha}{x}u(x,cx),$$

whose solution is given by $u(x,cx) = kx^\alpha$. Here $k = k(c)$ depends on c which may differ from characteristic to characteristic. Thus, we have the general solution $u(x,y) = k\left(\frac{y}{x}\right)x^\alpha$, where k is an arbitrary function. Now applying the condition $u = \phi(x)$ at $y = 1$, we get

$$\phi(x) = k\left(\frac{1}{x}\right)x^\alpha \text{ or } k(x) = \phi\left(\frac{1}{x}\right)x^\alpha$$

and hence, the required solution is

$$u(x,y) = \phi\left(\frac{x}{y}\right) y^{\alpha}.$$

There will be difficulties if instead we prescribe the initial condition on the x-axis (why?).

10.2 Quasi-linear Equations

The general form of a quasi-linear equation in two variables can be written as

$$a(x,y,u)u_x + b(x,y,u)u_y - c(x,y,u) = 0. \tag{10.2.1}$$

As in the linear case, we may prefer to introduce the characteristic curves by solving the ODE system as in (10.1.4). But in this case, the right hand side in (10.1.4) has to be replaced by $a(x(t),y(t),u(x,t)),b(x(t),y(t),$ $u(x,t))$ which depends on the unknown function u as well. Hence, the system is not complete and we need to adjoin the ODE for u as well to get a complete system. Thus, we need to introduce space curves, instead of just plane curves.

Definition 10.2.1

A solution $u(x,y)$ of (10.2.1) represented by a surface $z = u(x,y)$ in the three-dimensional (x,y,z) space is known as an *integral surface*.$\square$

The equation (10.2.1) can be written as a dot product, namely,

$$(a,b,c) \cdot (u_x,u_y,-1) = 0. \tag{10.2.2}$$

The vector $(u_x,u_y,-1)$ is the normal to the integral surface $z = u(x,y)$ at the point (x,y,z) on the surface. Thus, (10.2.1) can be interpreted as the condition that at each point on the integral surface, the vector (a,b,c) is tangent to the surface. Thus, the PDE (10.2.1) defines a vector (direction) field (a,b,c) in $\mathbb{R}^3$, called the *characteristic direction*, having the property that a surface $z = u(x,y)$ is an integral surface if and only if at each point on the surface, the tangent plane contains the characteristic direction.

This motivates us to look at an integral surface as a family of space curves whose tangent at any point on the curve coincides with the characteristic direction. These curves are known as *characteristic curves*. Thus, introduce the family of space curves given by

$$\frac{dx}{a(x,y,z)} = \frac{dy}{b(x,y,z)} = \frac{dz}{c(x,y,z)}. \tag{10.2.3}$$

In the parametric form, these equations can be written as

$$\left.\begin{aligned}
\frac{dx}{dt}(t) &= a(x(t),y(t),z(t)), \\[2mm]
\frac{dy}{dt}(t) &= b(x(t),y(t),z(t)), \\[2mm]
\frac{dz}{dt}(t) &= c(x(t),y(t),z(t)).
\end{aligned}\right\} \tag{10.2.4}$$

Note that in the linear case, a and b were independent of z and hence, the solution $(x(t),y(t))$ of (10.1.4) defined plane curves in the $x-y$ plane. In relation to (10.2.4), when a and b were independent of z, these plane curves are nothing but the projection of the space curves given by the solutions of (10.2.4). Using the results of Chapter 4, under suitable conditions on a,b,c, we see that through each point (x_0,y_0,z_0), there passes a unique characteristic curve

$$x(t) = x(x_0,y_0,z_0,t), y = y(x_0,y_0,z_0,t), z = z(x_0,y_0,z_0,t),$$

defined for small t. Now, it is trivial to see that if a surface is generated by a family of characteristic curves, then, it is an integral surface as both have the same tangential directions. Conversely, if $z = u(x,y)$ is an integral surface Σ and $(x_0,y_0,z_0 = u(x_0,y_0))$ is a point on Σ, then, the integral curve through (x_0,y_0,z_0) will lie completely on Σ and thus, Σ is generated by a family of characteristic curves. To see this, consider the solution of

$$\frac{dx}{dt} = a(x,y,u(x,y)), \frac{dy}{dt} = b(x,y,u(x,y))$$

with $x = x_0, y = y_0$ at $t = 0$. Then, the corresponding curve

$$x = x(t), y = y(t), z = z(t) = u(x(t),y(t))$$

satisfies

$$\frac{dz}{dt} = u_x \frac{dx}{dt} + u_y \frac{dy}{dt} = au_x + bu_y = c.$$

Thus, the curve $(x(t), y(t), z(t)) = u(x(t)), y(t))$ satisfies the system (10.2.4) and hence, it is the characteristic through (x_0, y_0, z_0). Moreover, it lies on Σ by definition. Further, if two integral surfaces intersect at a point, then the characteristic curve through the point would lie on both the surfaces and hence, they intersect along the whole characteristic through this common point. With this detailed discussion, we can now formulate the initial value problem as follows:

Initial Value Problem (IVP): Let Γ_0 be a given initial curve as in the linear case and the initial value u_0 be given. Since this data is insufficient to solve (10.2.4), we 'lift' the initial curve Γ_0 to a space curve $\bar{\Gamma}_0$ by adjoining the initial values as

$$\bar{\Gamma}_0 = \{(x_0(s), y_0(s), u_0(s)) : 0 \le s \le 1\}. \tag{10.2.5}$$

Now the IVP reads as follows: find an integral surface Σ passing through the lifted initial curve $\bar{\Gamma}_0$.

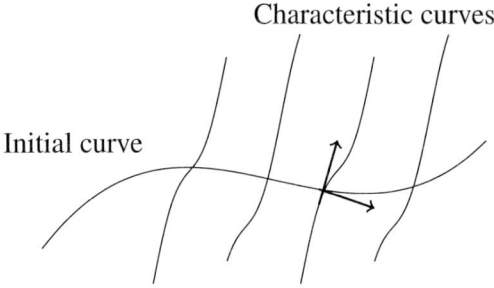

Characteristic curves

Initial curve

Fig. 10.2   Characteristic and initial curves

Theorem 10.2.2

Consider the PDE (10.2.1) and let a, b, c have continuous partial derivatives with respect to x, y, u. Suppose that along the initial curve Γ_0, the initial values $u = u_0(s)$ are prescribed, where x_0, y_0, u_0 are continuously differentiable functions for $0 \le s \le 1$. Define the initial space curve $\bar{\Gamma}_0$ by (10.2.5) and assume that the transversality condition holds:

$$a(x_0(s),y_0(s),u_0(s))\frac{dy_0}{ds} - b(x_0(s),y_0(s),u_0(s))\frac{dx_0}{ds} \neq 0 \quad (10.2.6)$$

for all $0 \leq s \leq 1$. Then, there exists a unique solution $u(x,y)$ defined in some neighborhood of the initial curve Γ_0, which satisfies the PDE (10.2.2) and the initial conditions

$$u(x_0(s),y_0(s)) = u_0(s). \tag{10.2.7}$$

□

The theorem thus, confirms that there is an integral surface through the space curve $\bar{\Gamma}_0$ in some neighborhood.

Proof: Consider the system of ODE

$$\frac{dx}{dt} = a(x,y,u), \quad \frac{dy}{dt} = b(x,y,u), \quad \frac{du}{dt} = c(x,y,u).$$

For any fixed $(x_0(s),y_0(s),u_0(s)) \equiv (x_0,y_0,z_0)$ on the initial space curve, by solving this system, we obtain a unique family of characteristics

$$x = x(x_0,y_0,u_0,t) \equiv x(t,s)$$

$$y = y(x_0,y_0,u_0,t) \equiv y(t,s) \tag{10.2.8}$$

$$u = u(x_0,y_0,u_0,t) \equiv u(t,s)$$

for small t, with $x(0,s) = x_0(s), y(0,s) = y_0(s), u(0,s) = u(x_0(s), y_0(s)) = u_0(s)$. Note that the derivatives of x,y,u with respect to s and t are continuous. Thus, we can solve for u along the characteristic curve. But we need to do more because we need to solve for u in any arbitrary point in the neighborhood of the initial space curve as we discussed earlier in the linear case; in other words, we need to answer the question: does there exist a characteristic curve passing through any arbitrary point in the neighborhood and meeting the initial space curve?   Since the Jacobian

$$\left.\frac{\partial(x,y)}{\partial(s,t)}\right|_{t=0} = \left.\begin{vmatrix} \dfrac{\partial x}{\partial s} & \dfrac{\partial x}{\partial t} \\[2mm] \dfrac{\partial y}{\partial s} & \dfrac{\partial y}{\partial t} \end{vmatrix}\right|_{t=0} = b\,\frac{dx_0}{ds} - a\,\frac{dy_0}{ds} \neq 0,$$

we can invoke the inverse function theorem ([Apo11, Rud76]) to obtain s,t as functions of x,y in some neighborhood of the initial curve $t = 0$, say $s = s(x,y), t = t(x,y)$. Now define

$$\varphi(x,y) = u(s(x,y),t(x,y)).$$

One can verify that φ is the unique solution satisfying the initial conditions.   □

Example 10.2.3

Consider the PDE : $uu_x + u_y = 1$ with initial conditions $x = s, y = s, u = \frac{1}{2}s, 0 \le s \le 1$ that is, the value of u are given on the diagonal of the unit square in $\mathbb{R}^2$.

Here $a = u, b = 1, c = 1$. Thus, the transversality condition (10.2.6):

$$a(x_0(s),y_0(s),u_0(s))\frac{dy_0}{ds} - b(x_0(s),y_0(s),u_0(s))\frac{dx_0}{ds} = \frac{s}{2}\cdot 1 - 1\cdot 1 \neq 0,$$

is satisfied for $0 \le s \le 1$. Solving the systems of ODE with the initial conditions:

$$\frac{dx}{dt} = u,\ \frac{dy}{dt} = 1,\ \frac{du}{dt} = 1,\ x(0,s) = s, y(0,s) = s, u(0,s) = \frac{s}{2},$$

we get the family of characteristic curves

$$x(t,s) = \frac{t^2}{2} + \frac{st}{2} + s,\ y(t,s) = t + s,\ u(t,s) = t + \frac{s}{2}$$

Solving s and t, we get $s = \dfrac{x - y^2/2}{1 - y/2},\ t = \dfrac{y - x}{1 - y/2}$ and finally, the solution is given by

$$u(x,y) = \frac{2(y-x) + (x - y^2/2)}{2 - y}.$$

10.3 General First Order Equation in Two Variables

Substituting $z = u(x,y), p = u_x = z_x, q = u_y = z_y$, a general first order equation in two variables can be written as

$$F(x,y,z,p,q) = 0, \tag{10.3.1}$$

where F is a given function with continuous second derivatives with respect to its variables x, y, z, p, q. The interesting and surprising fact is that, once again, we can reduce the study of the problem (10.3.1) to that of a system of ODE. But the geometry is more complicated than the case of quasi-linear equations and require much more complicated geometrical objects known as *strips* and *cones*.

Let (x_0, y_0, z_0) be a point in space and consider an integral surface $z = u(x,y)$ through (x_0, y_0, z_0). The direction numbers $(p, q, -1)$ define the normal direction to the tangent plane at (x_0, y_0, z_0) of the integral surface. Then, (10.3.1) states that there is a relation

$$F(x_0, y_0, z_0, p, q) = 0 \tag{10.3.2}$$

between the direction numbers p and q, that is, the differential equation (one relation with two numbers) will restrict its solutions to those surfaces having tangent planes belonging to a one parameter family.

In general, this one parameter family of planes will envelope a cone called the *Monge cone*. Thus, the differential equation (10.3.1) describes a field of cones having the property that a surface will be an integral surface if and only if it is tangent to a cone at each point.

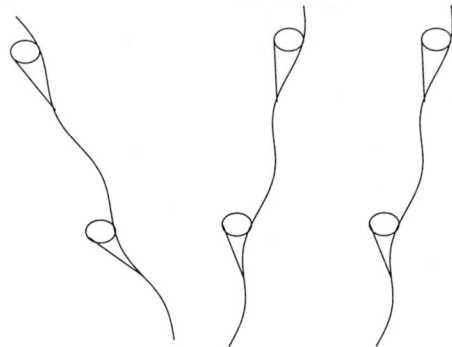

Fig. 10.3 Monge cone

Remark 10.3.1

In the quasi-linear case, the cone degenerates into a straight line whose direction is given by (a, b, c). ☐

At each point, the surface will be tangent to a Monge cone. The line of contact of the surface and the cones define a field of directions on the surface called the *characteristic directions* and the integral curves of this field define a family of characteristic curves. The Monge cone at (x_0, y_0, z_0) is the envelope of the one parameter family of planes (whose normal is $(p, q, -1)$) which can be written as

$$z - z_0 = p(x - x_0) + q(y - y_0), \tag{10.3.3}$$

where p, q solves (10.3.2). By solving (10.3.2) for q in terms of p as $q = q(x_0, y_0, z_0, p)$, we can write (10.3.3) as

$$z - z_0 = p(x - x_0) + q(x_0, y_0, z_0, p)(y - y_0)$$

which is a one parameter family of planes describing the Monge cone. Differentiating with respect to p, we get

$$0 = (x - x_0) + (y - y_0)\frac{dq}{dp}.$$

From (10.3.2), we have

$$\frac{dF}{dp} = F_p + F_q\frac{dq}{dp} = 0. \tag{10.3.4}$$

Eliminating $\dfrac{dq}{dp}$, the equations describing the Monge cone can be written as

$$\left. \begin{array}{l} F(x_0, y_0, z_0, p, q) = 0 \\[2mm] z - z_0 = p(x - x_0) + q(y - y_0) \\[2mm] \dfrac{x - x_0}{F_p} = \dfrac{y - y_0}{F_q}. \end{array} \right\} \tag{10.3.5}$$

Given p and q, the last two equations give the line of contact between the tangent plane and the cone. The last two equations can be written as

$$\frac{x-x_0}{F_p} = \frac{y-y_0}{F_q} = \frac{z-z_0}{pF_p+qF_q}. \tag{10.3.6}$$

Thus, on the given integral surface, at each point $p_0 = p(x_0,y_0), q_0 = q(x_0,y_0)$ are known, the tangent plane

$$z-z_0 = p_0(x-x_0) + q_0(y-y_0)$$

together with the third equation in (10.3.5) determines the line of contact with the Monge cone given by (10.3.6) or the characteristic direction. Thus, the characteristic curves are given by the system of ODE

$$\frac{dx}{F_p} = \frac{dy}{F_q} = \frac{dz}{pF_p+qF_q}$$

or

$$\frac{dx}{dt} = F_p, \quad \frac{dy}{dt} = F_q, \quad \frac{dz}{dt} = pF_p+qF_q. \tag{10.3.7}$$

As there are five unknowns $x(t), y(t), z(t), p(t), q(t)$, we need two more equations to complete the system (10.3.7). But along a characteristic curve on the given integral surface, we have

$$\left. \begin{array}{l} \dfrac{dp}{dt} = p_x\dfrac{dx}{dt} + p_y\dfrac{dy}{dt} = p_xF_p + p_yF_q \\[3mm] \dfrac{dq}{dt} = q_xF_p + q_yF_q. \end{array} \right\} \tag{10.3.8}$$

However, p_x, p_y, q_x, q_y are second derivatives of u which are undesirable; we need to eliminate them. Differentiating the given PDE with respect to x and y, we obtain

$$F_x + F_z p + F_p p_x + F_q q_x = 0,$$

$$F_y + F_z q + F_p p_y + F_q q_y = 0$$

so that (10.3.8) becomes

$$\left. \begin{array}{l} \dfrac{dp}{dt} = -F_x - F_z p \\[3mm] \dfrac{dq}{dt} = -F_y - F_z q, \end{array} \right\} \tag{10.3.9}$$

where we have used $p_y = \dfrac{\partial^2 u}{\partial y \partial x} = q_x$. Thus, on the integral surface $z = u(x,y)$, we have a family of characteristic curves with coordinates $x(t), y(t), z(t)$ along with the numbers $p(t), q(t)$ and which is given by the system (10.3.7), (10.3.9). Moreover along the curve, we have

$$\frac{dF}{dt} = F_x \frac{dx}{dt} + F_y \frac{dy}{dt} + F_z \frac{dz}{dt} + F_p \frac{dp}{dt} + F_q \frac{dq}{dt}$$

and we readily see that $\frac{dF}{dt} = 0$ using (10.3.7) and (10.3.9), showing that $F = constant$, is an integral of ODE. Thus, if $F = 0$ is satisfied at an initial point x_0, y_0, z_0, p_0, q_0 for $t = 0$, then (10.3.7), (10.3.9) will determine a unique solution $x(t), y(t), z(t), p(t), q(t)$ passing through this point and along which $F = 0$ will be satisfied for all t.

Hence, a solution can be interpreted using these five numbers and is called a *strip*, that is, a space curve $x = x(t), y = y(t), z = z(t)$ and, along it, a family of tangent planes whose normal directions are $(p(t), q(t), -1)$.

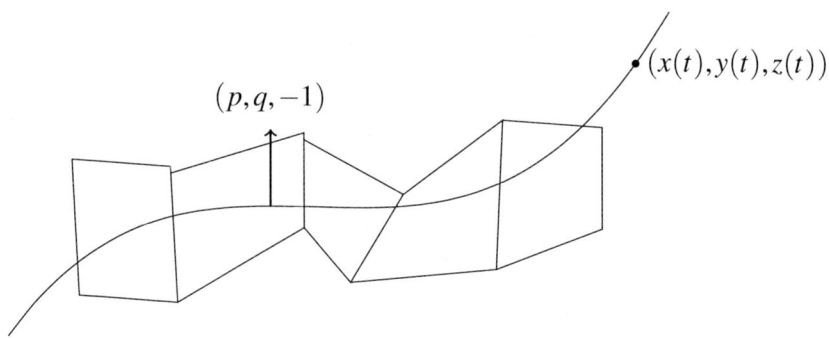

Fig. 10.4   Characteristic strips

For fixed t_0, the five numbers x_0, y_0, z_0, p_0, q_0 are said to define an *element* of the strip. That is, a point on the curve together with the tangent plane. From (10.3.7), we get

$$\frac{dz(t)}{dt} = p(t) \frac{dx(t)}{dt} + q(t) \frac{dy(t)}{dt}. \tag{10.3.10}$$

This is the condition that the planes are tangent to the curve and is called the *strip condition*. The strips which are solutions to (10.3.7), (10.3.9) are respectively called *characteristic strips* and the curves, *characteristic curves*.

Furthermore, as in the case of quasi-linear equations, if a characteristic strip has one element x_0, y_0, z_0, p_0, q_0 in common with an integral surface $z = u(x,y)$, then it lies completely on the surface. To see this, solve the ODE system

$$\frac{dx}{dt} = F_p(x,y,u(x,y),u_y(x,y)), \quad \frac{dy}{dt} = F_q(x,y,u(x,y),u_y(x,y))$$

to obtain a curve $x = x(t), y = y(t)$ satisfying the initial conditions $x(0) = x_0, y(0) = y_0$. Then, by defining $z(t) = u(x(t),y(t))$, $p(t) = u_x(x(t),y(t))$ $q(t) = u_y(x(t),y(t))$, we see that

$$\frac{dz(t)}{dt} = p(t)F_p + q(t)F_q, \quad \frac{dp(t)}{dt} = -F_x - F_z u_x, \quad \frac{dq(t)}{dt} = -F_y - F_z u_y.$$

Therefore, $x(t), y(t), z(t), p(t), q(t)$ determine a characteristic strip and by definition, they lie on the surface. But, by uniqueness of the characteristic strip with the initial element x_0, y_0, z_0, p_0, q_0, this is the given strip.

Initial Value Problem: Suppose now an initial curve $\Gamma : x = x_0(s), y = y_0(s), z = z_0(s)$ be given. Further, assume that we can assign functions $p_0(s)$ and $q_0(s)$ such that they together form a family of appropriate initial strips. That is, they satisfy the equation

$$F(x_0(s), y_0(s), z_0(s), p_0(s), q_0(s)) = 0 \tag{10.3.11}$$

and the strip condition

$$\frac{dz_0}{ds} = p_0 \frac{dx_0}{ds} + q_0 \frac{dy_0}{ds}. \tag{10.3.12}$$

So, by fixing s and taking an initial element, the idea is to construct the characteristic strip starting from the given initial strip. As s varies, we get a family of characteristic strips (see Figure 10.5). This, in turn will give the integral surface satisfying the initial conditions. Again, this requires the initial curve Γ_0 to be a non-characteristic curve; see the following theorem. We now state the theorem without proof.

Characteristic elements

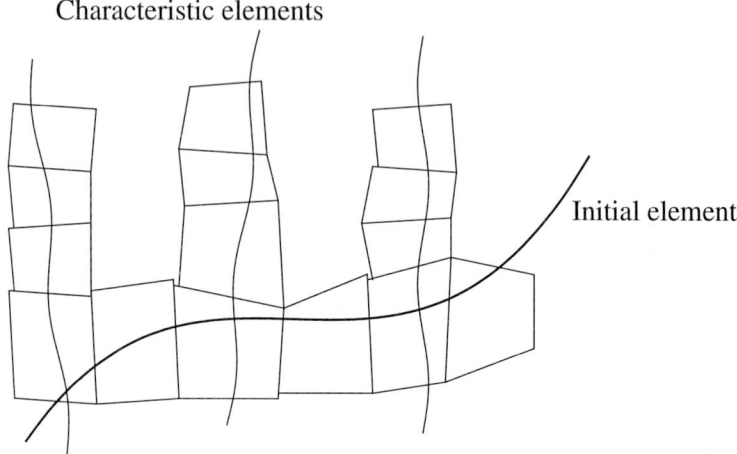

Initial element

Fig. 10.5   Characteristic and initial strips

Theorem 10.3.2

Let x_0, y_0, z_0, p_0, q_0 be as in (10.3.11) and (10.3.12). Assume x_0, y_0, z_0 have continuous derivatives and p_0, q_0 are continuously differentiable. Moreover, assume the following non-characteristic condition along the initial curve:

$$\frac{dx_0}{ds}F_q(x_0, y_0, z_0, p_0, q_0) - \frac{dy_0}{ds}F_p(x_0, y_0, z_0, p_0, q_0) \neq 0.$$

Then, in some neighborhood of the initial curve, there exists a solution $z = u(x, y)$ of (10.3.1) containing the initial strip; that is,

$$z(x_0(s), y_0(s)) = z_0(s), z_x(x_0(s), y_0(s)) = p_0(s), z_y(x_0(s), y_0(s)) = q_0(s).$$

In general, there is no uniqueness of solution. See the following example.

Example 10.3.3

Consider the equation $p^2 + q^2 = 1$ with initial condition $u(x, y) = 0$ on the line $x + y = 1$. Then, there are two solutions given by $u(x, y) = \pm \frac{1}{\sqrt{2}}(x + y - 1)$. The details are left as an exercise.   □

Remark 10.3.4

We can extend the method of characteristics to the first order equations in n variables, namely $F(x_1,\ldots,x_n,z,p_1,\ldots,p_n) = 0$, where $z = u(x_1,\ldots,x_n)$ and $p_i = \dfrac{\partial u}{\partial x_i}, 1 \leq i \leq n$. Here, the characteristic equations are given by

$$\frac{dx_i}{dt} = F_{p_i}, \frac{dz}{dt} = \sum_{i=1}^{n} p_i F_{p_i}, \frac{dp_i}{dt} = -F_{x_i} - p_i F_z, 1 \leq i \leq n$$

which is a system of $2n+1$ equations in $2n+1$ unknowns. $\qquad\square$

Remark 10.3.5

One can easily deduce the quasi-linear and linear case from the general equations. In the quasi-linear case

$$F(x,y,z,p,q) = a(x,y,z)p + b(x,y,z)q - c(x,y,z) = 0.$$

Thus, $F_p = a, F_q = b$ and $pF_p + qF_q = c$ and hence, the equations in (10.3.7) are independent of p and q and so it can be solved to determine the characteristic curves $(x(t),y(t),z(t))$. But, in the nonlinear case, one has to solve for $(x(t),y(t),z(t))$ together with the direction numbers p and q. Moreover, in the quasi-linear case, the Monge–Cone equations (10.3.6), reduces to

$$\frac{x - x_0}{a} = \frac{y - y_0}{b} = \frac{z - z_0}{c}$$

which represents the equation of a line in the space showing that the Monge cone degenerates to a line.

In the linear case, a and b are independent of z as well, so that the first two equations in (10.3.7) form a complete system for x and y; the characteristic curves are plane curves, that is, the curves lie on the (x,y) plane. Moreover, the third equation reduces to

$$\frac{du}{dt}(x(t),y(t)) = \frac{dz}{dt}(t) = c(x(t),y(t))$$

which can be solved to obtain u. $\qquad\square$

10.4   Hamilton–Jacobi Equation

Our main motivation in this section is to introduce the reader to an important equation, namely the Hamilton–Jacobi equation (HJ). This is given by

$$u_t + H(\mathbf{x}, Du) = 0. \tag{10.4.1}$$

Our presentation here is deliberately vague, essentially because even the explanation of the various terminology involved goes much beyond the scope of the present book. However, because of the importance of the HJ equation in numerous applications, including optimal control problems, where it is referred to as the Hamilton–Jacobi–Bellman(HJB) equation, we have included this brief introduction so that the interested reader can look into more advanced texts on the subject.

In the HJ equation, $u = u(t, \mathbf{x})$ is the unknown function of $n + 1$ variables, $t > 0$ and $\mathbf{x} \in \mathbb{R}^n$; the *Hamiltonian H* is a real valued smooth function defined on $\mathbb{R}^n \times \mathbb{R}^n$ and $Du = (u_{x_1}, \cdots, u_{x_n})$ is the gradient vector of u. Time dependent Hamiltonian, that is, $H = H(t, \mathbf{x}, Du)$ can also be considered. This equation arises in the classical *calculus of variations* and more generally in optimal control problems (here, it is called the Hamilton–Jacobi–Bellman(HJB) equation). Further, H may be of the general form $H(t, \mathbf{x}, Du)$. In a smooth set up, the minimum value of an associated cost functional or energy functional, namely *value function* is known to satisfy the HJ or HJB equation. Equation (10.4.1) is a first order PDE in $n + 1$ dimensions, and thus, the characteristic equations form a system of $2n + 3$ equations. But, due to the special structure of (10.4.1), the system of characteristic equations can be written as $2n$ equations known as *Hamilton's ODE system*

$$\left.\begin{array}{l} \dfrac{d\mathbf{x}}{dt} = D_{\mathbf{p}}H(\mathbf{x}, \mathbf{p}) \\[3mm] \dfrac{d\mathbf{p}}{dt} = -D_{\mathbf{x}}H(\mathbf{x}, \mathbf{p}) \end{array}\right\} \tag{10.4.2}$$

together with an ODE equation for the unknown u. Further, the variable t itself can be used as the parameter. See Example 1.2.8.

Calculus of Variations: Given a *Lagrangian* $L = L(\mathbf{x}, \mathbf{q})$, $\mathbf{x}, \mathbf{q} \in \mathbb{R}^n$, introduce the functional

$$J(\mathbf{w}) := \int_0^T L(\mathbf{w}(s), \mathbf{w}'(s)) ds, \qquad (10.4.3)$$

where $\mathbf{w} \in \mathscr{A}$, the space of admissible trajectories, say for example,

$$\mathscr{A} = \{\mathbf{w} \in C^2([0, T]; \mathbb{R}^n) : \mathbf{w}(0) = \mathbf{a}, \mathbf{w}(T) = \mathbf{b}\}.$$

Here $T > 0$ is the terminal time and the vectors $\mathbf{a}$ and $\mathbf{b}$ are the given vectors in $\mathbb{R}^n$. The basic problem is to find a curve $\mathbf{x} \in \mathscr{A}$ which solves the minimization problem

$$J(\mathbf{x}) = \min_{\mathbf{w} \in \mathscr{A}} J(\mathbf{w}). \qquad (10.4.4)$$

One can extend the elementary result on finite dimensional optimization that extremal points occur at critical points, to the present minimization problem which is infinite dimensional, since we are working on trajectories belonging to the infinite dimensional space $\mathscr{A}$. If $\mathbf{x} \in \mathscr{A}$ is a solution to the problem (10.4.4), known as *optimal trajectory*, then, $\mathbf{x}$ satisfies the *Euler–Lagrange(EL)* equations

$$\frac{d}{ds}\left(D_q L(\mathbf{x}(s), \dot{\mathbf{x}}(s))\right) + D_{\mathbf{x}} L(\mathbf{x}(s), \dot{\mathbf{x}}(s)) = 0. \qquad (10.4.5)$$

This is a system of n second order equations. Thus, any minimizer $x \in \mathscr{A}$ of (10.4.4) solves EL equations, but the converse need not be true. A solution to the EL equations is called a *critical point* of the functional J.

Since EL is a system of n second order equations, we can convert it into a system of $2n$ first order system as follows. Introduce $\mathbf{p}(s) = D_q L(\mathbf{x}(s), \dot{\mathbf{x}}(s))$, called the *generalized momentum* corresponding to the position $\mathbf{x}(\cdot)$ and velocity $\dot{\mathbf{x}}(\cdot)$. These terminologies were motivated from classical mechanics which we will see soon. We need an important hypothesis to get Hamilton's ODE.

Assumption: Suppose that, for given $\mathbf{x}, \mathbf{p} \in \mathbb{R}^n$, the equation $D_q L(\mathbf{x}, \mathbf{q}) = \mathbf{p}$ can be uniquely and smoothly solved for $\mathbf{q}$ as $\mathbf{q} = \mathbf{q}(\mathbf{x}, \mathbf{p})$. Now introduce the Hamiltonian

$$H(\mathbf{x}, \mathbf{p}) = \mathbf{p} \cdot \mathbf{q}(\mathbf{x}, \mathbf{p}) + L(\mathbf{x}, \mathbf{q}(\mathbf{x}, \mathbf{p})). \qquad (10.4.6)$$

Theorem 10.4.1

Let $\mathbf{x}$ solve the EL equations and $\mathbf{p}$ be the corresponding generalized momentum. Then, under this assumption, $\mathbf{x}$ and $\mathbf{p}$ satisfy Hamilton's ODE system with the Hamiltonian defined as in (10.4.6). Further, the mapping $s \to H(\mathbf{x}(s), \mathbf{p}(s))$ is constant.

We have earlier seen that Hamilton's ODE are characteristic curves corresponding to a PDE. Thus, the characteristic curves are nothing but the level curves of the energy. There are many results in this direction and more general analysis in the context of HJB equations. The analysis for HJB is not a direct generalization from classical HJ equations and it took almost two centuries to come up with a theory. Two theories emerged after the 1950s due to Pontryagin from USSR and Bellman from USA. The former one is based on Hamiltonian ODE whereas Bellman's theory is based on PDE known as HJB equations. We now give the classical example from Newtonian mechanics.

Example 10.4.2

Consider the motion of a particle of mass m under the influence of a force field $\mathbf{f}$ given by a potential ϕ, that is, $\mathbf{f} = \nabla \phi$.

Define the Lagrangian $L(\mathbf{x}, \mathbf{q}) = \frac{m}{2}|\mathbf{q}|^2 - \phi(\mathbf{x})$, the difference in kinetic and potential energy. Then, the corresponding EL equations describe Newton's second law of motion, namely $m\ddot{\mathbf{x}}(s) = \mathbf{f}(\mathbf{x}(s)) = \nabla\phi(s)$. Here, the assumption that $\mathbf{p} = D_{\mathbf{q}}L(\mathbf{x}, \mathbf{q}) = m\mathbf{q}$ is trivially solvable. It is easy to see that the Hamiltonian is the total energy $H(\mathbf{x}, \mathbf{p}) = \frac{1}{2m}|\mathbf{p}|^2 + \phi(\mathbf{x})$. In Newton's theory, position and velocity formulation is given (that is, Lagrangian formulation), whereas position and momentum (Hamiltonian formulation) are the unknowns given through the Hamiltonian system. It is possible to go from one formulation to the other.

10.5   Exercises

1. Find and sketch some sample characteristic curves of the PDE

$$(x+2)u_x + 2yu_y = 2u$$

in the $x - y$ plane. Write the ODE for u along a characteristic curve with x as parameter and then, solve the PDE with the initial condition $u(-1, y) = \sqrt{|y|}$.

2. Consider the PDE

$$xu_x + yu_y = 2u, \quad x > 0, y > 0.$$

Plot the characteristic curves and then, solve the equation with the following initial conditions: (a) $u = 1$ on the hyperbola $xy = 1$; and (b) $u = 1$ on the circle $x^2 + y^2 = 1$.

Can you solve the equation in general, if certain initial data is prescribed on the initial curve $y = e^x$? Justify with reasons.

3. Write down the characteristic ODE system for the equation $u_y = u_x^3$ and then, solve with the initial condition $2x^{3/2}$ on the positive x-axis.

4. Sketch the characteristic curve, the initial curve and solve the following problems

(a) $xu_x + yu_y = ku$, $x \in \mathbb{R}$, $y \geq \alpha > 0$; $u(x, \alpha) = F(x)$, where k, α are fixed and F is a given smooth function.

(b) $(x + 2)u_x + 2yu_y = \alpha u$; $u(-1, y) = \sqrt{y}$.

(c) $yu_x - xu_y = 0$; $u(x, 0) = x^2$.

(d) $x^2 u_x - y^2 u_y = 0$; $u(1, y) = F(y)$.

5. Find the characteristic curves of the following equations

(a) $(x^2 - y^2 + 1)u_x + 2xyu_y = 0$.

(b) $2xyu_x - (x^2 + y^2)u_y = 0$.

6. Solve the quasi-linear problem and verify the transversality condition.

(a) $uu_x + u_y = 0$, $u(x, 0) = x$.

(b) $uu_x + u_y = 1$, $u(x, x) = x/2, x \in (0, 1]$.

7. Find and sketch the characteristic curves of $uu_x + u_y = 0$ with the following initial conditions

(a) $u(x, 0) = \begin{cases} 0 & \text{if } x < 0 \\ 1 & \text{if } x \geq 0 \end{cases}$

(b) $u(x,0) = \begin{cases} 1 & \text{if } x < 0 \\ 0 & \text{if } x \geq 0 \end{cases}$

(c) $u(x,0) = \begin{cases} 0 & \text{if } x < 0 \\ 1 & \text{if } x \geq 1 \end{cases}$ and $u(x,0)$ is smooth and increasing.

8. Find the integral surface of the equation $x\left(\dfrac{\partial u}{\partial x}\right)^2 + y\dfrac{\partial u}{\partial y} = u$ passing through the line $y = 1$, $x + z = 0$.

9. Consider the equation $p^2 + q^2 = 1$ with initial condition $u(x,y) = 0$ on the line $x + y = 1$. Show that there are two solutions given by $u(x,y) = \pm\dfrac{1}{\sqrt{2}}(x+y-1)$ using the method of characteristics.

10.6 Notes

The purpose of this chapter is not to give an expository introduction to PDE, but to show how the ODE play an important role in the analysis of first order PDE. The reader can refer to [Eva98, Joh75, RR04, PR96] for further discussion and more details.

Appendix A

Poincarè–Bendixon and Leinard's Theorems

A.1 Introduction

In this appendix (see ([CL72])), we present a proof of the Poincarè–Bendixon theorem concerning the existence of periodic orbits to two- dimensional autonomous systems. We also discuss Leinard's theorem. First we discuss some basic notions of limit sets. Consider an n-dimensional autonomous system

$$\dot{\mathbf{x}} = \mathbf{f}(\mathbf{x}), \tag{A.1.1}$$

where $\mathbf{f} : \mathbb{R}^n \to \mathbb{R}^n$ is a continuous, locally Lipschitz function. Denote by $\phi_t(\mathbf{x}_0)$, the unique solution $\mathbf{x}$ of (A.1.1) with $\mathbf{x}(0) = \mathbf{x}_0$, for $t \in I_{\mathbf{x}_0}$, where $I_{\mathbf{x}_0}$ is the corresponding maximal interval of existence.

Recall the definition of an *invariant* set. A set $A \subset \mathbb{R}^n$ is said to be invariant with respect to (A.1.1), if $\phi_t(\mathbf{x}) \in A$ for every $\mathbf{x} \in A$ and $t \in I_{\mathbf{x}}$.

Next, recall the definitions of orbit $\mathscr{O}(\mathbf{x})$, positive (semi) orbit $\mathscr{O}^+(\mathbf{x})$ and negative (semi) orbit $\mathscr{O}^-(\mathbf{x})$ through a given point $\mathbf{x} \in \mathbb{R}^n$:

$$\mathscr{O}(\mathbf{x}) = \{\phi_t(\mathbf{x}) : t \in I_{\mathbf{x}}\};$$

$$\mathscr{O}^+(\mathbf{x}) = \{\phi_t(\mathbf{x}) : t \in I_{\mathbf{x}}, t \geq 0\};$$

$$\mathscr{O}^-(\mathbf{x}) = \{\phi_t(\mathbf{x}) : t \in I_{\mathbf{x}}, t \leq 0\}.$$

It is easy to verify that a set $A \subset \mathbb{R}^n$ is invariant if and only if

$$A = \bigcup_{\mathbf{x} \in A} \mathscr{O}(\mathbf{x}).$$

We now introduce the notions of α-*limit* and ω-*limit* sets (observe that α and ω are, respectively, the first and the last letters of the Greek alphabet).

Defintion A.1.1

Given $\mathbf{x} \in \mathbb{R}^n$, the α-*limit* set and the ω-*limit* set of $\mathbf{x}$, with respect to (A.1.1) are defined, respectively, by

$$\alpha(\mathbf{x}) = \alpha_{\mathbf{f}}(\mathbf{x}) = \bigcap_{\mathbf{y} \in \mathscr{O}(\mathbf{x})} \overline{\mathscr{O}^-(\mathbf{y})}$$

and

$$\omega(\mathbf{x}) = \omega_{\mathbf{f}}(\mathbf{x}) = \bigcap_{\mathbf{y} \in \mathscr{O}(\mathbf{x})} \overline{\mathscr{O}^+(\mathbf{y})}. \qquad\qquad \square$$

We have the following result concerning ω-limit sets.

Theorem A.1.2

Suppose the positive orbit $\mathscr{O}^+(\mathbf{x})$ of a point $\mathbf{x} \in \mathbb{R}^n$ is bounded. Then, the following hold:

1. the ω-limit set $\omega(\mathbf{x})$ is non-empty, compact and connected;

2. a point $\mathbf{y} \in \omega(\mathbf{x})$ if and only if there is a sequence $t_k \nearrow +\infty$ such that $\phi_{t_k}(\mathbf{x}) \to \mathbf{y}$ as $k \to \infty$;

3. the ω-limit set $\omega(\mathbf{x})$ is positively invariant, that is, $\phi_t(\mathbf{y}) \in \omega(\mathbf{x})$ for every $\mathbf{y} \in \omega(\mathbf{x})$ and for all $t > 0$;

4. $\inf\{\|\phi_t(\mathbf{x}) - \mathbf{y}\| : \mathbf{y} \in \omega(\mathbf{x})\} \to 0$ as $t \to +\infty$.

Proof: Put $K = \overline{\mathscr{O}^+(\mathbf{x})}$. By hypothesis, K is compact. Being the arbitrary intersection of closed sets, $\omega(\mathbf{x})$ is closed. Since $\omega(\mathbf{x}) \subset K$, it follows that $\omega(\mathbf{x})$ is compact.

Since $\mathcal{O}^+(\mathbf{x})$ is bounded, it follows that $(0,\infty) \subset I_{\mathbf{x}}$; see Chapter 4 on extension of the interval of existence of a solution. Therefore,

$$\omega(\mathbf{x}) = \bigcap_{t>0} A_t, \tag{A.1.2}$$

where

$$A_t = \overline{\{\phi_s(\mathbf{x}) : s > t\}}.$$

Using (A.1.2), we now establish the second property. If $\mathbf{y} \in \omega(\mathbf{x})$, then, there exists a sequence $t_k \nearrow \infty$ such that $\mathbf{y} \in A_{t_k}$ for $k \in \mathbb{N}$. Then, there is also a sequence $s_k \nearrow \infty$, $s_k \geq t_k$ such that $\phi_{s_k}(\mathbf{x}) \to \mathbf{y}$. Conversely, if there exists a sequence $t_k \nearrow \infty$ such that $\phi_{t_k}(\mathbf{x}) \to \mathbf{y}$, then, $\mathbf{y} \in A_{t_k}$ for $k \in \mathbb{N}$, and hence,

$$\mathbf{y} \in \bigcap_{k=1}^{\infty} A_{t_k} = \bigcap_{t>0} A_t,$$

since $A_t \subset A_{t'}$ for $t > t'$.

Next, we show that $\omega(\mathbf{x}) \neq \emptyset$. Consider the sequence $\{\phi_k(\mathbf{x})\}$ contained in K. By compactness, there is a subsequence $\{\phi_{t_k}(\mathbf{x})\}$ with $t_k \nearrow \mid \infty$, converging to a point in K. The second property therefore that $\omega(\mathbf{x}) \neq \emptyset$.

We now show that $\omega(\mathbf{x})$ is connected. Suppose not. Then, since $\omega(\mathbf{x})$ is closed, there exist two closed, disjoint subsets A, B such that $\omega(\mathbf{x}) = A \cup B$. By compactness, A, B are at a positive distance apart, that is,

$$\delta \equiv \inf\{\|\mathbf{a} - \mathbf{b}\| : \mathbf{a} \in A, \mathbf{b} \in B\} > 0.$$

Since A is a subset of $\omega(\mathbf{x})$, it follows that $d(\phi_{t_k}(\mathbf{x}), A) \leq \delta/2$ for some sequence $t_k \nearrow \infty$, and hence, by triangle inequality, $d(\phi_{t_k}(\mathbf{x}), B) \geq \delta/2$, for sufficiently large k. By compactness of K, there is a subsequence such that $\phi_{t_k}(\mathbf{x}) \to \mathbf{y} \in K$. Therefore, $\mathbf{y} \in \omega(\mathbf{x})$. By continuity of the distance function and the solution, it follows that $d(\mathbf{y}, A) = \delta/2$. But then, $d(\mathbf{y}, B) \geq \delta/2$, by triangle inequality. Thus, $\mathbf{y} \notin \omega(\mathbf{x})$, which is a contradiction.

We now proceed to prove the third property. Suppose $\mathbf{y} \in \omega(\mathbf{x})$. Then, by the second property, there is a sequence $t_k \nearrow +\infty$ such that $\{\phi_{t_k}(\mathbf{x})\} \to \mathbf{y}$ as $k \to \infty$. Recall from the continuous dependence of solutions on initial data (Chapter 4), the function $\mathbf{y} \mapsto \phi_t(\mathbf{y})$ is continuous for each fixed t. Therefore, for given $t > 0$, we have

$$\phi_{t_k+t}(\mathbf{y}) = \phi_t(\phi_{t_k}(\mathbf{y})) \to \phi_t(\mathbf{y})$$

as $k \to \infty$. Since $t_k + t \nearrow +\infty$ also, we conclude that $\phi_t(\mathbf{y}) \in \omega(\mathbf{x})$.

For the last property, assume the contrary. There would then exist a sequence $t_k \nearrow +\infty$ and a constant $\delta_1 > 0$ such that

$$\inf_{\{\mathbf{y} \in \omega(\mathbf{x})\}} \|\phi_{t_k}(\mathbf{x}) - \mathbf{y}\| \geq \delta_1 \qquad (A.1.3)$$

for $k \in \mathbb{N}$. But then, by compactness of K, there is a subsequence $\phi_{t'_k}(\mathbf{x})$ of $\phi_{t_k}(\mathbf{x})$ converging to some $\mathbf{p} \in \omega(\mathbf{x})$. The inequality (A.1.3) then implies that $\|\mathbf{p} - \mathbf{y}\| \geq \delta_1$ for all $\mathbf{y} \in \omega(\mathbf{x})$. This contradiction proves the last property and the proof is complete.   □

Analogous statements may be made and proved for negative orbits. In fact, these properties follow if we apply Theorem A.1.2 to the system $\dot{\mathbf{x}} = -\mathbf{f}(\mathbf{x})$.

A.2   Poincarè–Bendixon Theorem

We now restrict the analysis to $\mathbb{R}^2$ and establish one of the important results in the qualitative theory of autonomous systems, namely the Poincarè–Bendixon Theorem.

A.2.1   Intersection with Transversals

Let $\mathbf{f} : \mathbb{R}^2 \to \mathbb{R}^2$ be a C^1 function; $\mathbf{f}$ may be defined in an open set in $\mathbb{R}^2$. A point in $\mathbb{R}^2$ which is not an equilibrium point will be called a *regular point*. A finite line segment L is said to be *transversal to* $\mathbf{f}$, if each point on L is a regular point and at each point $\mathbf{x}$ on L, the directions of L and $\mathbf{f}(\mathbf{x})$ generate $\mathbb{R}^2$. In other words, the directions of L and $\mathbf{f}$ are linearly independent at every point on L. We also say that L is *transverse* to $\mathbf{f}$. A point on L is said to be an interior point if it is not an end point of the line segment L.

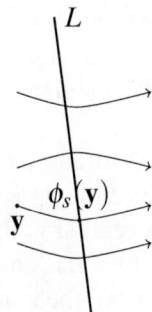

Fig. A.1   Intersection of orbits with a transversal

Lemma A.2.1

The following statements hold:

1. Every regular point **x** is an interior of some transversal, which may have any direction except that of **f**(**x**).

2. Every orbit which meets a transversal L must cross it, and all such orbits cross L in the same direction.

3. Let x_0 be an interior point of a transversal L. Then, for each $\varepsilon > 0$, there is a circle C_ε with x_0 as center such that every orbit passing through a point inside C_ε at $t = 0$ crosses L at some t, $|t| < \varepsilon$.

Proof: By continuity of **f**, there is a circle C centered at **x** such that $\mathbf{f}(\mathbf{y}) \neq \mathbf{0}$ for all **y** in C and its interior. Any diameter of C, which is not in the direction of **f**(**x**) serves as a transversal. This proves the first statement.

Let L be a transversal which is represented by the equation $\mathbf{a} \cdot \mathbf{x} + b = 0$ for some non-zero vector $\mathbf{a} \in \mathbb{R}^2$. The meaning of the first part of the assertion is the following:

If there is a $t_0 < \infty$ (t_0 cannot be $\pm\infty$; why?) such that $\mathbf{a} \cdot \mathbf{x}(t_0) + b = 0$ for an orbit $\mathbf{x}(t)$, then, for small t, $\mathbf{a} \cdot \mathbf{x}(t) + b < 0$ for $t < t_0$ and $\mathbf{a} \cdot \mathbf{x}(t) + b > 0$ for $t > t_0$, or vice versa. Since L is a transversal, it follows that $\mathbf{a} \cdot \mathbf{f}(\mathbf{x}(t)) \neq 0$ for any orbit **x**. If the said assertion were not true, then, we must have $\mathbf{a} \cdot \mathbf{f}(\mathbf{x}(t_0)) = 0$, in case L is tangent to the orbit at $t = t_0$, or $\mathbf{a} \cdot \mathbf{f}(\mathbf{x}(t)) = 0$, in case the orbit moves along L for t closer to t_0. In either case, we arrive at a contradiction to the transversality condition.

For the other assertion, let the orbits **x**,**y** meet L at $t = t_1$ and $t = t_2$ respectively. Suppose $\mathbf{a} \cdot \mathbf{f}(\mathbf{x}(t_1)) > 0$ and $\mathbf{a} \cdot \mathbf{f}(\mathbf{y}(t_2)) < 0$. As $t_1 \neq t_2$, by continuity, we then have $\mathbf{a} \cdot \mathbf{f}(\mathbf{z}) = 0$ for some $\mathbf{z}$ along the line segment joining $\mathbf{x}(t_1)$ and $\mathbf{y}(t_2)$ along L. This again contradicts the transversality condition, proving the second statement. Some of such crossings are depicted in Fig. A.1.

For the last statement in the theorem, let the equation for L be given by $\mathbf{a} \cdot \mathbf{x} + b = 0$ for some non-zero vector $\mathbf{a} \in \mathbb{R}^2$. By continuity of $\mathbf{f}$, there is a circle around $\mathbf{x}_0$ containing only regular points. The solution $\phi_t(\mathbf{x})$ passing through any $\mathbf{x}$ inside this circle at $t = 0$ is continuous in $(t, \mathbf{x})$ in an open set containing $(0, \mathbf{x}_0)$; this follows from the continuous dependence on initial data. Put $L(t, \mathbf{x}) = \mathbf{a} \cdot \phi_t(\mathbf{x}) + b$. Then, $L(0, \mathbf{x}_0) = 0$ and $\frac{\partial L}{\partial t}(0, \mathbf{x}_0) \neq 0$, by transversality. Hence, by the implicit function theorem, for any $\varepsilon > 0$, there is a circle C, centered at $\mathbf{x}_0$ and a continuous function $t = t(x)$ defined inside C satisfying $t(\mathbf{x}_0) = 0$ and $L(t(x), \mathbf{x}) = 0$ for all $\mathbf{x}$ inside C_ε. By continuity of t at $\mathbf{x}_0$, it now follows that, given any $\varepsilon > 0$, there is a circle C_ε, centered at $\mathbf{x}_0$ such that $|t(\mathbf{x})| < \varepsilon$ for all $\mathbf{x}$ inside C_ε. Hence, the orbit passing through any $\mathbf{x}$ inside C_ε at $t = 0$ crosses L at time $t(\mathbf{x})$ and $|t(\mathbf{x})| < \varepsilon$. $\square$

Proposition A.2.2

Given $\mathbf{x} \in \mathbb{R}^2$, the intersection $\omega(\mathbf{x}) \cap L$ contains at most one point, where L is transversal to $\mathbf{f}$.

Proof: Assume $\omega(\mathbf{x}) \cap L$ is non-empty and let $\mathbf{q} \in \omega(\mathbf{x}) \cap L$. By Theorem A.1.2, there is a sequence $t_k \nearrow \infty$ such that $\phi_{t_k}(\mathbf{x}) \to \mathbf{q}$ as $k \to \infty$. Since L is a transversal to $\mathbf{f}$, it follows from Lemma A.2.1 that for each $\mathbf{y} \in \mathbb{R}^2$ sufficiently close to $\mathbf{q}$, there exists a unique time s such that $\phi_s(\mathbf{y}) \in L$ and $\phi_t(\mathbf{y}) \notin L$ for $t \in (0, s)$ if $s > 0$, or $t \in (s, 0)$ if $s < 0$; in particular, for each $k \in \mathbb{N}$, sufficiently large, there is an $s = s_k$ as earlier such that $\mathbf{x}_k = \phi_{t_k + s_k}(\mathbf{x}) \in L$.

We now consider two cases: either the sequence $\{\mathbf{x}_k\}$ is a constant sequence or is a non-constant sequence. In the first case, the orbit of $\mathbf{x}$ is periodic and hence, the ω-limit set $\omega(\mathbf{x}) = \mathcal{O}(\mathbf{x})$ only intersects L at the constant value of the sequence $\{\mathbf{x}_k\}$, and thus, $\omega(\mathbf{x}) \cap L = \{\mathbf{q}\}$. In the second case, let us consider two successive points of intersection $\mathbf{x}_k$ and $\mathbf{x}_{k+1}$; these intersections occur in one of the two ways as shown in Fig. A.2. By Lemma A.2.1, we know that along L, the vector field $\mathbf{f}$ always points to the same side. Next, note that the segment of the orbit between $\mathbf{x}_k$ and $\mathbf{x}_{k+1}$ together with the line segment between these two points form a continuous curve C. Hence, by Jordan's curve theorem, its complement $\mathbb{R}^2 \setminus C$ has two connected components, one bounded and another unbounded. The bounded component is the shaded region in Fig. A.2. Due to the direction of $\mathbf{f}$ on the line segment between $\mathbf{x}_k$ and $\mathbf{x}_{k+1}$, the positive orbit $\mathscr{O}^+(\mathbf{x}_k)$ is contained in the unbounded component. This implies that the next point of intersection $\mathbf{x}_{k+2}$ does not belong to the line segment between $\mathbf{x}_k$ and $\mathbf{x}_{k+1}$. Therefore, the points $\mathbf{x}_k, \mathbf{x}_{k+1}$ and $\mathbf{x}_{k+2}$ are ordered on the transversal L as shown in Fig. A.3. Due to the monotonicity of the sequence $\{\mathbf{x}_k\}$ along L, it has at most one limit point in L and hence, $\omega(\mathbf{x}) \cap L = \{\mathbf{q}\}$. $\qquad\square$

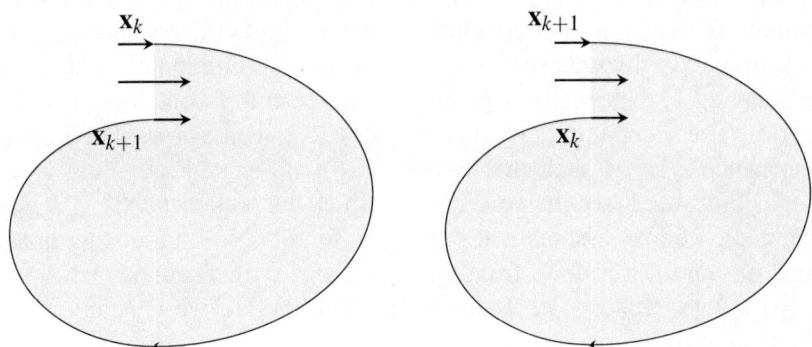

Fig. A.2   The direction of an orbit while crossing a transversal

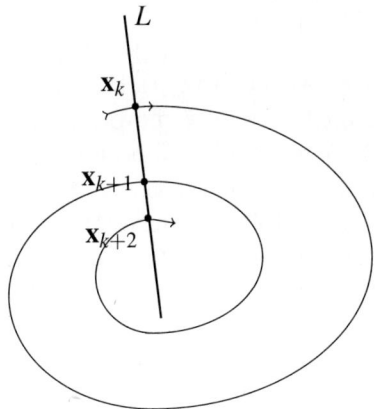

Fig. A.3   Crossings of an orbit on a transversal

We now come to the Poincarè–Bendixon theorem.

Theorem A.2.3

[Ponicarè–Bendixon] Let $\mathbf{f} : \mathbb{R}^2 \to \mathbb{R}^2$ be a C^1 function. For the equation (A.1.1), if the positive orbit $\mathcal{O}^+(\mathbf{x})$ of a point $\mathbf{x} \in \mathbb{R}^2$ is bounded and $\omega(\mathbf{x})$ contains no equilibrium points of $\mathbf{f}$, then, $\omega(\mathbf{x})$ is a periodic orbit.

Proof: Since the positive orbit $\mathcal{O}^+(\mathbf{x})$ is bounded, it follows from Theorem A.1.2 that $\omega(\mathbf{x})$ is non-empty. Let $\mathbf{p} \in \omega(\mathbf{x})$. It follows from the first and third properties in Theorem A.1.2 and the definition of ω-limit set, that $\omega(\mathbf{p})$ is non-empty and $\omega(\mathbf{p}) \subset \omega(\mathbf{x})$. Now consider any $\mathbf{q} \in \omega(\mathbf{p})$. By hypothesis, $\mathbf{q}$ is not an equilibrium point. Thus, by Lemma A.2.1, there a line segment L transversal to $\mathbf{f}$ containing $\mathbf{q}$ in its interior. The second property in Theorem A.1.2, then, shows that there is a sequence $t_k \nearrow \infty$ such that $\phi_{t_k}(\mathbf{p}) \to \mathbf{q}$ as $k \to \infty$. Proceeding as in Proposition A.2.2 (see the procedure in which the sequence $\{\mathbf{x}_k\} \subset L$ was obtained), we may assume that $\phi_{t_k}(\mathbf{p}) \in L$ for $k \in \mathbb{N}$. On the other hand, since $\mathbf{p} \in \omega(\mathbf{x})$, it follows from the third property in Theorem A.1.2 that $\phi_{t_k}(\mathbf{p}) \in \omega(\mathbf{x})$ for $k \in \mathbb{N}$. Since $\phi_{t_k}(\mathbf{p}) \in \omega(\mathbf{x}) \cap L$, we conclude from Proposition A.2.2 that

$$\phi_{t_k}(\mathbf{p}) = \mathbf{q}, \text{ for every } k \in \mathbb{N}.$$

This implies that $\mathcal{O}(\mathbf{p})$ is a periodic orbit. In particular, $\mathcal{O}(\mathbf{p}) \subset \omega(\mathbf{x})$.

It remains to show that $\mathcal{O}(\mathbf{p}) = \omega(\mathbf{x})$. Suppose not. Since $\omega(\mathbf{x})$ is connected, in each neighborhood of $\mathcal{O}(\mathbf{p})$, there exist points of $\omega(\mathbf{x})$ that are not in $\mathcal{O}(\mathbf{p})$. The continuity of $\mathbf{f}$ shows that in a sufficiently small neighborhood of $\mathcal{O}(\mathbf{p})$, there are no equilibrium points of $\mathbf{f}$. Then, Lemma A.2.1 shows that there exists a transversal L' to $\mathbf{f}$ containing in its interior, a point of $\omega(\mathbf{x})$ and a point of $\mathcal{O}(\mathbf{p})$. Since $\mathcal{O}(\mathbf{p}) \subset \omega(\mathbf{x})$, it follows that $\omega(\mathbf{x}) \cap L'$ contains at least two points, a contradiction to Proposition A.2.2. Therefore, $\mathcal{O}(\mathbf{p}) = \omega(\mathbf{x})$ and $\omega(\mathbf{x})$ is a periodic orbit. □

As an extension of the Poincarè–Bendixon theorem, to include the case of $\omega(\mathbf{x})$ containing equilibrium points, we have the following theorem.

Theorem A.2.4

Let $\mathbf{f} : \mathbb{R}^2 \to \mathbb{R}^2$ be a C^1 function. Assume that the positive orbit $\mathcal{O}^+(\mathbf{x})$ for (A.1.1) of a point $\mathbf{x} \in \mathbb{R}^2$ is contained in a compact set having at most finite number of equilibrium points of $\mathbf{f}$. Then, one of the following alternatives hold: the ω-limit set $\omega(\mathbf{x})$ is

(i) an equilibrium point;

(ii) a periodic orbit;

(iii) a union of a finite number of equilibrium points and a homoclinic or heteroclinic orbits.

A *homoclinic orbit* is a solution $\mathbf{x}(t)$ of (A.1.1) such that $\lim\limits_{t \to \pm\infty} \mathbf{x}(t) = \xi$ for some equilibrium point ξ of (A.1.1). For Duffing's equation without damping, we have seen that the orbit 'connecting' the equilibrium point at the origin is a homoclinic point.

A *heteroclinic orbit* is a solution $\mathbf{x}(t)$ of (A.1.1) such that $\lim\limits_{t \to \infty} \mathbf{x}(t) = \xi_1$ and $\lim\limits_{t \to -\infty} \mathbf{x}(t) = \xi_2$ for some equilibrium points ξ_1 and ξ_2 of (A.1.1). For the pendulum equation $\ddot{x} + k\sin x = 0 \,(k > 0)$, we have seen that the orbit 'connecting' the equilibrium points, for example, $(-\pi, 0)$ and $(\pi, 0)$ is a heteroclinic orbit.

Proof (of Theorem A.2.4) Since $\omega(\mathbf{x}) \subset \overline{\mathcal{O}^+(\mathbf{x})}$, the set $\omega(\mathbf{x})$ contains at most finite number of equilibrium points. If $\omega(\mathbf{x})$ consists of only equilibrium points, then it is necessarily a single equilibrium point because of connectedness.

Now, assume $\omega(\mathbf{x})$ contains some regular points as well and at least one periodic orbit $\mathcal{O}(\mathbf{p})$. We claim that $\omega(\mathbf{x})$ is the periodic orbit. For otherwise, by connectedness, there would exist a sequence $\{\mathbf{x}_k\} \subset \omega(\mathbf{x}) \setminus \mathcal{O}(\mathbf{p})$ and a point $\mathbf{x}_0 \in \mathcal{O}(\mathbf{p})$ such that $\mathbf{x}_k \to \mathbf{x}_0$ as $k \to \infty$. Next consider a transversal L to $\mathbf{f}$ such that $\mathbf{x}_0 \in L$. It follows from Proposition A.2.2 that $\omega(\mathbf{x}) \cap L = \{\mathbf{x}_0\}$. On the other hand, proceeding as in Proposition A.2.2, we infer that $\mathcal{O}^+(\mathbf{x}_k) \subset \omega(\mathbf{x})$ intersects L for sufficiently large k. Since $\omega(\mathbf{x}) \cap L = \{\mathbf{x}_0\}$, it follows that $\mathbf{x}_k \in \mathcal{O}(\mathbf{x}_0) = \mathcal{O}(\mathbf{p})$ for sufficiently large k, which contradicts the choice of the sequence $\{\mathbf{x}_k\}$. Therefore, $\omega(\mathbf{x})$ is a periodic orbit.

Finally, assume that $\omega(\mathbf{x})$ contains regular points, but no periodic orbit. We show that for any regular $\mathbf{p} \in \omega(\mathbf{x})$, the sets $\omega(\mathbf{p})$ and $\alpha(\mathbf{p})$ are equilibrium points.

If $\mathbf{p} \in \omega(\mathbf{x})$ is a regular point, notice that $\omega(\mathbf{p}) \subset \omega(\mathbf{x})$. If $\mathbf{q} \in \omega(\mathbf{p})$ is a regular point and L is a transversal to $\mathbf{f}$ containing $\mathbf{q}$ in its interior, then by Proposition A.2.2, we have

$$\omega(\mathbf{p}) \cap L = \omega(\mathbf{x}) \cap L = \{\mathbf{q}\};$$

in particular, the orbit $\mathscr{O}^+(\mathbf{p})$ intersects L at a point $\mathbf{x}_0$. Since $\mathscr{O}^+(\mathbf{p}) \subset \omega(\mathbf{x})$, we have $\mathbf{x}_0 = \mathbf{q}$, and thus, $\mathscr{O}^+(\mathbf{p})$ and $\omega(\mathbf{x})$ have the point $\mathbf{q}$ in common. Then, proceeding as in Proposition A.2.2, we conclude that $\mathscr{O}(\mathbf{p}) = \omega(\mathbf{p})$ is a periodic orbit. This contradiction shows that $\omega(\mathbf{p})$ contains only equilibrium points and by connectedness, it consists of a single equilibrium point. The proof for $\alpha(\mathbf{p})$ is similar. $\quad\square$

A.3   Leinard's Theorem

We now present Leinard's theorem concerning the existence of periodic orbits of a second order nonlinear equation and a proof of the same (see [CL72, Per01, Sim91]).

Theorem A.3.1

[Leinard's Theorem] Consider Leinard's equation

$$\ddot{x} + f(x)\dot{x} + g(x) = 0, \tag{A.3.1}$$

where $f, g : \mathbb{R} \to \mathbb{R}$ are C^1 functions satisfying the following conditions:

1. The function g is odd and satisfies $g(x) > 0$ for $x > 0$; f is an even function.

2. The odd function $F(x) = \int_0^x f(\xi) d\xi$ has exactly one positive zero at $x = a$, is negative in $(0, a)$, is positive and non-decreasing for $x > a$, and $F(x) \to \infty$ as $x \to \infty$.

Then, (A.3.1) has a unique periodic orbit surrounding the origin in the phase plane, and this orbit is approached spirally by every other (nontrivial) orbit as $t \to \infty$.

Proof: First let us write (A.3.1) as a first order system

$$\left.\begin{array}{l} \dot{x} = y \\ \dot{y} = -g(x) - f(x)y. \end{array}\right\} \tag{A.3.2}$$

Since f, g are assumed to be of class C^1, the basic theorem on existence and uniqueness of solutions holds. By condition (1), it follows that $g(0) = 0$ and $g(x) \neq 0$ for $x \neq 0$. Therefore, origin is the only equilibrium point for the system (A.3.2). Thus, any periodic orbit must surround the origin. Now

$$\ddot{x} + f(x)\dot{x} = \frac{d}{dt}\left[\frac{dx}{dt} + \int_0^x f(\xi)\,d\xi\right]$$

$$= \frac{d}{dt}[y + F(x)],$$

which suggests that we introduce a new variable

$$z = y + F(x).$$

Thus, (A.3.2) can be written in the equivalent form

$$\left.\begin{array}{l} \dot{x} = z - F(x) \\[1mm] \dot{z} = -g(x), \end{array}\right\} \tag{A.3.3}$$

in the $x - z$ plane. Again, the origin is the only equilibrium point for (A.3.3) and the usual existence and uniqueness result holds. Because of assumption (2), the correspondence $(x,y) \leftrightarrow (x,z)$ between the points in the two planes is one–one and continuous both ways. Therefore, periodic orbits correspond to periodic orbits and the configurations of orbits in the two planes are qualitatively similar. The orbits of (A.3.3) satisfy the differential equation

$$\frac{dz}{dx} = \frac{-g(x)}{z - F(x)}. \tag{A.3.4}$$

The orbits of (A.3.3) may easily be described using the hypothesis in the theorem and (A.3.4). We will now make the following observations which will help in understanding the directions of the orbit in Fig. A.4.

First note that since g and F are odd functions, (A.3.3) and (A.3.4) are unchanged when x, z are replaced by $-x, -z$. This means that any curve symmetric to an orbit with respect to the origin is also an orbit. Therefore, if we know an orbit in the right half plane $(x > 0)$, it is also known in the left half plane $(x < 0)$ by reflection through the origin.

Next, if an orbit starts on the z-axis with positive z coordinate (denoted by P in Fig. A.4), then, the orbit is horizontal (parallel to the x-axis) at this point. The x coordinate of the orbit is increasing and thus, moves into the right half plane, until the orbit meets the curve $z = F(x)$ (denoted by Q in Fig. A.4), where the orbit becomes vertical, that is, parallel to the z-axis; after crossing the curve $z = F(x)$, the x coordinate of the orbit starts decreasing, which continues up to the time when the orbit meets the z-axis again (denoted by R in Fig. A.4). As long as the orbit is in the right half plane, the z coordinate is decreasing. Let b be the abscissa (the x coordinate) of the point Q and denote by C_b the orbit described previously.

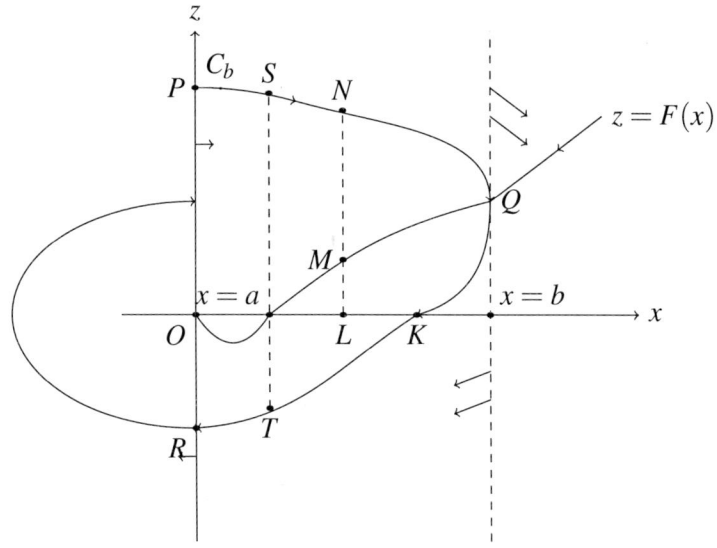

Fig. A.4   Proof of Leinard's theorem

It is not hard to see that when the orbit is continued beyond P and R into the left half plane, the result will be a periodic orbit if and only if the distances OP and OR are equal, by using the reflection through the origin; O is the origin. Therefore, to show that there is a unique periodic orbit, it suffices to show that there is a unique value of b which gives $OP = OR$.

To this end, we introduce the function

$$G(x) = \int_0^x g(\xi)\, d\xi$$

and consider the function

$$E(x,z) = \frac{z^2}{2} + G(x).$$

Note that $E(0,z) = \dfrac{z^2}{2}$. Along any orbit, we have

$$\dot{E} = g(x)\dot{x} + z\dot{z}$$

$$= -[z - F(x)]\dot{z} + z\dot{z} \text{ (using (A.3.4))}$$

$$= F(x)\dot{z},$$

which may be written as $dE = F\,dz$. Now evaluate the line integral of F along the orbit C_b from P to R, to obtain

$$I(b) = \int_{PR} F\,dz = \int_{PR} dE = E_R - E_P = \frac{1}{2}(OR^2 - OP^2).$$

Thus, it suffices to show the existence of a unique b such that $I(b) = 0$.

For $b \le a$ (see the hypothesis), we have $F < 0$ and $\dot{z} < 0$. Hence, $I(b) > 0$. For $b > a$, write $I(b) = I_1(b) + I_2(b)$, where

$$I_1(b) = \int_{PS} F\,dz + \int_{TR} F\,dz \text{ and } I_2(b) = \int_{ST} F\,dz.$$

See Fig. A.4. Since $F < 0$ and $\dot{z} < 0$ as we move along C_b from P to S and from T to R, we have $I_1(b) > 0$. On the other hand, when we move from S to T along C_b, we have $F > 0$ and $\dot{z} < 0$, so $I_2(b) < 0$. Therefore, we need to find a b such that $I_1(b) = -I_2(b) > 0$.

We now show that $I(b)$ is a decreasing function of b for $b \ge a$ and tends to $-\infty$ as $b \to \infty$. Since $I(a) > 0$, this gives, by continuity, a unique b_0 such that $I(b_0) = 0$. We, then, have the unique periodic orbit C_{b_0}.

From (A.3.4), it follows that

$$F\,dz = F\frac{dz}{dx}dx = \frac{-g(x)F(x)}{z - F(x)}dx.$$

Hence, the effect of increasing b is to raise arc PS (the arc PS is part of the orbit) and to lower arc TR, which decreases the magnitude of $\dfrac{-g(x)F(x)}{z - F(x)}$ for a given $x \in (0,a)$. Since the limits of integration for $I_1(b)$ are fixed, the result is a decrease in $I_1(b)$. Furthermore, since F is positive and non-decreasing for $x > a$, we see that an increase in b gives an increase in the positive number $-I_2(b)$, and hence, decrease in $I_2(b)$. This proves the first assertion that $I(b)$ is a decreasing function in $b \geq a$. To show that $I_2(b) \to -\infty$ as $b \to \infty$, which in turn implies that $I(b) \to -\infty$ as $b \to \infty$, consider a point L as shown in Fig. A.4. fiFix it and let K be a point to the right of L. Then, using $F(x) > 0$ for $x > a$, we have

$$I_2(b) = \int_{ST} F \, dz < \int_{NK} F \, dz \leq -(LM) \cdot (LN);$$

and since $LN \to \infty$ as $b \to \infty$, we have $I_2(b) \to -\infty$. This proves the first assertion in the theorem. For the second part, observe that $OR > OP$ for $b < b_0$ and by symmetry, we conclude that the orbits which start inside C_{b_0} spiral out to C_{b_0}. Similarly, using $OR < OP$ for $b > b_0$, we see that the orbits which start outside C_{b_0} spiral in to C_{b_0}. This completes the proof. $\square$

Bibliography

[AMR95]   Ascher, U. M., R. M. M. Mattheij, and R.D. Russel. 1995. *Numerical Solution of Boundary Value Problems for Ordinary Differential Equations.* Philadelphia: SIAM.

[AO12]   Agarwal, Ravi P. and Donal O'Regan. 2012. *An Introduction to Ordinary Differential Equations.* NewDelhi: Springer.

[Apo11]   Apostol, Tom M. 2011. *Calculus, Vol 1 and 2.* New Delhi: Wiley.

[Arn98]   Arnold, V. I. 1998. *Ordinary Differential Equations.* India: Prentice-Hall.

[AS72]   Abramowitz, M. and A. Stegun. 1972. *Handbook of Mathematical Functions.* New York: Dover.

[BR03]   Birkhoff, G. and G-C. Rota. 2003. *Ordinary Differential Equations.* New York: John Wiley and Sons.

[Bra75]   Braun, Martin. 1975. *Differential Equations and Their Applications.* New York: Springer-Verlag.

[Bra78]   Braun, Martin. 1978. *Differential Equations and Their Applications.* New Delhi: Springer International.

[BS05]   Bartle, R. G. and D. R. Sherbert. 2005. *Introduction to Real Analysis.* New Delhi: John Wiley and Sons.

[CL45]   Cartwright, M. L. and J. E. 'Littlewood. 1945. On the nonlinear differential equations of the second order. i. the equation $\ddot{y} + k(1 - y^2)\dot{y} + y = b\lambda k\cos(\lambda t + a); k$ large'. *J.Lond.Math.Soc.* 20: 180–189.

[CL72]   Coddington, E. A. and N. Levinson. 1972. *Ordinary Differential Equations.* New Delhi: Tata McGraw-Hill.

[Eva98]   Evans, L. C. 1998. *Partial Differential Equations.* Providence, Rhode Island: American Mathematical Society.

[GH83]   Guckenheimer, John and Philip Holmes. 1983. *Nonlinear Oscillations, Dynamical Systems and Bifurcation of Vector Fields.* New York: Springer Verlag.

[Hao84]   Hao, B. L. 1984. *Chaos.* Singapore: World Scientific.

[HK97]   Hoffman, Kenneth and Ray Kunze. 1997. *Linear Algebra.* New Delhi: Prentice Hall.

[HSD04]   Hirsch, M. W., S. Smale, and R. L. Devaney. 2004. *Differential Equations, Dynamical System and an Introduction to Chaos.* New Delhi: Academic Press.

[Inc26]   Ince, E. L. 1926. *Ordinary Differential Equations.* New York: Dover, Inc.

[Joh75]   John, F. 1975. *Partial Differential Equations, 2 ed..* New York: Springer-Verlag.

[JS03]   Jordan, D. W. and P. Smith. 2003. *Nonlinear Differential Equations, third edition (reprint).* Oxford: Oxford University Press.

[Kel90]   Keller, H. B. 1990. *Numerical Solution of Two-point Boundary Value Problems.* New York: Dover.

[Kum00]   Kumaresan, S. 2000. *Linear Algebra: A Geometric Approach.* New Delhi: Prentice Hall.

[Lef77]   Lefschetz, Solomon. 1977. *Differential Equations: Geometric Theory.* New York: Dover.

[Lev49]   Levinson, N. 1949. 'A second order differential equation with singular solutions.' *Ann.Math.* 50: 127–153.

[Mer97]   Merkin, David R. 1997. *Introduction to the Theory of Stability.* New York: Springer.

[MU78]   Myint-U, T. 1978. *Ordinary Differential Equations.* New York: North-Holland.

[Per01]   Perko, Lawrence. 2001. *Differential Equations and Dynamical Systems.* New York: Springer.

[PR96]   Prasad, P. and R Ravindran. 1996. *Partial Differential Equations, 3 ed.* New Delhi: New Age International.

[RR04]   Renardy, M. and R. C. Rogers. 2004. *Partial Differential Equations.* New York: Springer-Verlag.

[Rud76]   Rudin, Walter. 1976. *Principles of Mathematical Analysis.* London: McGraw-Hill.

[Sim91]   Simmons, G. F. 1991. *Differential Equations with Applications and Historical Notes.* New Delhi: McGraw-Hill International Edition.

[SK07]   Simmons, G. F. and S. G. Krantz. 2007. *Differential Equations: Theory, Technique and Pratice.* New Delhi: Tata McGraw-Hill.

[Str06]   Strang, Gilbert. 2006. *Linear Algebra and its Applications.* New Delhi: Cenagage Learning.

[Tay11]   Taylor, Michael E. 2011. *Introduction to Differential Equations, Indian edition (2013).* Providence, Rhode Island: American Mathematical Society.

[TS86]   Thomson, J.M.T. and H. B. Stewart. 1986. *Nonlinear Dynamics and Chaos.* New York: John Wiley and Sons.

[Wig90]   Wiggins, S. 1990. *Introduction to Applied Nonlinear Dynamical Systems and Chaos.* New York: Springer.

Index